Basic Biochemistry
for Medical Students

P. N. Campbell (editor)

J. B. C. Findlay

H. Hassall

R. P. Hullin

A. J. Kenny

B. A. Kilby (editor)

J. H. Parish

Department of Biochemistry
University of Leeds,
England

1975

Academic Press

London New York San Francisco

A Subsidiary of Harcourt Brace Jovanovich, Publishers

ACADEMIC PRESS INC. (LONDON) LTD.
24/28 Oval Road,
London NW1

United States Edition Published by
ACADEMIC PRESS INC.
111 Fifth Avenue,
New York, New York 10003

Library of Congress Catalog Card Number: 75 543
ISBN: 0 12 158150 0

Printed in Great Britain by William Clowes & Sons Limited,
London, Colchester and Beccles

Basic Biochemistry

for Medical Students

Preface

In the last few years Medical Schools in the U.K. and elsewhere have generally adopted a much more flexible approach to the medical curriculum. Although various patterns of medical education have emerged from this process, the general tendency has been for greater integration of the basic Medical Sciences with clinical teaching. Biochemistry is no less relevant to clinical studies than is anatomy or physiology, although it is more demanding of the student in the sense that it requires an understanding of the language of chemistry for it to be intelligible. Without some knowledge of this basic grammar the subject can degenerate into a string of unconnected jargon. Our experience has convinced us that a course in biochemistry, placed early in the curriculum, is an essential basis on which the later integrated and applied teaching can be built. The pattern at Leeds is not unlike that in other medical schools both in the U.K. and overseas. The first Medical Year contains courses of about equal weight in anatomy, physiology and biochemistry together with short courses in cell biology and on man in society. In the second and, to diminishing degrees in subsequent years, biochemistry teaching is integrated with that of other departments who collaborate in courses on endocrinology, energy, nutrition, genetics, haematology and gastroenterology.

This book has its origin in the material given by teachers in the Department of Biochemistry in the first medical year. This course comprises about 70 lectures in addition to laboratory classes and small-group teaching in Seminars. In this book we have attempted to cover the ground in mammalian biochemistry in a concise fashion. We have assumed that students are familiar with the essential principles of organic and physical chemistry—a justifiable assumption in view of the high standard of entry now achieved in Medical Schools. Nevertheless we believe that the majority of students benefit from further instruction in at least some aspects of chemistry and so we include the principles of pH, dissociation and elementary thermodynamics and the chemistry of sugars and nucleotides. We then proceed to the structure of proteins which leads naturally into enzymes. The scene is set for carbohydrate metabolism which is followed by nitrogen and lipid metabolism. These form the basis of a study in depth of the bioenergetics of mitochondria. Owing to their increasing importance considerable attention is given to membranes both from the structural viewpoint and their role in the transport of materials. There follows a discussion of the biosynthesis of nucleic acids and proteins. The book ends with an introduction to the integration and regulation of metabolism, which reminds the reader of the way in which the biochemistry of the various tissues helps to explain the physiology of man.

The book is the work of seven teachers who differ in their styles of lecturing and of writing. The editors have not attempted to foist any uniformity of style upon their colleagues. Nonetheless each contribution has been read by at least two other authors and each of us has benefited from these critical comments in revising the final drafts. We hope this critical process has reduced both errors and any unnecessary duplication to a minimum.

In developing a lecture course over a number of years, we, like any other group of teachers, have drawn heavily on the published work of others. Where a figure or table has been reproduced more or less in the original form, permission has been sought and grateful acknowledgement made. But there are, no doubt, other figures or concepts that were at one time borrowed for inclusion in a lecture and have then undergone revision, amalgamation or simplification to the point where the ancestry has become obscure or even forgotten. If any reader can trace such material to something of their own work we hope they will be more flattered by the imitation than angered by our unintended discourtesy.

Note on Formulae, Equations and Units

Most acids and bases exist in living systems as an equilibrium mixture of ionized and un-ionized forms and it is not always known in which form the molecule reacts. Biochemists often use terms such as pyruvate and pyruvic acid quite arbitrarily. We also have not been consistent, but have employed whichever appeared most convenient in a particular context. It is usually clearer when writing equations to employ the un-ionized form, but α-amino acids are frequently shown in the zwitterion form as a reminder that this inner salt structure is the one that predominates at physiological pH.

A phosphate group in a molecule is conveniently represented as —P, the symbol standing for —PO(OH)$_2$ or written as ~P when it is desirable to indicate a "high energy" phosphate compound, as in ATP or phosphoenolpyruvic acid. When inorganic orthophosphate or pyrophosphate occur in an equation, the anions or acids are represented as P$_i$ and PP$_i$ respectively.

SI units are now being adopted increasingly in biochemistry, especially in Europe and have therefore been used in this book. For example, small lengths are expressed in nanometers (nm) instead of Angstroms (Å) and energy in kilojoules (kJ). But since the kilocalorie (kcal) is still so familiar and its use so widespread, both nomenclatures have been used in this case. A kilocalorie is about 4·2 kilojoules.

The cover illustrations show (from top to bottom): an enzyme complex–fatty acid synthetase; bacteriophage; pancreatic cell; and kidney cell visualized by freeze-etching.

Contents

1

Introduction

What is biochemistry all about and what is it that biochemists strive to achieve? The short answer can be suggested by the alternative form of the question. "How is it that living things are composed of lifeless molecules?" Two of the most characteristic properties of all living organisms are, (1) their ability to extract and transform energy from their environment, and (2) their ability to reproduce themselves, i.e., their capacity for precise self-replication. There is, therefore, something very special about the phenomenon of life.

In attempting to answer the question, the biochemist breaks open the living cells and isolates their constituents. When he does this he finds that while he is impressed with the structural complexity of many of the cellular components, the isolated substances do seem to conform to the laws which he has learnt from his training in chemistry and physics. In other words, the living organism is radically different from the inanimate parts from which it is composed.

The answer to the problem given by the medieval philosophers was vitalism. They believed that living tissues were endowed with some mysterious life-force. While we do not to-day believe in vitalism, we have to admit that we, as biochemists, have not yet reached our goal of explaining all the properties of living cells in terms of the laws of chemistry and physics. A recurring problem which we also hope to answer eventually concerns the origin of life, i.e., how did living things come to this earth?

1

We have said that the first experimental approach of the biochemist was to determine the chemical structure of the components of cells. Let us see, therefore, where this approach leads us. The chemical substances in cells are in the main compounds of carbon, and since this has long been recognized to be so, such substances are known as organic. Apart from carbon they also contain a considerable proportion of nitrogen. So far as the earth's crust is concerned, carbon and nitrogen are only found in substances with comparatively simple structures, CO_2, N_2, carbonates and nitrates, so-called inorganic compounds. The situation can be depicted as:

Intra cellular	Extra cellular
Organic compounds	CO_2 N_2 Carbonates, nitrates

For the purposes of an example, we might consider the substances to be found in one of the unicellular organisms known as bacteria. A particular example might be the bacterium known as *Escherichia coli* which is found in the intestine. *E. coli* contains about 5000 different organic substances and 3000 of these might be different kinds of proteins and 1000 different kinds of nucleic acids. A multicellular organism such as man, is composed of many different kinds of differentiated cells. While many of the same substances will be shared between the different types of cells, other substances will be peculiar to certain cells. If we add up all the different kinds of substances in man, therefore, we find that the situation is infinitely more complex than in *E coli*, and man probably contains about 5 million different proteins. None of these proteins will be the same as those in *E. coli* or other kinds of bacteria. Indeed each species of organism has its own chemically distinct proteins and nucleic acids. Since there are about 1,200,000 species of organism there must be about 10^{10} different proteins and a further 10^{10} different kinds of nucleic acids. To put matters in perspective we recall that the total number of compounds that have so far been synthesized by organic chemists is about 10^6. It follows that if the answer to the question posed by the biochemists depends on their ability to unravel the structure of all the different compounds found in living organisms, the task would appear hopeless. Fortunately for us things are not as black as this, for we now know enough to realize that the basic pattern of the complex organic molecules found in living cells is rather simple.

Proteins are essentially polymers made up of about 24 different monomeric units known as amino acids. Similarly nucleic acids are polymers made up of 4 different nucleotides. It is now clear that the monomeric units in both proteins and nucleic acids are identical, irrespective of the organism. We see, therefore, that the different proteins arise simply by ringing the changes on the order in which the same 24 different amino acids are linked together to form the polymers which are the basic structures in the proteins. The same considerations apply for the nucleic acids.

Another interesting point is that we see the amino acids and nucleotides not only as the essential ingredients of proteins and nucleic acids but also in other roles. Thus short chains of amino acids may serve as hormones and certain amino acids may be a little modified to serve as the pigment of the skin. The nucleotides have an important role in the form of coenzymes.

The versatility of the organic compounds serves to emphasize another point. Living cells are only just as complex as is required for them to fulfil their particular function. Thus a cell contains no useless molecules without a function, and, moreover, there is a great economy in the use of the substances for many different purposes.

The fact that the amino acids and nucleotides are common to all forms of living organism suggests that we may presume that all living things came from a common ancestor. One of the important characteristics of living things that we previously mentioned concerned the transformation of energy from the environment to a form that was useful to them.

Input energy	Chemical engine	Cellular energy
Sunlight	isothermal using enzymes	chemical energy used for synthesis of compounds,
Food	⟶	transport of compounds, mechanical work.

In the case of plants, the energy of the environment is sunlight, and for animals it is food. The transformation process is very efficient and in this respect is much better than that of man-made machines such as we find in the conversion of oil into electrical energy. The cellular machine has to operate at a given temperature and cannot use heat itself as a source of energy. It is, therefore, said to be isothermal. The energy that is transformed is converted into chemical energy which is then available for a variety of purposes, such as the synthesis of organic compounds e.g., proteins, and the transport of substances into and out of the cell, or for mechanical work. The reason why cells can function as chemical engines is that they possess enzymes. These are highly specialized protein molecules which can greatly enhance the rate of specific chemical reactions. Well over 1000 different enzymes have now been characterized, each of which is able to assist a specific chemical reaction.

The presence of an enzyme in a reaction increases its speed and ensures that no by-products are formed i.e., the reaction is 100% efficient. In a normal chemical reaction in the test tube, one never gets such efficiency and nearly always there are by-products.

The reason why enzymes are such efficient catalysts is that there is a very precise interaction between the substrate (the substance to engage in the reaction) with the enzyme protein. This introduces a new principle known as *structural complementarity* which underlies the specificity of many different types of molecular interactions in cells. Such interactions depend on rather weak non-covalent bonds; since these bonds are weak a large number are required to give the complex stability, and for this to happen, the two substances have to come together in close juxtaposition over a large area. This demands a very specific "fit" or structural complementarity. For the cells to function properly, the individual enzymic reactions have to be linked as are a series of traffic lights controlling the flow of traffic through a city. The control of the enzymic reactions, which is again effected by the interaction of small molecules with the protein enzymes, serves to ensure that there are no pile-ups which would result in the accumulation of intermediates. A common arrangement is for the first reaction in a series to be slowed down thus preventing the completion of the whole operation.

It is fascinating to note that once again we find that there is a common thread in the method whereby energy is transformed throughout all living cells. Thus while some cells utilize food, and others sunlight as energy input, we observe that in all cases the energy is recovered in the form of the same substance, adenosinetriphosphate, commonly known as ATP.

The other characteristic of living cells we mentioned was their capacity to reproduce themselves with near perfect fidelity. This process goes on for hundreds of thousands of generations. The immense amount of information contained in the single sperm cell and the single egg resides in the deoxyribose nucleic acid, DNA, which they contain. Once again structural complementarity plays a role in ensuring that the information is handed on from one generation to another by the fact that a new molecule of DNA is formed by using the old molecule as a template. No new DNA can be made without a DNA template and this ensures that the new molecules are faithful copies of the old.

In the words, therefore, of Lehninger,* we can say that a living cell is a self-assembling, self-adjusting, self-perpetuating system. The machinery of living cells functions, we believe, within the same set of laws that governs the operation of man-made machines, but the chemical reactions have been refined far beyond the capabilities of chemical engineering. The important conclusion is that biochemistry has a set of organizing

* See "Biochemistry: The molecular basis of cell structure and function". Lehninger, A. L. (1970); a book which has much influenced this introduction.

principles and is not merely a collection of unselected facts about living matter. We endeavour in this book to emphasize wherever possible these principles.

The Place of Biochemistry in the Medical Course

Because of the great advances in recent years towards an understanding of the principles that we have been discussing, there is now no doubt about the importance of biochemistry in medical research. Biochemists, and the technology which they have created, ultimately hold the clue to our understanding of such subjects as the mechanism of action of antibiotics, ageing, transplantation immunity, all aspects of nutrition, endocrinology, hereditary diseases and cancer. Thus those who intend to enter upon a career in medical research will undoubtedly need an understanding of biochemistry.

In the United Kingdom, as in most countries, medical education is organized so that all students attend the same basic course. This is irrespective of whether the student intends to work after qualification as a medical scientist, a surgeon, a physician or as a general practitioner. In the event, perhaps only 10% of any group of students will eventually do medical research. The biochemists will no doubt do their best to interest and stimulate that 10% but what of the other 90%, what is the relevance of the subject to them? It could be said that a general practitioner does not need a very detailed knowledge of biochemistry and it would be hard to argue against this viewpoint. We do, however, refuse to take too narrow a view of the needs of a general practitioner believing, as we do, that we should attempt to provide a reasonably broad education that will serve the student for perhaps 40 years of medicine. We do not wish to cram in facts but to give the opportunity to grasp the general principles on which the subject is based whilst acknowledging that principles are only understandable when founded on some factual material. We want the student to be able to share with us the excitement, not only of the new developments in research that are happening to-day, but also to get enough of the "feel" of the subject so that he can tune into and assess the developments that arise in the future. In the chapters that follow we do our best to indicate wherever possible the relevance of each subject to medicine. This matter of relevance can, however, be overdone, so that the teacher not only becomes intellectually dishonest but the principles are either omitted or obscured; we will do our best to avoid these pitfalls.

2

Acid-base Dissociations and their Relevance to Biological Systems

It is not easy to justify to the new student the inclusion of a chapter on pH and dissociation in an introductory textbook of biochemistry. It is even more difficult to do this when biochemistry is being studied as a subsidiary subject in a larger course such as medicine. Much of the justification has to be taken on trust, and to some extent any acquired knowledge has to be held in cold-storage until its relevance is more easily appreciated.

We can begin by pointing out that most biologically occurring compounds, with the exception of certain lipids and storage macromolecules, are characterized by one or more acidic or basic groups, and exist in aqueous solution as charged species. The degree of dissociation or extent of ionization of a particular chemical group, and hence the biochemical reactivity of the molecule itself, are strongly influenced by the hydrogen ion concentration of the solution. This applies not only to intermediates of metabolic pathways but also to the biological catalysts, the enzymes, that control the reactions taking place in the cell. Enzymes, being proteins, have a large number of weakly acidic or weakly basic groups which play an important role both in maintaining the molecular structure or conformation and in determining the properties of the active centre.

This in itself might be sufficient justification for ensuring that the student has at least an elementary grasp of the principles of acid-base chemistry. However, there are other reasons too. The subject is fundamental to an understanding of buffer systems in general, electrolyte balance and renal transport, acid secretion in the stomach and gas transport at the lungs and tissues. In addition, it is perhaps not surprising that many biochemical separation methods exploit the differences in acidic and basic properties shown by most metabolites and cell constituents. These differences, usually small, can frequently be amplified by deliberately altering the pH of the solution so that separation is more readily achieved.

The techniques relying most heavily upon this kind of approach are ion-exchange chromatography and electrophoresis, where the degree of separation is directly influenced by pH. A number of other methods, particularly in the field of protein chemistry, also depend indirectly upon changes in the extent of ionization of various acidic or basic groups. This applies to salt fractionation and isoelectric precipitation where in each case the effect of charge on solubility is the underlying principle of the fractionation procedure.

2.1. Water

(a) Structural properties

Water is in many ways unique. Compared with other related hydrides, such as H_2S, it has a high boiling point so that it is a liquid at physiological temperatures and atmospheric pressure; it also has a relatively high freezing point.

These properties are a consequence of the strong interactions between individual water molecules. This in turn arises because of the strong dipole moment of the HOH structure. By this we simply mean that the hydrogen atoms exhibit partial positive charges because the electron distribution of the molecule is shifted towards the heavier nucleus of the oxygen atom. The latter, as a result, has a net negative charge associated with it.

Interactions of the type shown in Fig. 2.1 will then occur between adjacent H_2O molecules. In the liquid state, these interactions will be random and will fluctuate rapidly; in ice they are much more ordered and permanent.

The attractive force between the hydrogen atom of one molecule and the oxygen atom of another is known as a hydrogen bond. Compared with the normal intramolecular covalent bond it is a rather weak interaction but nevertheless, in a system where many hydrogen bonds are possible, their total contribution may be quantitatively large.

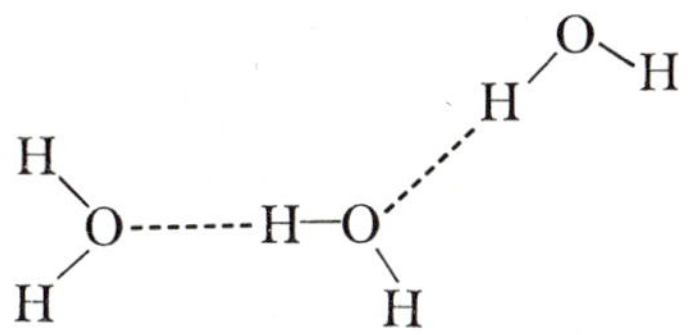

Fig. 2.1 Hydrogen bonding between water molecules.

Hydrogen bonds are not only important when we consider the dissociation properties of water but they are also found in many compounds of biological interest. Here, they may be relevant to the structure and solution properties of individual molecules, to the interaction between molecules and frequently to the biological function of the particular compound. For example, they are formed between $>C=O$ and $H-N<$ groups of the polypeptide chain of proteins (See Chapter 4.4) and between similar groups (and between $N-H$ and $N\!\!<$) of complementary bases of nucleic acids (See Chapter 3.18). Hydrogen bonds are one type of the non-covalent bonds mentioned in the Introduction.

(b) Dissociation properties

A small proportion of the water molecules are dissociated; at body temperatures, the extent of this dissociation is one molecule in approximately every 5×10^8. Although it is convenient for us to represent the dissociation of water by Eqn 1, this is an oversimplification.

$$H_2O \rightleftharpoons H^+ + OH^- \qquad (1)$$

Because of the high reactivity of the hydrogen ion (or proton), and the dipole moment of the water molecule, H^+ does not exist as such in aqueous solution but reacts with a second molecule of H_2O to form the hydronium ion, H_3O^+; this in turn is hydrogen bonded to three other molecules of water to form the hydrated species, $H_9O_4^+$, as shown in Fig. 2.2.

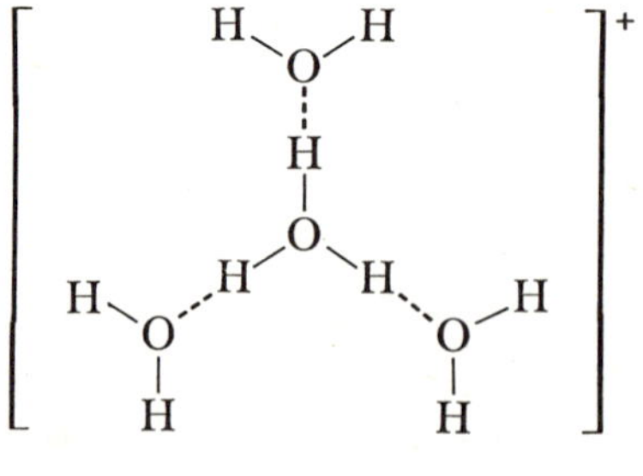

Fig. 2.2 The hydrated hydronium ion, $H_9O_4^+$.

It should, perhaps, be remembered that this is the structure that we are really talking about when we refer to a proton (H^+) in aqueous solution.

The dissociation constant for the reaction shown in Eqn 1 can be written as

$$K_a = \frac{[H^+][OH^-]}{[H_2O]} \qquad (2)$$

where square brackets are used to denote concentration. At 25°C, the dissociation constant for water has a value of $1\cdot8 \times 10^{-16}$. From it, we are able to derive a much more useful term, K_w (the ionic product of water), where

$$K_w = [H^+][OH^-] \qquad (3)$$

Since the dissociation of water is so slight, the concentration of undissociated water molecules can be assumed to be constant at 100/18 or 55·5 M. It follows that (at any given temperature) K_w is also a constant which is related to K_a as follows:

$$K_w = K_a \times 55\cdot5$$
$$K_w = 1\cdot8 \times 10^{-16} \times 55\cdot5 = 10^{-14}$$

Therefore at all times, in aqueous solution at 25°C

$$[H^+][OH^-] = 10^{-14} \qquad (4)$$

and if either $[H^+]$ or $[OH^-]$ is known then we can readily calculate the other.

In pure water, for example, $[H^+]$ must be equal to $[OH^-]$ and therefore each must have a value of $\sqrt{10^{-14}}$ or 10^{-7}. The units used are moles per litre (moles/l) or molar (M). Aqueous solutions having this concentration of hydrogen ion are said to be neutral. At 37° C (body temperature), the ionic product of water is slightly larger ($10^{-13\cdot6}$) and $[H^+] = 10^{-6\cdot8}$ M. We shall, however, use the former value in all calculations and derivations.

2.2. Acids and bases

(a) Definitions

For centuries, acids were recognized as a group of substances that reacted with metals, liberated a gas (CO_2) from chalk, and (presumably in weak solution), had a bitter taste. They were also seen to change the colour of certain vegetable dyes such as litmus. Alkalis (bases) were simply compounds which abolished or "neutralized" the properties of acids.

Davy was the first to recognize that all acids contained hydrogen, and Arrhenius between 1880 and 1890, showed that the hydrogen responsible for acidity must be ionizable in solution. The definition of an acid that serves us best was developed directly from this concept by Brønsted and Lowry in 1923.* It states that an acid is a substance that loses, or tends to

---

* The definition by Lewis, although chemically more comprehensive, is not used, since in our experience, it leads to confusion for the non-specialist and has no particular advantages. According to this definition, an acid is any substance that can accept electrons and a base is a substance that can donate them. Thus, in the reaction $Ag^+ + 2CN^- = Ag(CN)_2^-$, Ag^+ would be the acid and CN^- would be the base.

lose, a proton. For example, acetic acid

$$CH_3COOH \rightleftharpoons CH_3COO^- + H^+$$

An extension of this definition says that a base is a substance that accepts, or tends to accept, a proton. For example, ammonia

$$NH_3 + H^+ \rightleftharpoons NH_4^+$$

It is important to note that a base is not defined as a substance producing hydroxyl ions since this is not necessarily true except in aqueous solution. Here, hydroxyl ions are produced even by weak bases such as ammonia because hydrogen ions are removed from the water thus causing more of the latter to dissociate.

The net reaction can then be represented as:

$$NH_3 + H_2O \rightleftharpoons NH_4^+ + OH^-$$

As we shall see, strong bases, for example sodium hydroxide, do produce hydroxyl ions directly.

$$NaOH \rightleftharpoons Na^+ + OH^-$$

The hydroxyl ion is basic (by definition) since it reacts with a proton

$$OH^- + H^+ \rightleftharpoons H_2O$$

It can be seen from the Brønsted-Lowry definition that water is amphoteric, that is, it exhibits both acidic and basic properties. In the reaction:

$$H_2O + H_2O \rightleftharpoons OH^- + H_3O^+$$
$$\quad A \qquad\ B$$

molecule A behaves as an acid because it loses a proton; molecule B behaves as a base because it accepts one.

(b) Conjugate pairs

If we consider the dissociation of acetic acid, the following equilibrium exists

$$CH_3COOH \rightleftharpoons H^+ + CH_3COO^-$$

From the definitions of acids and bases given above, and considering the reaction in the forward direction, CH_3COOH is clearly an acid since it donates (or at equilibrium *tends* to donate) a proton. Less obviously perhaps, is the fact that when we look at the reverse reaction, the acetate ion (CH_3COO^-), by accepting or tending to accept a proton, is a base. In a reaction of this kind, which is typical of all acid-base equilibria, the acid (CH_3COOH) and the base (CH_3COO^-) are termed a conjugate pair.

(c) The strength of an acid

It follows axiomatically from the Brønsted-Lowry definition that strong

acids will be substances which have a high tendency to lose hydrogen ions.
If the acid HA dissociates as follows:

$$HA \rightleftharpoons A^- + H^+$$

then the acid dissociation constant is given by

$$K_a' = \frac{[A^-][H^+]^*}{[HA]} \tag{5}$$

It follows that the larger the value of K_a' then the greater is the strength
of the acid.

Strong acids are those compounds which have K_a' values high enough
for the acid to be effectively fully dissociated even in relatively strong
solution. On the other hand, the dissociation constants of weak acids are
much lower so that the extent (or degree) of dissociation is much more
concentration-dependent and the acid is only fully dissociated at infinite
dilution.

The degree of dissociation is given the symbol α which is equal to the
fraction of the molecules that are in the dissociated form at a given
concentration, c. It follows that

$$[H^+] = \alpha c \tag{6}$$

In Table 2.1, the degree of dissociation of a number of strong and weak
acids is shown at two arbitrary concentrations. The mineral acids, HCl,
HNO_3 and H_2SO_4, are all strong acids whereas the organic acids are weak.
Halogen substitution of organic acids, as shown by trichloroacetic acid,
greatly increases their strength.

TABLE 2.1. The effect of concentration on the degree of
dissociation of strong and weak acids

	Degree of dissociation, α, at	
	0·5 M	10^{-3} M
Strong acids		
HCl	0·862	0.993
HNO_3	0·862	0·997
CCl_3COOH	0·760	0·990
Weak acids		
HCOOH	0·020	0·386
CH_3COOH	0·006	0·126
HCN	0·00005	0·0011

* Strictly speaking, the acid dissociation constant, K_a, for the reaction shown is equal to

$$\frac{\text{activity of } A^- \times \text{activity of } H^+}{\text{activity HA}}$$

However, it is more convenient for us to use the apparent acid dissociation constant, K_a', which is
the one expressed above, in Eqn 5, in terms of concentrations. This approach is valid as long as one
is dealing with dilute solutions where the activity coefficients are very close to 1.

(d) The strength of a base

Although a base is defined as a substance which accepts a proton, it is perhaps more convenient for us to have a slightly different working definition, since we are concerned only with bases in aqueous solution. In water, bases are compounds which bring about an increase in $[OH^-]$. As stated earlier, they can do this in one of two ways. Either they dissociate to give OH^- directly or they remove H^+ ions from the water thus causing a net increase in $[OH^-]$. Strong bases such as KOH, NaOH, and $Ba(OH)_2$ are in the first group while weak bases, for example ammonia and methylamine, are in the second.

For strong bases we can still use the term degree of dissociation (α) but for weak bases we need to introduce another concept, that of hydrolysis. For methylamine, which establishes the following equilibrium:

$$CH_3NH_2 + H_2O \rightleftharpoons CH_3NH_3^+ + OH^-$$

the hydrolysis constant, K_h, is given by

$$K_h = \frac{[CH_3NH_3^+][OH^-]}{[CH_3NH_2][H_2O]}$$

If we use α both for the degree of dissociation of a strong base and for the degree of hydrolysis of a weak base, then

$$[OH^-] = \alpha c$$

Examples of the way in which α is dependent upon concentration are shown for a number of bases in Table 2.2. As we might expect, strong bases are those for which α is large even at high concentrations.

TABLE 2.2. The effect of concentration on the degree of dissociation or hydrolysis of strong and weak bases

	Degree of dissociation or hydrolysis (α), at	
	0·5 M	10^{-3} M
Strong bases		
KOH	0·826	0·981
NaOH	0·795	0·996
Weak bases		
CH_3NH_2 (+H_2O)	0·0313	0·500
NH_3 (+H_2O)	0·0068	0·141
$C_6H_5NH_2$ (+H_2O)	0·00005	0·00068

(e) The pH scale

Because of the range of hydrogen ion concentration that is encountered, a way of expressing this other than in absolute terms is desirable. The pH scale was introduced therefore largely as a matter of convenience by

Sørensen in 1909. The pH of a solution is defined as the negative logarithm to the base 10 of the hydrogen ion concentration:

$$pH = -\log_{10}[H^+]$$

Thus, for a solution of 0·1 M HCl, which we can assume to be fully dissociated,

$$[H^+] = 0·1 \text{ M} = 10^{-1} \text{ M}$$
$$\log[H^+] = -1$$
$$pH = -\log[H^+] = 1$$

and similarly for 0·1 M NaOH,

$$[OH^-] = 10^{-1} \text{ M}$$

since

$$[H^+][OH^-] = 10^{-14} \quad \text{(see Eqn 4)}$$
$$[H^+] = \frac{10^{-14}}{10^{-1}} = 10^{-13}$$
$$\log[H^+] = -13$$
$$pH = 13$$

This type of calculation is normally accomplished with a high degree of accuracy by all students. A significant failure rate, however, is found in the next type of calculation where the pH is not a whole number. The student is at first advised to include all the intermediate steps in calculations of the kind shown below.

Example (1) The hydrogen ion concentration of a sample of urine is 2×10^{-6} M. What is its pH?

$$[H^+] = 2 \times 10^{-6}$$
$$\therefore \log[H^+] = \bar{6}·301$$

Since this is the same as $(\bar{6}) + (\overset{+}{0}·301)$

$$\log[H^+] = -5·699$$
$$\therefore pH = 5·7$$

And when calculating from pH to $[H^+]$:

Example (2) The pH of a sample of serum is 7·4. What is the hydrogen ion concentration?

$$pH = 7·4$$
$$\therefore -\log[H^+] = 7·4$$
$$\therefore \log[H^+] = -7·4 = \bar{8}·6$$
$$\therefore [H^+] = 4 \times 10^{-8} \text{ M}$$

The pH of a number of biological fluids and foodstuffs is shown in Table 2.3. The convenience of the pH scale is apparent when we remember that a decrease in pH of one unit represents a tenfold increase in hydrogen ion concentration. A doubling in hydrogen ion concentration occurs for every decrease in pH of 0·301 (i.e., $\log_{10} 2$).

TABLE 2.3. The pH of some biological fluids and foodstuffs

	pH		pH
Gastric juice	1·2-3·0	Citrus fruits	2-4
Urine	5-8	Wines	2-4
Saliva	6·5-7·5	Beer	4-5
Milk	6·4-6·6	Eggs	7·6-8·0
Blood	7·4		
Pancreatic secretion	7·8-8·0		

(f) pK_w and pK_a

Just as it is convenient to use pH as a measure of $[H^+]$ so we can use pOH as an index of $[OH^-]$ and, more importantly, we can introduce the concept of pK.
Thus:

$$pOH = -\log_{10} [OH^-]$$

and

$$pK = -\log_{10} K$$

For water, therefore, we can derive the following relationship.
Since

$$[H^+] [OH^-] = K_w = 10^{-14} \qquad \text{(Eqns 3 and 4)}$$

$$(-\log H^+) + (-\log OH^-) = -\log K_w = 14$$

$$\therefore pH + pOH = pK_w = 14 \qquad (7)$$

(g) The Henderson-Hasselbalch Equation

For a weak acid, which dissociates as follows:

$$HA \rightleftharpoons H^+ + A^-$$

$$K_a' = \frac{[H^+] [A^-]}{[HA]} \qquad \text{(Eqn 5)}$$

An interesting and extremely useful relationship between pH and pK_a' can be obtained simply by taking logs of the above:

$$\log K_a' = \log[H^+] + \log[A^-] - \log[HA]$$

$$\therefore -\log[H^+] = -\log K_a' + \log[A^-] - \log[HA]$$

or

$$\mathrm{pH} = \mathrm{p}K_a' + \log \frac{[A^-]}{[HA]} \quad *$$

The most convenient form of this equation, the Henderson-Hasselbalch equation, is

$$\mathrm{pH} = \mathrm{p}K_a' + \log \frac{[\mathrm{conjugate\ base}]}{[\mathrm{conjugate\ acid}]} \quad (8)$$

It is the equation which is used for the calculation of the pH or composition of a buffer solution (See Section 2.3).

By using $\mathrm{p}K_a'$ values, we are able to express the strength of an acid (i.e. its tendency to dissociate) with reference to the pH scale. If K_a', the dissociation constant, is large, then $\mathrm{p}K_a'$ will have a low numerical value. Perhaps it is useful to look at this in another way: if we consider the situation where the acid is one half dissociated, in other words where $[A^-]$ is equal to $[HA]$, then, substituting in Eqn 8:

$$\mathrm{pH} = \mathrm{p}K_a' + \log 1$$

$$\therefore \mathrm{pH} = \mathrm{p}K_a' + 0$$

$$\therefore pH = pK_a'$$

This means that an acid is half dissociated when the pH of the solution is numerically equal to the $\mathrm{p}K_a'$ of the acid. Therefore acids with the lowest $\mathrm{p}K_a'$ values are able to dissociate in solutions of low pH, i.e., where the hydrogen ion concentration is high.

Table 2.4 gives some examples of K_a' and $\mathrm{p}K_a'$ values for a number of acids which are listed in decreasing strength.

TABLE 2.4. K_a' and $\mathrm{p}K_a'$ values for some acids

Acid	K_a'	$\mathrm{p}K_a'$
Trichloroacetic	2×10^{-1}	0·7
Dichloroacetic	5×10^{-2}	1·3
Monochloroacetic	$1·55 \times 10^{-3}$	2·81
Formic	$2·1 \times 10^{-4}$	3·68
Benzoic	$7·8 \times 10^{-5}$	4·11
Acetic	$1·86 \times 10^{-5}$	4·73
H_2CO_3	$2·9 \times 10^{-7}$	6·54
H_2S	$5·8 \times 10^{-8}$	7·24
HCN	$1·3 \times 10^{-9}$	8·89

* If this derivation presents any difficulty then the student is reminded of the following relationships:

(i) $\log(x \times y) = \log x + \log y$

(ii) $\log \dfrac{x}{y} = \log x - \log y$

(iii) $\log\left(\dfrac{1}{x}\right) = -\log x$

(iv) $\log \sqrt{x} = \frac{1}{2} \log x$.

(h) The dissociation of a weak acid in aqueous solution: The relationship
     between pH, pK_a', the degree of dissociation (α) and concentration (c)

Since the degree of dissociation of a weak acid is concentration-
dependent, we need a simple method of calculating its value for any given
set of conditions.

If we begin with a weak acid, HA, of concentration c, then if α is the
degree of dissociation

$$[H^+] = \alpha c \quad \text{(see Eqn 6)}$$

and at equilibrium

$$HA \rightleftharpoons H^+ + A^-$$

$$c(1-\alpha) \quad \alpha c \quad \alpha c$$

$$\therefore K_a' = \frac{[H^+][A^-]}{[HA]} = \frac{(\alpha c)^2}{c(1-\alpha)} = \frac{\alpha^2 c}{(1-\alpha)}$$

Where α is much less than 1, as it is for a weak acid at the
concentrations that we encounter, $(1-\alpha)$ approximates to 1,
and

$$K_a' = \alpha^2 c$$

or

$$\alpha = \sqrt{\frac{K_a'}{c}} \tag{9}$$

Since

$$[H^+] = \alpha c$$

$$[H^+] = c\sqrt{\frac{K_a'}{c}}$$

$$\therefore [H^+] = \sqrt{K_a' \cdot c} \tag{10}$$

This in itself is a useful equation but by taking logs, we obtain

$$\log[H^+] = \tfrac{1}{2}\log K_a' + \tfrac{1}{2}\log c$$

$$\therefore -\log[H^+] = -\tfrac{1}{2}\log K_a' - \tfrac{1}{2}\log c$$

or

$$pH = \tfrac{1}{2}pK_a' - \tfrac{1}{2} log\ c \tag{11}$$

We can apply either Eqn 10 or 11 to obtain answers to the following
example:

Example What is the hydrogen ion concentration and pH of a
solution of 0·1 M acetic acid if the dissociation constant of acetic acid is
1·86 x 10^{-5} (pK = 4·73)?

Using Eqn 10:

$$[H^+] = \sqrt{1 \cdot 86 \times 10^{-5} \times 0 \cdot 1}$$
$$[H^+] = \sqrt{1 \cdot 86} \times 10^{-3}$$
$$[H^+] = 1 \cdot 36 \times 10^{-3} \ M$$

from which, pH = 2·87.

OR Using Eqn 11

$$pH = 2 \cdot 37 - \tfrac{1}{2}(-1 \cdot 0)$$
$$pH = 2 \cdot 87$$

from which, $[H^+] = 1 \cdot 36 \times 10^{-3} \ M$.

(i) Titration curves of acids and bases

Although we are primarily interested in the behaviour of weak acids and bases it is useful for us to consider, initially, the titration of a strong acid with a strong base.

If we begin with 0·1 M HCl, then since this is fully dissociated, the pH of the solution will be 1. As 0·1 M NaOH is added, the pH will rise slowly and will reach pH 2 only after neutralization of 90% of the acid has occurred (i.e., when the HCl concentration is effectively 0·01 M); after 99% neutralization, the pH will be 3. Only after this point, near to the end point, will the pH rise quickly. This is shown in curve (a) in Fig. 2.3.

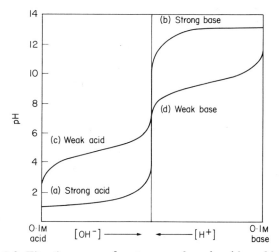

Fig. 2.3 Titration curves for strong and weak acids and bases.

If the reverse titration is carried out by adding 0·1 M HCl to 0·1 M NaOH then curve (b) is obtained. Again we are able to calculate reference pH values, this time of 13, 12 and 11 which correspond to the addition of 0, 0·9 and 0·99 of an equivalent of acid.

By comparison, when we titrate 0·1 M acetic acid, a weak acid, with 0·1 M NaOH, a strong base, then a completely different relationship is seen (curve (c) in Fig. 2.3). First, as we have already calculated (See Chapter 2.2(h)), the pH of the acid solution is 2·87. Secondly, when base is progressively added, there is initially a relatively steep rise in pH, followed by a flattening of the titration curve, and then a final steep rise at the equivalence point. If base is added beyond the equivalence point then curve (b) is followed (i.e., the addition of OH^- to a solution of Na^+ and CH_3COO^- ions, with respect to pH, is the same as adding OH^- to water).

The titration of a weak base, such as methylamine, with a strong acid (HCl) follows curve (d) and continues along curve (a) after the addition of one equivalent. The titration of a weak base with a weak acid gives the composite (c)-(d) curve.

We are particularly interested in knowing how to calculate the pH of points along curve (c), that is, for mixtures of a weak acid and its salt; any expression that we derive for this curve is also applicable to curve (d).

If we take the specific example of 0·1 M acetic acid, titrated with 0·1 M NaOH, then initially we have

$$CH_3COOH \rightleftharpoons CH_3COO^- + H^+$$

where, as we have seen, the concentration of H^+ and CH_3COO^- is small ($1·36 \times 10^{-3}$ M) compared with that of undissociated CH_3COOH (10^{-1} M). As strong base (OH^-) is added, more acetic acid will dissociate to replace H^+ ions which are removed by OH^- ions. The overall reaction becomes:

$$CH_3COOH + OH^- \rightleftharpoons CH_3COO^- + H_2O$$

Since the amount of CH_3COO^- present initially is so small, the total amount present *during* the titration can be taken as equal to the amount of strong base added. We can then calculate the pH by introducing the appropriate values into the Henderson-Hasselbalch equation, Eqn 8.

Example   What is the pH of a solution of 30 ml of 0·1 M NaOH and 100 ml of 0·1 M acetic acid, given that the pK_a' for acetic acid is 4·73?

Since the variable term in the Henderson-Hasselbalch equation is a *ratio* of concentration values, then it follows that the units of concentration are irrelevant. Concentration can be taken as being proportional to the number of ml of each solution added. Thus $[CH_3COO^-]$ is proportional to the 30 ml of base added and the concentration of the *remaining* undissociated acid is proportional to $(100 - 30)$.

Therefore

$$pH = pK_a' + \log \frac{30}{(100 - 30)}$$
$$pH = 4·73 \log 0·429$$
$$pH = 4·73 + \bar{1}·63$$
$$pH = 4·36$$

Note that the same pH would be obtained if we mixed (i) 30 ml of 0·1 M sodium acetate with 70 ml of 0·1 M acetic acid, or (ii) added 70 ml of 0·1 M HCl to 100 ml of 0·1 M sodium acetate solution.

Once again, an interesting situation is obtained where $[CH_3COO^-]$ = $[CH_3COOH]$, i.e., after the addition of half of an equivalent of NaOH, in that $pH = pK'_a$. This is of importance since it allows us to determine the pK'_a of a given acidic (or basic) group directly from the titration curve.

Finally, a word of caution should be added, in that it must be emphasized that it is only an assumption that the amount of acetate ion (CH_3COO^-) present is equal to the amount of base (OH^-) added. Since some CH_3COOH is dissociated before the addition of base then this cannot be true throughout the whole of the titration curve. For the region where small quantities of NaOH have been added, calculations based on this assumption will give only approximate answers. For similar reasons, accurate pH values close to the equivalence point cannot be calculated by the method described above.

(j) The pH of a solution of a pure salt formed from a weak acid and a strong base

We have seen how the pH of a weak acid and also that of a solution containing both the acid and its salt may be calculated. Briefly, we must now consider how to calculate the pH of a solution of the pure salt (e.g. sodium acetate). Since this is the pH obtained at the end-point of the titration of a weak acid with a strong base there is a tendency to accept that such solutions are neutral. This is, of course, erroneous. End-points (equivalence-points) can be obtained, for different compounds, over a very wide range of pH. For this reason the term titration (and titration curve) is preferred to neutralization (or neutralization curve). Only in the case of the titration of a strong acid and a strong base can the two safely be taken to be synonymous.

The derivation shown below is the final one that we shall be concerned with and it is undoubtedly of far less use to us, as biologists, than either of the two that we have already discussed. It does, however, illustrate some important concepts.

If we add sodium acetate to water there is complete dissociation into CH_3COO^- and Na^+ ions and some of the former will combine with H^+ to form CH_3COOH. Water then dissociates to maintain the constancy of the ionic product (K_w) so that the overall reaction is effectively

$$CH_3COO^- + H_2O \rightleftharpoons CH_3COOH + OH^-$$

The equilibrium constant for this reaction can be written as:

$$K = \frac{[CH_3COOH][OH^-]}{[CH_3COO^-][H_2O]}$$

or, more simply (since water is present at fixed concentration) as the hydrolysis constant

$$K_h = \frac{[CH_3COOH][OH^-]}{[CH_3COO^-]}$$

However, in aqueous solution:

$$[OH^-] = \frac{K_w}{[H^+]} \quad \text{(see Eqn 3)}$$

$$\therefore \quad K_h = \frac{[CH_3COOH][K_w]}{[CH_3COO^-][H^+]}$$

or

$$K_h = \frac{K_w}{K_a}^* \tag{12}$$

We can redefine K_h for a system where the initial concentration of salt (CH_3COO^-) is c and where α is the degree of hydrolysis

$$K_h = \frac{\alpha^2 c}{c(1-\alpha)} = \alpha^2 c$$

or

$$\alpha = \sqrt{\frac{K_h}{c}} \quad \text{(see Eqn 9 for assumptions)}$$

Since

$$[OH^-] = \alpha c$$

$$[OH^-] = c\sqrt{\frac{K_h}{c}} = \sqrt{K_h c}$$

and substituting for K_h (Eqn 12) we obtain

$$[OH^-] = \sqrt{\frac{K_w \cdot c}{K_a}}$$

but

$$[H^+] = \frac{K_w}{[OH^-]} \quad \text{(see Eqn 3)}$$

$$\therefore [H^+] = K_w\sqrt{\frac{K_a}{K_w \cdot c}} = \sqrt{\frac{K_w \cdot K_a}{c}}$$

By taking negative logarithms of this equation

$$-\log[H^+] = -\tfrac{1}{2}\log K_w - \tfrac{1}{2}\log K_a + \tfrac{1}{2}\log c$$

$$\therefore pH = \tfrac{1}{2}pK_w + \tfrac{1}{2}pK_a + \tfrac{1}{2}\log c \tag{13}$$

---

* This can be expressed as $K_a \cdot K_h = K_w$ which is a constant. Since K_a is a measure of the acidity of an acid and K_h is a measure of the basicity of its conjugate base, it follows that the stronger the acid, the weaker is the conjugate base. Thus Cl^-, the conjugate base of a strong acid, HCl, is itself a very weak base. Conversely the conjugate base (CH_3COO^-) of a weak acid (CH_3COOH) is relatively strong.

The way in which we might use this equation is as follows:

Example   What is the pH of 0·1 M sodium acetate given that the pK_a of acetic acid is 4·73?

By substitution in equation

$$pH = \tfrac{1}{2}\cdot 14 + \tfrac{1}{2} 4\cdot 73 + \tfrac{1}{2}(-1)$$

$$pH = 8\cdot 87$$

2.3. Buffer Solutions

By definition, buffered systems are those which show very small changes in pH on addition of either acid or base. They are of interest to us for two reasons. Firstly they are normally included in reaction mixtures when one is studying the metabolism of isolated biological preparations where it is desirable to maintain the pH constant. Their inclusion is made necessary by the fact that many reactions lead to the production of acid or base which would, within a short time, inactivate the enzymes in the pre-paration being studied.

Secondly, the buffer systems within the body are vital to the maintenance of a constant internal environment (homeostasis) in which the various cells and tissues can carry out their functions. Of particular interest are the mechanisms which are used to maintain the pH of the blood as it transports CO_2 from the tissues to the lungs. Other biological fluids, such as the urine and various digestive secretions, also have their characteristic buffer components.

There are many clinical conditions in which altered pH occurs and in which compensatory mechanisms operate to keep these changes to a minimum; a lowered pH is referred to as acidosis and a raised one as alkalosis. A discussion of these conditions is beyond the scope of this book and a simple analysis of them is made difficult by the fact that—to quote one authority*—"there is a bewildering variety of pseudoscientific jargon in medical writing on this subject". However, lowered blood pH tends to be found during respiratory failure where there is a raised partial pressure of CO_2 in the plasma, and also in cases of diabetic ketosis where there is an accumulation of the acidic ketone bodies (See Chapter 8.6). In the latter case there is a greatly increased acidity of the urine due to the presence of the same compounds. Renal failure also leads to a disturbance in the blood buffer systems while repeated vomiting, which causes loss of gastric HCl, can cause a rise in blood pH.

Even in the conditions referred to above, the pH changes which occur do not exceed a few tenths of a pH unit. The way in which the body is

---

* Creese, Niel, Ledingham and Vere, (1962) *The Lancet* Vol. 1, p. 419. This short paper is relevant to this chapter as a whole since it has some very pertinent comments to make about the teaching of acid-base chemistry to medical students.

able to accomplish this buffering, even under adverse conditions, is a fascinating area of study shared by the physiologist, the biochemist and the clinician.

(a) The composition of buffer systems

Considering the subject from a more theoretical viewpoint, we must first define what constitutes a buffer system. Quite simply, it is either a solution of a weak acid and its conjugate base, or a weak base and its conjugate acid. An example of each would be the $CH_3COOH : CH_3COO^-$ system or $CH_3NH_2 : CH_3NH_3^+$ system, respectively.

(b) Buffer capacity

The resistance to pH change of a buffer (its buffer capacity or buffer value) depends upon two factors:

  (i) the *concentration* of the buffer (the total molarity of conjugate acid and conjugate base).
 (ii) the *ratio* of the concentration of conjugate base to that of conjugate acid.

By reference to the Henderson-Hasselbalch equation (8), another way of putting this is to say that the buffering capacity of a given system depends upon its pH as well as upon its strength.

If we look at the titration curve of the acetic acid-sodium acetate system (Fig. 2.4), then we have visual justification for such a statement.

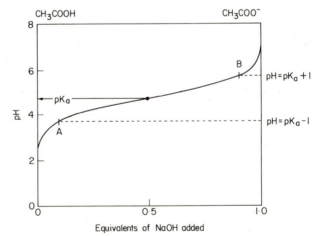

Fig. 2.4 The titration curve of acetic acid.

Here, we can see that the change in pH for the addition of the same amount of OH^- or H^+ is smaller at some stages of the titration than at others. Maximum buffering (represented by the most horizontal portion of the curve) is obtained at the mid-point of the titration, which is where $[CH_3COOH] = [CH_3COO^-]$ or when $pH = pK_a$.

To see why this is so we must first understand what happens to the H^+ and OH^- ions that are introduced into such a system.

(i) On addition of acid:

$$H^+ + CH_3COO^- \rightleftharpoons CH_3COOH$$

(ii) On addition of base:

$$OH^- + CH_3COOH \rightleftharpoons CH_3COO^- + H_2O$$

Since the final pH is given by

$$pH = pK_a + \log\frac{[CH_3COO^-]}{[CH_3COOH]} \quad \text{(see Eqn 8)}$$

it is determined entirely by the ratio of [conjugate base] to [conjugate acid], a ratio of two relatively large concentrations compared with the amount of OH^- or H^+ added. The larger these reservoirs of conjugate base and conjugate acid, and the more nearly equal that they are, the greater will be the buffer capacity of the solutions. Normally and somewhat arbitrarily, a buffer system is considered to be effective only within one pH unit of the pK_a value, i.e. within the limits of a [conjugate base] to [conjugate acid] ratio from 1 : 10 to 10 : 1. These ratios correspond to points A and B on the titration curve of acetic acid (Fig. 2.4).

The actual buffer capacity is defined as the number of moles of strong acid or strong base that are required to change the pH of one litre of solution by one pH unit. It should be noted that, except for a system where pH is equal to pK_a, definition in terms of both acid and base would not give the same numerical value. For example a buffer is more efficient for bases below the pK_a value, and for acids above it.

(c) Titratable acidity

Because of the considerations discussed above, when dealing with bio-logical fluids, particularly urine, it is not enough to know the pH in order to determine "acidity". Since the pH of a buffered solution is determined by the ratio [conjugate base] : [conjugate acid], and effectively is independent of the absolute concentrations of these two species, it follows that a well-buffered urine of pH 5·5 may contain more acidic compounds than a weakly buffered one at pH 5·0. A useful measure of the true or "hidden" acidity is the titratable acidity. This refers to the amount of strong base (NaOH, KOH) required to titrate a particular urine "back" to the blood pH of 7.4.

2.4. Indicators

With the ready availability of pH meters, the practical applications of indicator theory are of limited value. It is included here because a knowledge of the principles involved in the use of indicators is relevant to an understanding of acid-base dissociations in general.

Indicators are themselves weak acids and weak bases with conjugate species that have different coloured forms. Since they undergo titration in any given system to which they are added, their concentration must always be kept as low as possible. Broadly speaking we use them to determine the pH of a solution but the accuracy with which we need to know this pH, and whether this is of primary or secondary consideration, governs our choice of indicator. The two purposes for which we use them are:

(a) *Titration*   Here we are interested only in finding the end-point of a titration, the point at which one equivalent of acid or base has been added. Since at this point there is a rapid change in pH we need not know its actual value accurately: all that is required is for us to use an indicator

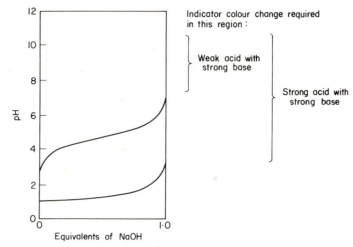

Fig. 2.5 Indicator ranges for the titration of strong and weak acids with a strong base.

which changes colour at a pH value in this region. This means that we must use one that has a pK value (pK_I) at a pH which is representative of the end-point of the titration.

For the titration of a given acid or base we can calculate the pH range that is acceptable as being indicative of the end-point. If, for example, sodium hydroxide is added to a weak acid, then the titration is 50% complete when pH = pK_a. When pH = pK_a + 1, the titration is approximately 90% complete and at pH = pK_a + 3, only 0·1% of the acid remains (i.e., [conjugate base] to [conjugate acid] = 1000). Ideally, therefore, for a titration of this kind we need an indicator that changes colour (has a pK_I) at least three pH units above the pK_a of the acid being titrated. Reference to Fig. 2.5 may make this simpler to understand. Table 2.5 gives examples of some of the indicators that are suitable for various systems.

By using an unsuitable indicator, very misleading results may be obtained. If methyl red were used for the titration of a weak acid and

TABLE 2.5. Choice of indicator for various titrations

System titrated	Indicator	pK$_I$
Strong acid-strong base	Any	—
Weak acid-strong base	Phenol red	7·9
	Thymol blue	8·9
	Phenolphthalein	9·7
Weak base-strong acid*	Methyl red	5·1
	Bromocresol green	4·7
	Bromophenol blue	4·0
Weak base-weak acid*	None suitable	—

* Refer to Fig. 2.3 for these titration curves.

strong base, not only would an indistinct end-point be obtained, some-where in the region of pH 5 to pH 5·5, but at this point, only some 60-70% of the acid would have been titrated.

(b) *Determination of pH*   A comparator method is used in that the colour obtained with an indicator and the sample is compared either with those obtained using reference buffers of known pH or, more commonly, with a colour chart. The latter is the principle of pH indicator papers which frequently contain a number of different indicators, thus spanning a wide pH range.

For this method, we require an indicator which shows maximum colour change (with pH) at a pH as near as possible to that which we are measuring. The reasons for this can be explained by reference to Fig. 2.6.

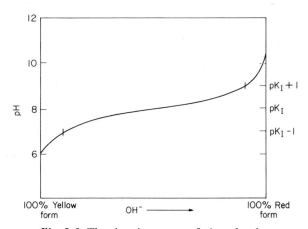

Fig. 2.6 The titration curve of phenol red.

When pH = pK$_I$ + 1 the ratio of the red form of the indicator to the yellow will be 10 : 1, so that at this pH, and above, further colour changes will be difficult to detect. Similarly, at pH = pK$_I$ − 1, the solution will be decidedly yellow and would change colour only slightly on further lowering of the pH. The maximum sensitivity of colour change is found in the region of pK_I.

A useful guide when choosing an indicator for comparator purposes is that it will only be satisfactory for measuring pH within ±1 pH unit of its pK_I.

2.5. Multiple Dissociations

So far, we have used in our examples only compounds which have a single dissociating group. Substances of this kind probably represent only a small proportion of naturally occurring compounds since many of the latter have additional acidic or basic groups. The principles of complex or multiple dissociations are exactly the same as those that we have relied upon when describing the simple dissociation of acetic acid.

A good example to take initially is the tribasic (or triprotic) acid, phosphoric acid. Later, we shall see how the multiple dissociations of the amino acids and proteins can be considered in the same way.

Phosphoric acid has the following structure with each of the hydrogens being dissociable

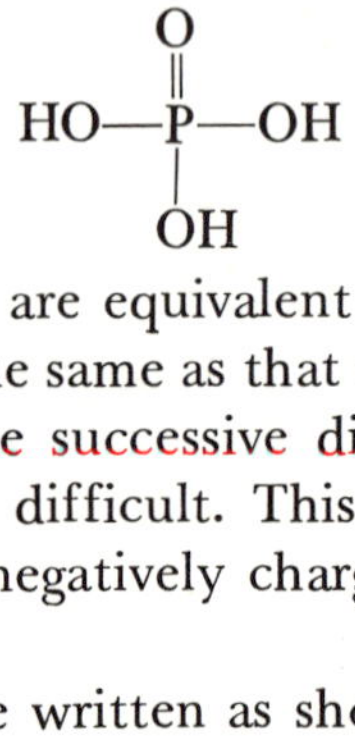

$$\begin{array}{c} O \\ \parallel \\ HO—P—OH \\ \mid \\ OH \end{array}$$

The three hydrogen atoms are equivalent in that the probability of any one of them dissociating is the same as that of the other two. However, as the first hydrogen is lost, the successive dissociation of the second and third protons becomes more difficult. This we might expect since these hydrogens are lost from the negatively charged ions, $H_2PO_4^-$ and HPO_4^{2-}, respectively.

The overall reaction can be written as shown below with the constants K_1', K_2' and K_3' being the apparent dissociation constants for the successive dissociations.

$$H_3PO_4 \underset{K_1'}{\overset{H^+}{\rightleftharpoons}} H_2PO_4^- \underset{K_2'}{\overset{H^+}{\rightleftharpoons}} HPO_4^{2-} \underset{K_3'}{\overset{H^+}{\rightleftharpoons}} PO_4^{3-}$$

The dissociation constants have the following values:

$$K_1' = \frac{[H_2PO_4^-][H^+]}{[H_3PO_4]} = 1{\cdot}1 \times 10^{-2}; \quad (pK_1' = 1{\cdot}96)$$

$$K_2' = \frac{[HPO_4^{2-}][H^+]}{[H_2PO_4^-]} = 1{\cdot}6 \times 10^{-7}; \quad (pK_2' = 6{\cdot}8)$$

$$K_3' = \frac{[PO_4^{3-}][H^+]}{[HPO_4^{2-}]} = 10^{-12}; \quad (pK_3' = 12{\cdot}0)$$

The titration curve obtained when a strong base, such as NaOH, is added to H_3PO_4 is shown in Fig. 2.7. Distinct end-points are recognizable for the titration of the first two dissociations since each is indicated by a relatively sharp rise in pH. However, because pK_3 is quite high (12·0) and close to the pH of the sodium hydroxide being added (pH 13 for 0·1 M NaOH), the end point for the addition of the third equivalent of base cannot be found easily in this way. At different stages of the titration, the various ionic species will predominate as shown.

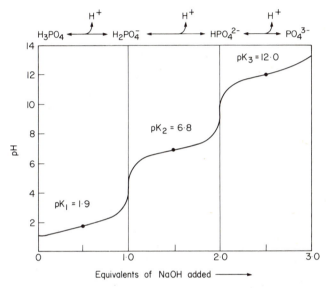

Fig. 2.7 The titration curve of phosphoric acid with NaOH.

Because phosphoric acid has three dissociable hydrogens there are three regions of pH, around the three pK values, where phosphate systems may be used as effective buffers. The most important of these physiologically is the $H_2PO_4^- \rightleftharpoons HPO_4^{2-} + H^+$ system since the pH of many biological fluids lies within ±1 pH unit of the pK' for this dissociation.

If we need to calculate the relative concentrations of phosphoric acid and the various phosphate species present at any pH, we again apply the Henderson-Hasselbalch equation (See Chapter 2.2(g)) using, in turn, the pK' values for the three dissociations. In practice, we are usually concerned only with the determination of the concentrations of two species at any given pH.

For example, at pH 7·4, we can deduce from the titration curve that the major dissociation is that of dihydrogen phosphate to hydrogen phosphate, viz.,

$$H_2PO_4^- \xrightleftharpoons{} HPO_4^{2-} + H^+$$

conjugate acid   conjugate base   ($pK_2' = 6·8$)

since

$$pH = pK + \log \frac{[\text{conjugate base}]}{[\text{conjugate acid}]} \quad \text{(Eqn 8)}$$

$$7 \cdot 4 = 6 \cdot 8 + \log \frac{[HPO_4^{2-}]}{[H_2PO_4^-]}$$

$$\therefore \log \frac{[HPO_4^{2-}]}{[H_2PO_4^-]} = 0 \cdot 6$$

$$\therefore \frac{[HPO_4^{2-}]}{[H_2PO_4^-]} = 4$$

Therefore, at pH 7·4, the ratio of the concentration of hydrogen phosphate to that of dihydrogen phosphate is 4.

However, although these are the major, or predominant, species present at pH 7·4, there will also be small amounts of H_3PO_4 and PO_4^{3-}, since the dissociation of phosphoric acid is written as an *equilibrium* reaction between all species (see above). As one might expect, the concentrations of H_3PO_4 and PO_4^{3-} present at pH 7·4 are very low compared with those of $H_2PO_4^-$ and HPO_4^{2-}. We may calculate them as follows:

For $[H_3PO_4]$, relative to $[H_2PO_4^-]$, using pK_1'

$$7 \cdot 4 = 1 \cdot 96 + \log \frac{[H_2PO_4^-]}{[H_3PO_4]}$$

$$\therefore \log \frac{[H_2PO_4^-]}{[H_3PO_4]} = 5 \cdot 44$$

$$\therefore \log \frac{[H_3PO_4]}{[H_2PO_4^-]} = -5 \cdot 44 \quad (= \bar{6} \cdot 56)$$

$$\therefore \frac{[H_3PO_4]}{[H_2PO_4^-]} = 3 \cdot 6 \times 10^{-6}$$

and for $[PO_4^{3-}]$ relative to $[HPO_4^{2-}]$

$$7 \cdot 4 = 12 \cdot 0 + \log \frac{[PO_4^{3-}]}{[HPO_4^{2-}]}$$

$$\therefore \log \frac{[PO_4^{3-}]}{[HPO_4^{2-}]} = -4 \cdot 6 \quad (= \bar{5} \cdot 4)$$

$$\therefore \frac{[PO_4^{3-}]}{[HPO_4^{2-}]} = 2 \cdot 5 \times 10^{-5}$$

We are now able to give the relative concentrations of all species at pH 7·4:

$$[H_3PO_4] = 3 \cdot 6 \times 10^{-6}$$

$$[H_2PO_4^-] = 1$$

$$[HPO_4^{2-}] = 4$$

$$[PO_4^{3-}] = 1 \times 10^{-4}$$

From this distribution we can see why it is possible, for most purposes, to ignore the concentrations of the minority species.

2.6. The Dissociation Properties of Amino Acids

α-Amino acids are of great importance biochemically, because they are the component "building blocks" of proteins. They are of interest to us, therefore, because their metabolism is an integral part of protein metabolism, and because we must explain all the properties of proteins, including their biological functions, in terms of their amino acid compositions and amino acid sequences.

There are twenty amino acids that are commonly found in proteins and four others that occur less frequently (Chapter 4). They are amphoteric in that each has at least one acidic and one basic group and all of them, with the exception of proline and hydroxyproline (which are imino acids) can be represented by the general formula:

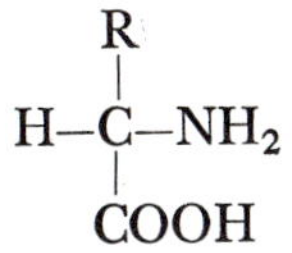

$$
\begin{array}{c}
\text{R} \\
| \\
\text{H-C-NH}_2 \\
| \\
\text{COOH}
\end{array}
$$

Since this structure indicates the presence of an asymmetric carbon atom (Chapter 3) it follows that amino acids, with the exception of glycine where R = H, exist in optically active forms. It is the L form of each amino acid that occurs predominantly in nature and these are the only ones found in proteins; D amino acids are restricted in occurrence mainly to plants and the cell walls of fungi and bacteria, but here again, the L-amino acids are the predominant ones.

From the generalized structure shown above, it can be seen that it is solely the nature of the R group that differentiates one amino acid from another. The functional significance of this group is emphasized still further when the amino acid is part of a protein because the α-carboxyl and α-amino groups form peptide bonds and no longer have their characteristic acid-base properties.

There are a number of ways in which amino acids can be classified depending upon the nature of the R group. This may be aliphatic (linear or branched), aromatic or heterocyclic; in some amino acids it contains sulphur or a hydroxyl group and in others it may have an amino or carboxyl group. At this stage, it is convenient for us to classify amino acids as either neutral, acidic or basic. Thus if R is neutral, the amino acid, having only one carboxyl group and one amino group, is also neutral. However, if R contains an acidic or basic group, the amino acid then has an excess of one type of group over the other and is consequently acidic or basic.

TABLE 2.6. The amino acids found in protein

Neutral	Acidic	Basic
Glycine	Aspartic acid	Lysine
Alanine	Glutamic acid	
Valine		
Leucine		
Isoleucine	Cysteine	Arginine
Serine		
Threonine		
Asparagine		
Glutamine		
Methionine		
Proline	Tyrosine	Histidine
Phenylalanine		
Tryptophan		

Aspartic acid / Glutamic acid } —COOH (carboxyl)

Lysine: —NH₂ (amino) with H$^+$

Cysteine: —SH (thiol or sulphydryl)

Arginine: $^+$H$_2$N=C with NH₂ and NH (guanidino)

Tyrosine: (phenolic —OH)

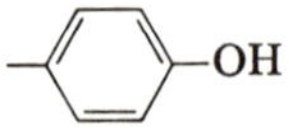

Histidine: (imidazole)

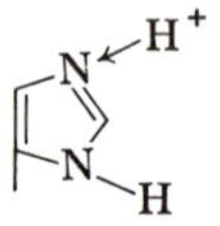

TABLE 2.7. Apparent pK_a values and isoionic points (pl) for some amino acids.
(For each amino acid, the pK_a values are placed in the order pK_1, pK_2 and,
where relevant, pK_3)

Amino acid	α Carboxyl	R group	α Amino	R group	Isoionic point
Neutral					
Glycine	2·34	—	9·60	—	5·97
Alanine	2·34	—	9·69	—	6·01
Serine	2·21	—	9·15	—	5·68
Tryptophan	2·38	—	9·39	—	5·89
Acidic					
Aspartic	1·88	3·65 } (carboxyl)	9·60	—	2·77
Glutamic	2·19	4·35 }	9·67	—	3·22
Cysteine	1·96	8·18 (sulphydryl)	10·28	—	5·07
Tyrosine	2·20	—	9·11	10·07 (phenolic OH)	5·66
Basic					
Lysine	2·18	—	8·95	10·53 (amino)	9·74
Arginine	2·17	—	9·04	12·48 (guanidino)	10·76
Histidine	1·82	6·00 (imidazole)	9·17	—	7·59

(a) Neutral amino acids

When the R group of an amino acid is neutral its composition has little effect on the dissociation properties of either the α-carboxyl group or the α-amino group and the pK values for these dissociations remain relatively constant (Table 2.7). As a result, the titration curves of all the neutral amino acids are almost superimposable and we can consider one of them, alanine, as being representative of them all.

If alanine is dissolved in water, the structure adopted is that of the dipolar ion or zwitterion and the pH of the solution is approximately 6. In this form, each molecule of alanine is isoionic in that it has no net charge and it is also said to be isoelectric since it has no mobility when direct, electric current is passed through the solution.

$$\underset{\underset{H}{|}}{\overset{\overset{CH_3}{|}}{{}^+H_3N-C-COO^-}}$$

The isoionic form of alanine

The titration curve that we obtain when strong acid or strong base is added to a solution of alanine is shown in Fig. 2.8.

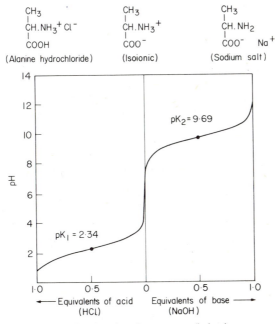

Fig. 2.8 The titration curve of alanine.

The pK'_a values for the α-COOH and α-NH$_3^+$ are pK_1 = 2·34 and pK_2 = 9·69. At a low pH, below pK_1, most of the alanine molecules exist as the fully protonated species carrying one net positive charge. Another way of putting this, is to say that when the hydrogen ion concentration of the solution is high then the tendency of hydrogen to dissociate from the structure is minimal.

The successive dissociations of the hydrogen ions from the fully protonated form can be expressed as shown below:

$$
\begin{array}{ccccc}
\mathrm{CH_3} & & \mathrm{CH_3} & & \mathrm{CH_3} \\
| & \xrightarrow{\;\mathrm{H^+}\;} & | & \xrightarrow{\;\mathrm{H^+}\;} & | \\
\mathrm{CH\cdot NH_3^+} & & \mathrm{CH\cdot NH_3^+} & & \mathrm{CH\cdot NH_2} \\
| & & | & & | \\
\mathrm{COOH} & & \mathrm{COO^-} & & \mathrm{COO^-}
\end{array}
$$

$$\mathrm{p}K_1 = 2\!\cdot\!34 \qquad\qquad \mathrm{p}K_2 = 9\!\cdot\!69$$

The carboxyl group, having the lower $\mathrm{p}K_a$ (i.e., being the more acidic group) dissociates before the protonated amino group. The pH at which a neutral amino acid is isoionic (the isoionic point) lies mid-way between the two $\mathrm{p}K_a$ values, i.e.,

$$\text{Isoionic point} = \frac{\mathrm{p}K_1 + \mathrm{p}K_2}{2}$$

$$\text{for alanine} \quad = \frac{2\!\cdot\!34 + 9\!\cdot\!69}{2} = 6\!\cdot\!01$$

(b) Acidic amino acids

The two most common acidic α-amino acids are aspartic and glutamic. Since both of these have R groups which contain a second carboxyl group, they are relatively strong acids when compared with the other acidic amino acids, the pK values of the —SH group of cysteine and the —OH of

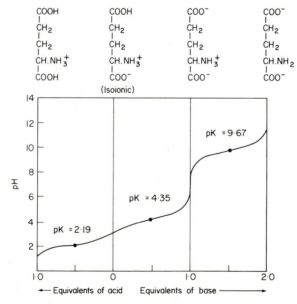

Fig. 2.9 The titration curve of glutamic acid.

tyrosine being much higher than those of carboxyl groups (Table 2.7). It is, therefore, impossible to give a completely "generalized" titration curve for all the acidic amino acids and a similar restriction also applies when we consider the basic amino acids.

The effect of having a dissociating group in the R moiety of the amino acid is that there is an increased lability of the carboxyl proton (an increased tendency for it to dissociate) and the pK_a value is lowered; the α-carboxyl group simply becomes more acidic.

The titration curve of glutamic acid and the dissociation of the fully protonated form is shown in Fig. 2.9.

The isoionic point of glutamic acid, and the other acidic amino acids, again lies mid-way between pK_1 and pK_2. Therefore, for glutamic acid,

$$\text{Isoionic point} = \frac{2{\cdot}19 + 4{\cdot}25}{2} = 3{\cdot}22$$

(c) Basic amino acids

The chemical groups that confer the basicity on the naturally occurring basic amino acids are the ϵ-NH$_2$ group of lysine, the guanidino group of

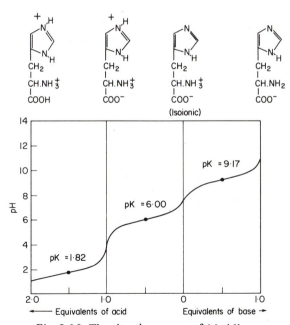

Fig. 2.10 The titration curve of histidine.

arginine and the imidazole group of histidine (Table 2.6, Table 2.7). The last of these three is of particular interest since it is only weakly basic, having a pK_a of 6·0, and therefore exists as a mixture of the protonated and dissociated forms in solutions at physiological pH. The imidazole

groups of the histidine residues of proteins are consequently of significance when we consider the buffering potential of the proteins of certain
body fluids, particularly blood.

As was the case with the acidic amino acids, since different groups are
involved in the R moiety, the titration curves of the basic amino acids are
not superimposable. That of histidine is shown in Fig. 2.10.

For basic amino acids, the isoelectric point lies between pK_2 and pK_3.
For histidine it is calculated as follows:

$$\text{Isoionic point} = \frac{6\cdot00 + 9\cdot17}{2} = 7\cdot59$$

2.7. The Dissociation Properties of Proteins

With some notable exceptions (see Chapter 4.6) protein molecules in
solution occupy the smallest possible volume and thus have a tendency to
be globular. This is shown by studies of viscosity, diffusion and rates of
sedimentation in the ultracentrifuge. For a number of proteins, such as
myoglobin, haemoglobin, lysozyme and ribonuclease, the concept of a
compact structure has been confirmed by X-ray studies of the crystalline
protein. Since protein crystals, unlike crystals of inorganic salts and
smaller compounds, may contain over 50% water, it is reasonable to
assume that the conformation of a protein molecule in a crystal is similar
to that adopted in aqueous solution.

The various forces that maintain protein conformation are described
fully in Chapter 4. The ones that interest us here are of two types; firstly,
those involving interactions between neighbouring R groups of the amino
acid residues and secondly, those resulting from interactions between
certain individual R groups and the aqueous environment. Broadly
speaking, the polar or charged groups on a protein (for example the R
groups of aspartic acid, glutamic acid, lysine, histidine and arginine) tend
to be situated on the outside of the molecule. Since they interact strongly
with water they are said to be hydrophilic. Other R groups which are
uncharged (for example those of alanine, valine, leucine, isoleucine and
phenylalanine) are repelled by the water molecules and tend to group
together in the centre of the protein structure; these groups are said to
be hydrophobic.

Many charged groups, however, because of the opposite polarity of the
charges carried, are able to interact with each other. For example, at
neutral pH the free carboxyl group of a glutamic acid or aspartic acid
residue will be ionized and carry a negative charge, while the basic residues
of lysine and arginine will carry positive charges. Interactions between
appropriately positioned acidic and basic residues can therefore exist as is
shown opposite.

$$\overset{\displaystyle NH}{\underset{\displaystyle CO}{HC}}-CH_2-COO^{-}\cdots{}^{+}H_3N-(CH_2)_4-\overset{\displaystyle CO}{\underset{\displaystyle NH}{CH}}$$

    Aspartic acid residue   Lysine residue

It must be emphasized again that the α-carboxyl and α-amino groups of the amino acids constituting a protein no longer have their acidic and basic properties since, with the exception of the terminal groups, they form the peptide bonds of the polypeptide chain (See Chapter 4.5).

Alteration of the hydrogen ion concentration of a protein solution will affect the degree of dissociation of charged residues and therefore alter the charge density and charge distribution on the outside of the molecule. Many of the original ionic interactions will cease to exist and new ones will be formed. At extremes of pH these may lead to a less ordered, more random structure with a decreased solubility and precipitation may occur.

When a curve is plotted showing the solubility of a given protein at different pH values, it is found that the solubility has a minimum value at a pH corresponding to the isoelectric point of the protein. At this pH, where there is no net charge on the protein, there is a minimum interaction of the charged groups with the solvent and maximum inter- and intra-molecular ionic bonding. These are factors which increase the tendency of the protein to precipitate.

Isoelectric precipitation is occasionally used during protein purification where the differential precipitation of proteins in a mixture can be obtained. The isoelectric points for a number of proteins are given in Table 2.8. The higher the proportion of acidic amino acid residues in a protein then the lower is the isoelectric point. Conversely, a protein such as cytochrome *c* with a high isoelectric point has an excess of basic over acidic residues.

TABLE 2.8. Isoelectric points of some common proteins

Blood Proteins		Miscellaneous Proteins	
Protein	Isoelectric point	Protein	Isoelectric point
α_1 Globulin	2·0	Pepsin	1·0
Haptoglobin	4·1	Ovalbumin	4·6
Serum albumin	4·7	Insulin	5·4
γ_1 Globulin	5·8	Histones	7·5-11·0
Fibrinogen	5·8	Ribonuclease	9·6
Haemoglobin	7·2	Cytochrome c	9·8
γ_2 Globulin	7·4	Lysozyme	11·1

As one might expect, the biological activity of a protein is particularly sensitive to changes in pH. This is perhaps best illustrated where the protein is an enzyme. Large changes in hydrogen ion concentration, affecting the overall conformation of the protein molecule, will have a correspondingly pronounced effect on activity. Smaller changes in pH may affect the state of ionization of functional groups at the active centre of the enzyme and also cause similar changes in the substrate molecule itself. The net result of these effects is that each enzyme exhibits a characteristic pH optimum (p. 152).

(a) Proteins as buffers

Because they are amphoteric and have the characteristics of weak acids and weak bases, proteins have a significant buffering capacity. When hydrogen ions are added to a protein solution, they will associate with either dissociated carboxyl groups or the basic $-NH_2$ or imidazole groups. At neutral pH, it is the imidazole groups of histidine residues that are particularly important since these are the groups with pK_a values in this region.

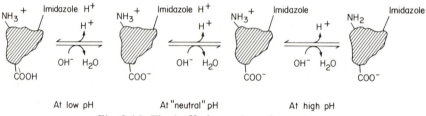

At low pH   At "neutral" pH   At high pH
Fig. 2.11 The buffering action of proteins.

(b) Haemoglobin-oxyhaemoglobin as a buffer system

Although the main function of haemoglobin can undoubtedly be seen to be the carriage of oxygen from the lungs to the tissues, it also plays a significant role in the transport of CO_2 in the reverse direction and in maintaining a constant blood pH as these processes occur.

Carbon dioxide which is released in the tissues during the terminal oxidation of foodstuffs reacts with water according to the reaction shown below:

$$CO_2 + H_2O \rightleftharpoons H_2CO_3$$

The reaction is catalyzed by carbonic anhydrase but the equilibrium position lies very much to the left so that little carbonic acid accumulates. Carbonic acid also dissociates as follows:

$$H_2CO_3 \rightleftharpoons HCO_3^- + H^+$$

Since the pK_a for this dissociation is 6·1, at pH 7·4 (the pH of blood) the ratio of the concentration of bicarbonate ion to that of undissociated H_2CO_3 is approximately 20. The effect of this dissociation is to allow more H_2CO_3 to be formed from CO_2 than would otherwise occur.

From the above, it can be seen that the release of CO_2 at the tissues is synonymous with the production of acid. It can be estimated that in the course of a day, each individual produces some 350-500 l of CO_2, which, taking an average value, is equivalent to 20 moles of H_2CO_3 (20 l of 1 M acid). Mainly owing to the unique properties of haemoglobin, this quantity of acid is transported from the tissues to the lungs with no significant change in blood pH.

Haemoglobin (See Chapter 4.7) is a conjugated protein of molecular weight 68,000 which consists of four subunits of almost identical size. These subunits are of two types, α and β, two of each comprising the haemoglobin molecule. Each subunit is associated with a haem group which has at its centre one atom of iron. This iron is in the ferrous state and does not undergo oxidation to the ferric form as oxygen is transported.

Each subunit of haemoglobin can react with one molecule of oxygen to form oxyhaemoglobin. It is convenient to represent the reaction as:

$$Hb + O_2 \rightleftharpoons HbO_2$$

but we should remember that four molecules of O_2 are carried by the complete haemoglobin molecule.

The oxygen saturation curve of haemoglobin is significant for two reasons, firstly its shape and secondly its response to the presence of carbon dioxide. Both these are shown in Fig. 2.12. The sigmoidal shape of

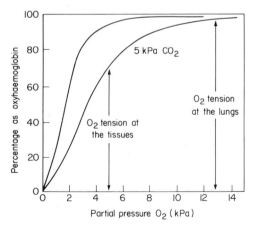

Fig. 2.12 The oxygen saturation curves of haemoglobin in the absence (top) and presence (bottom) of CO_2.

the oxygen saturation curve is indicative of co-operative interactions between subunits and in this respect it is similar to the velocity-substrate concentration curve obtained for an allosteric enzyme (See Chapter 5.11). It is possible to calculate the rate constants for the binding of the four separate O_2 molecules to haemoglobin and it is then apparent that the affinity for O_2 becomes progressively greater as successive O_2 sites are filled.

The effect of CO_2, most pronounced at low partial pressures of O_2, is to decrease the amount of haemoglobin which is present in the oxygenated form. The physiological significance of this has been appreciated for many years. The partial pressure of O_2 in the blood at the lungs is approximately 13 kPa (97 mm Hg) so that under these conditions haemoglobin becomes fully oxygenated before passing to the tissues. In the tissues the oxygen tension is about 5 kPa (35-40 mm Hg) and, because of this, oxyhaemoglobin will release O_2 simply as a result of mass action effects. However, this tendency of oxyhaemoglobin to become reduced is further increased by the presence of CO_2 (normally about 6·1 kPa, 46 mm Hg, in the tissues). In this way, both a low partial pressure of O_2 and a high partial pressure of CO_2 operate together to liberate O_2 so that it becomes available for tissue respiration. It should be noted that CO_2 has a minimal effect on the haemoglobin saturation curve in the regions where the oxygen tension is high. Thus the formation of oxyhaemoglobin at the lungs goes virtually to completion even in the presence of a significant concentration of CO_2.

The carbon dioxide released in the tissues, in addition to forming carbonic acid, reacts with amino groups of proteins to form carbamino derivatives according to the following equation:

$$\text{Protein-NH}_2 + CO_2 \rightleftharpoons \text{Protein-NH} \cdot COO^- + H^+$$

The reaction is a general one so that the plasma proteins, as well as haemoglobin, are involved. Haemoglobin, however, being the most abundant protein in blood is quantitatively the more significant. In venous blood, approximately 7% of the CO_2 is in the form of carbamino compounds and 5% is in the dissolved form. The remainder is present mainly as bicarbonate, most of which is in the plasma rather than the erythrocytes.

Haemoglobin is particularly important in CO_2 transport because both its ability to form carbamino compounds and its buffering capacity are very much dependent upon whether it is carrying oxygen. The carbamino capacity of reduced haemoglobin is about three times as high as that of oxyhaemoglobin. Because of this, and the consequent rapid release of CO_2 from carbamino groups as haemoglobin is oxygenated, it has been estimated that, at rest, as much as 25-30% of the respired CO_2 is transported from tissues to lungs in this form.

Of fundamental significance in blood buffering is the contribution made by haemoglobin itself. The pK_a value for the dissociation of the imidazole groups of certain histidine residues in the haemoglobin molecules increases from approximately 6·6 for oxyhaemoglobin to 8·2 when oxygen is released (Fig. 2.13). Haemoglobin (Hb), therefore, is less

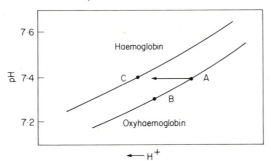

pKₐ = 8.2 for haemoglobin.

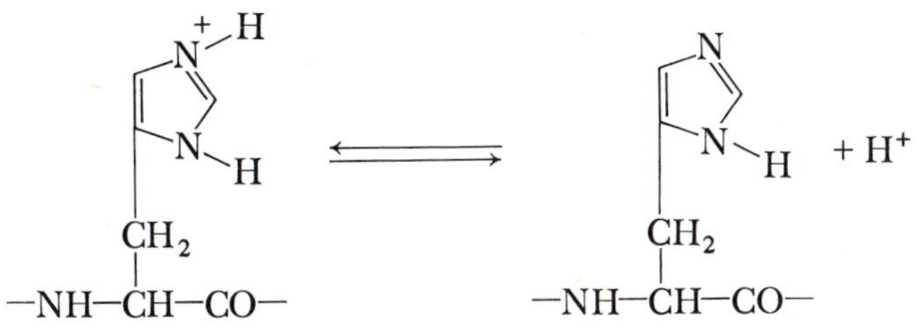

pK_a = 8.2 for haemoglobin.

pK_a = 6.6 for oxyhaemoglobin.

Fig. 2.13 Dissociation of key imidazole residues in haemoglobin and oxyhaemoglobin.

acidic than oxyhaemoglobin (HbO_2) and is consequently the better buffer of the two for the H^+ formed in the tissues both from carbonic acid and during the synthesis of carbamino compounds. The physiological importance of this transition can be appreciated by reference to Fig. 2.14 which shows, in simplified form, regions of the titration curves of haemoglobin and oxyhaemoglobin.

Fig. 2.14 Simplified titration curves of haemoglobin and oxyhaemoglobin in the region of pH 7·4.

If we consider point A on the HbO_2 curve, a point corresponding to pH 7·4, then as hydrogen ions are added to the system, instead of the pH falling in the direction of B, conversion of HbO_2 to Hb allows the pH to remain constant and effectively move along the line, shown by the arrow, towards C. This phenomenon is known as the isohydric shift. Under physiological conditions, a slight fall in pH will occur (from 7·40 to approximately 7·37) so that the actual pH change will follow a line which lies just below AC.

2.8. Electrophoresis and Ion Exchange Chromatography

(a) Electrophoresis

When a charged molecule (or ion) in solution is subjected to an electric current, it migrates to either the cathode or anode depending upon the nature of the charge carried. This phenomenon forms the basis of

electrophoresis, a technique that has proved to be invaluable in the
analysis of many cell constituents and derivatives, particularly amino
acids, peptides, proteins and nucleotides. It is also occasionally used on a
preparative scale for the purification of some of these substances.

To minimize diffusion and convections, the electrophoresis buffer is
normally held stationary in a support medium. The simplest of these is
buffered filter paper which will allow an excellent separation of amino
acids and which can be used with limited success for proteins. Cellulose
acetate, starch gel or acrylamide gel are undoubtedly better for the latter.
Acrylamide gel is particularly attractive since it is transparent and, because
an element of molecular sieving is introduced, the resolution that can be
obtained is extremely high.

In order to detect the presence of amino acids and proteins after
electrophoresis, they must be treated with an appropriate reagent so that
visible bands or spots are obtained. The principles of electrophoresis are
shown in Fig. 2.15 where the correlation between mobility and charge is
shown for a neutral, an acidic and a basic amino acid.

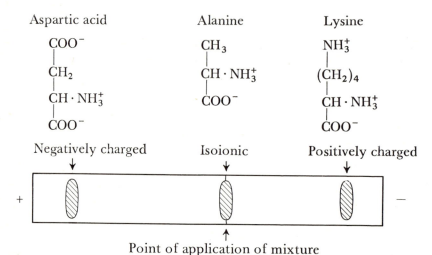

Fig. 2.15 An example of the separation of amino acids by paper electrophoresis at
pH 6·0.

At pH 6 it is possible to separate the basic and acidic amino acids
leaving the neutral ones stationary at the point of application. Mixtures of
neutral amino acids are best resolved by electrophoresis at a low pH (e.g.,
pH 1·8) where they all migrate towards the anode with slightly different
mobilities.

The electrophoretic analysis of plasma proteins is carried out routinely
in clinical laboratories for the screening of a number of pathological
conditions, particularly for multiple myeloma, agammaglobulinaemia,
macroglobulinaemia and the nephrotic syndrome. In addition, many
genetic variants of haemoglobin are detectable in this way.

(b) Ion-exchange chromatography

Charged molecules may frequently be separated from each other by their differential elution from a column of insoluble material to which they are ionically bound. For the separation of negatively charged ions (anions) a positively charged matrix is used and the process is said to be one of anion exchange. The elution of substances from a column of this kind is brought about either by lowering the pH of the eluting buffer, thereby decreasing the extent of the charge on the molecule, or by increasing the concentration of a competing ion (e.g., CH_3COO^-) which displaces other anions bound to the column. Positively charged ions (cations) are separated by cation exchange chromatography on a negatively charged matrix.

TABLE 2.9. The separation of amino acids and proteins by ion exchange chromatography; examples of commonly used materials

Material	Functional Group	Type and Uses
Cross-linked polystyrene carrying quaternary ammonium groups	$-\langle\ \rangle-\overset{+}{N}(CH_3)_3$	Strongly basic anion exchanger. Separation of amino acids as R-COO$^-$
Diethylaminoethyl (DEAE) cellulose. Aminoethyl (AE) and triethylaminoethyl (TEAE) cellulose also used.	$CH_2O \cdot C_2H_4\overset{+}{N}H(C_2H_5)_2$	Weakly basic anion exchanger. Separation of proteins at a pH above their isoelectric points.
Cross-linked sulphonated polystyrene	$-\langle\ \rangle-SO_3^-$	Strongly acidic cation exchanger. Separation of amino acids as R-NH$_3^+$
Carboxymethyl (CM) cellulose	$CH_2O \cdot CH_2COO^-$	Weakly acidic cation exchanger. Separation of proteins at a pH below their isoelectric points. Particularly useful for basic proteins.

Since amino acids and proteins may be either negatively or positively charged, depending on whether the pH of the solution is above or below the isoelectric point, they may be separated from mixtures either by cation or anion exchange chromatography. For proteins, however, which are predominantly negatively charged at neutral pH, anion exchange is the more frequently used method. In this respect, diethylaminoethyl (DEAE) cellulose has been invaluable; carboxymethylcellulose (CM-cellulose) is used for the separation of proteins at a pH below their isoelectric points. The properties of some useful exchange materials are given in Table 2.9.

3

Structure and Properties of Carbohydrates and Nucleotides

3.1. The Nature of Carbohydrates

Carbohydrates form a large class of substances which contain the elements carbon, hydrogen and oxygen, the last two frequently being in the ratio of $2 : 1$. They can be divided into two groups, the low-molecular-weight carbohydrates or sugars and the high-molecular-weight carbohydrates or polysaccharides, the latter consisting of simple sugar units linked together in chains. Additionally, there are numerous derivatives of carbohydrates that contain other elements such as nitrogen and phosphorus and which will be considered together with the carbohydrates proper.

The importance of the carbohydrates derives from their varied roles in metabolism. The sugars and polysaccharides provide a major source of dietary energy. The degradation of a particular monosaccharide— glucose—is one of the central metabolic pathways for producing energy in mammals. Polysaccharides are additionally important as stores of energy and their synthesis is a vital metabolic process. Other polysaccharides are essential structural macromolecules. Moreover, many of the co-factors and intermediates in metabolic pathways are sugar phosphates such as the nucleotides which contain a nitrogenous moiety attached to a sugar phosphate. The nucleotides are also the low-molecular-weight components of the nucleic acids.

Any discussion of the carbohydrates starts with the monosaccharides since these substances form the basis of all the other carbohydrates.

3.2. Monosaccharides

Monosaccharides are substances whose molecules consist of a short chain of carbon atoms (up to seven), with no carbon-carbon double bonds. The sugars can have an open chain structure which is in equilibrium with a cyclic form in which the ring contains one oxygen atom and four or five carbon atoms (see later). Although the cyclic form predominates, it is easier to appreciate the chemical properties of sugars if the open chain form is considered first. In this form, all the oxygen atoms are present as hydroxyl groups except one, which is present as a carbonyl group. There are two classes of monosaccharides, aldoses in which the carbonyl group is present on a terminal carbon atom, and the compounds are, therefore, aldehydes, and ketoses where the carbonyl group is not terminal and the compounds are ketones. In both types all the remaining oxygen atoms are in alcoholic groups. Examples of five-carbon monosaccharides are shown.

$$
\begin{array}{ll}
1 \quad C{<}^{H}_{O} & CH_2OH \quad 1 \\
2 \quad CHOH & C{=}O \quad 2 \\
3 \quad CHOH & CHOH \quad 3 \\
4 \quad CHOH & CHOH \quad 4 \\
5 \quad CH_2OH & CH_2OH \quad 5
\end{array}
$$

$$\text{Aldopentose} \qquad\qquad \text{Ketopentose}$$

In these formulae, we have introduced several conventions. The five-carbon sugars are called pentoses; the corresponding names for sugars containing different numbers of carbon atoms are trioses (for 3), tetroses (for 4) and hexoses (for 6). Moreover, the chain of carbon atoms has been written vertically with the CHO-group (in the case of the aldose) or the CH_2OH nearest the $C{=}O$ (in the case of the ketose) at the top. The carbon atoms are then numbered from the top downwards.

Thus the chemistry of the monosaccharides is the chemistry of alcohols and of carbonyl compounds. It is somewhat more complicated than the chemistry of such substances as ethanol, acetaldehyde and acetone for two reasons. One is that monosaccharides frequently contain more than one optically active (asymmetric) carbon atom in the same molecule, and the other is that they are substances with several functional groups in the same molecule. We shall consider first the isomerism exhibited by these substances by discussing the isomers of the aldoses systematically.

3.3. The Simple Aldoses

The simplest aldose theoretically is hydroxyacetaldehyde (or glycollic aldehyde) which we could regard as a "diose".

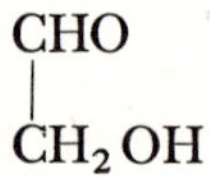

$$\begin{array}{c} CHO \\ | \\ CH_2OH \end{array}$$

Hydroxyacetaldehyde

In fact, this substance which is of no metabolic importance, is not regarded as a monosaccharide but rather as a chemical curiosity—it is extremely reactive and is a powerful reducing agent.

The first "real" aldose is glyceraldehyde (aldotriose):

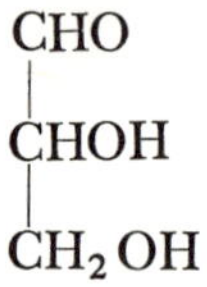

$$\begin{array}{c} CHO \\ | \\ CHOH \\ | \\ CH_2OH \end{array}$$

Glyceraldehyde

Glyceraldehyde exhibits the simple optical isomerism characteristic of, say, lactic acid. The middle carbon atom (C-2) is surrounded by four different groups (CHO, H, OH, and CH_2OH) and as a consequence of the tetrahedral configuration of the bonds around a saturated carbon atom, there are two possible structures that are non-superimposable mirror images of one another. Such pairs of substances are characterized by their solutions rotating the plane of polarized light in opposite directions. The substance that rotates the plane in a right-handed direction is called dextrorotatory (+) and the one which rotates it in a left-handed direction, laevorotatory (−). In the formulae, we have maintained the convention of drawing the carbon skeleton vertically, so the figures are a diagrammatic way of illustrating the view of the tetrahedron with the "H—OH edge" nearest the reader.

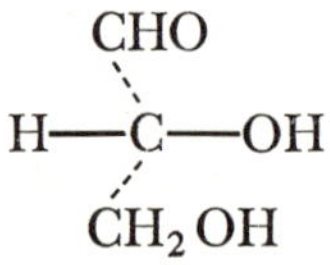

CHO	CHO
H—C—OH	HO—C—H
CH_2OH	CH_2OH
+ Glyceraldehyde	− Glyceraldehyde

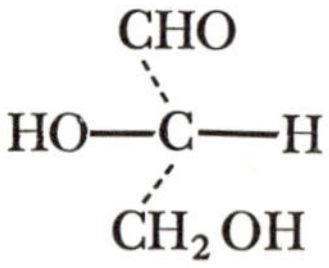

After glyceraldehyde, the next more complex aldoses are the tetroses. There are altogether four such sugars; their structures are as follows:

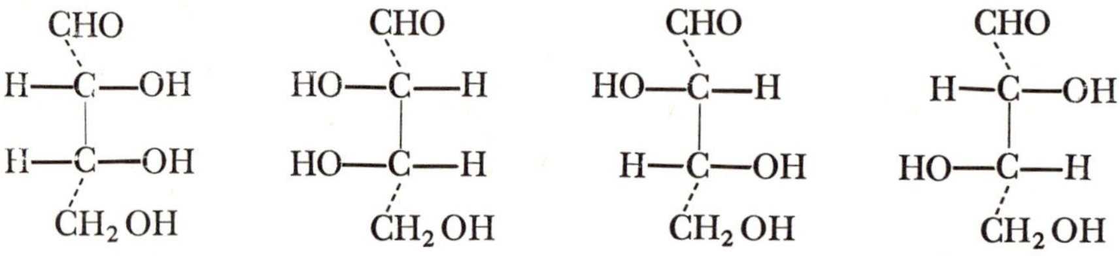

CHO	CHO	CHO	CHO
H—C—OH	HO—C—H	HO—C—H	H—C—OH
H—C—OH	HO—C—H	H—C—OH	HO—C—H
CH_2OH	CH_2OH	CH_2OH	CH_2OH
I	II	III	IV

Tetroses

These four structures introduce several new stereochemical features. First of all, let us consider the ways in which they are related to one another. Substance I is related to II in the same way that + glyceraldehyde is related to − glyceraldehyde. That is to say, the one structure is a non-superimposable mirror image of the other. Pairs of substances such as + and − glyceraldehyde, or I and II (and also III and IV) are said to be "enantiomers". Two enantiomers have chemical and physical properties identical to one another with one exception. All asymmetric molecules are optically active; therefore, in solutions of equal strength, I will rotate the plane of polarized light to an equal extent but in the opposite direction to II. As with all enantiomeric substances (including glyceraldehyde and lactic acid) an equimolar mixture of I and II will be optically inactive (will not rotate the plane of polarized light at all). Such mixtures are called "racemates" or "+/− mixtures". So much for I and II (and likewise III and IV). What is the structural relationship that exists between I and III? We could generate an image of the structure of III by looking at the mirror image of the top half of molecule I and keeping the bottom half the same. That is to say, C-3 in III has the same configuration as C-3 in I, but C-2 has the opposite (mirror-image) configuration. No matter how you twist the molecules around, III is non-superimposable on I and it is non-superimposable on its mirror image (II). Thus I and III are not the same substance, nor are they enantiomers of one another. Rather they demonstrate a special case of stereoisomerism which occurs only in substances which contain more than one optically active (asymmetric) carbon in the same molecule. I and III have different (but not equal and opposite) specific rotations; the relative configurations of the substituents are absolutely different and the other physical properties (such as solubility) differ slightly. Likewise the chemical reactivities of the hydroxyl groups vary slightly between I and III. Clearly II and IV are related in exactly the same way as I and III. Thus I, II, III and IV comprise two separate enantiomeric pairs.

All asymmetric molecules that occur in living cells are optically resolved; cells do not synthesize racemates. Moreover, among the mono-saccharides, the enantiomers that occur are invariably those that have the same configuration at the next-to-last carbon (C-3 in tetroses) as does + glyceraldehyde. Thus I and III are the two "natural" enantiomers among the four tetroses. We refer, therefore, to I and III as "D-sugars". In this context, D stands for dextro (+) glyceraldehyde and refers to the absolute configuration of the next-to-last carbon atom. The molecular parameters that determine whether an asymmetric molecule will be dextro- or laevo-rotatory are complex (and outside the scope of this book); it does not follow that every D-sugar will be dextrorotatory, except in the case of glyceraldehyde, where D-glyceraldehyde = (+)glyceraldehyde, by definition. The "unnatural" sugars (II and IV above) are referred to as L-sugars (L standing for laevo (−)glyceraldehyde).

We are now able to construct a table that illustrates the structures of the D-aldoses. Before doing so, the convention already described for glyceraldehyde can be simplified. The structure of (for example) D-glyceraldehyde can be represented in such a form that the molecule appears as a flat projection of its tetrahedral configuration thus:

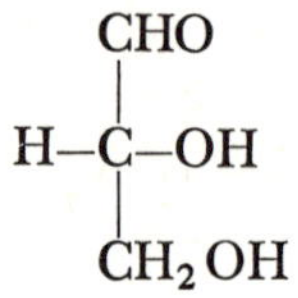

$$CHO$$
$$|$$
$$H-C-OH$$
$$|$$
$$CH_2OH$$

D-glyceraldehyde

In practice, the central carbon atom is not drawn in and the hydrogen is omitted. (It is the usual practice in drawing organic formulae to assume that the valency requirements of carbon are "satisfied" with hydrogen atoms that are usually left out of the structure.) Thus D-glyceraldehyde becomes:

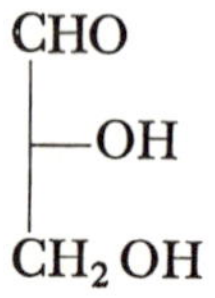

$$CHO$$
$$|$$
$$\!\!-OH$$
$$|$$
$$CH_2OH$$

D-glyceraldehyde

Structure I (for the tetrose; D-erythrose) becomes

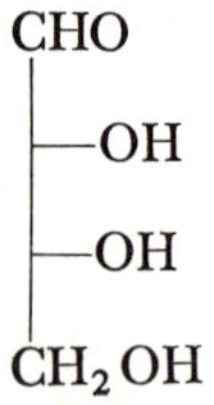

$$CHO$$
$$|$$
$$\!\!-OH$$
$$|$$
$$\!\!-OH$$
$$|$$
$$CH_2OH$$

D-erythrose

When looking at these conventional pictures of such structures, it is important to remember that we are "seeing" the molecule from a particular point of view and with the C–C–C groups in a particular orientation. This is particularly important in the case of a sugar with a chain longer than 3 (like the erythrose above). As the C–C single bonds are axes of free rotation, this latter point is the same as saying that the convention involves looking at only one of the infinite number of shapes that the molecule can assume (as a result of such rotation occurring). Thus

D-erythrose could theoretically be represented as:

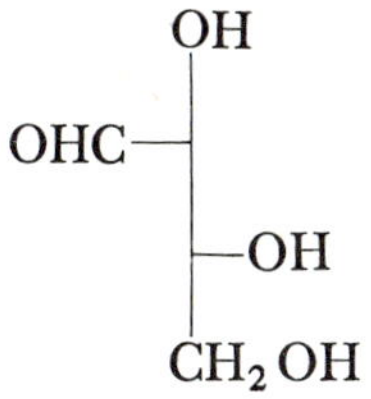

$$
\begin{array}{c}
\text{OH} \\
\mid \\
\text{OHC}\!\!-\!\!\!\!\!\begin{array}{c}\; \\ \text{} \\ \end{array} \\
\mid \\
-\text{OH} \\
\mid \\
\text{CH}_2\text{OH}
\end{array}
$$

D-erythrose

This structure is the "flattened" version of:

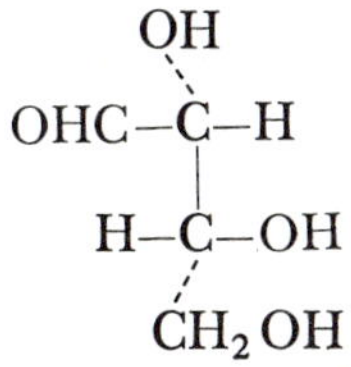

$$
\begin{array}{c}
\text{OH} \\
\mid \\
\text{OHC}-\text{C}-\text{H} \\
\mid \\
\text{H}-\text{C}-\text{OH} \\
\mid \\
\text{CH}_2\text{OH}
\end{array}
$$

which can be generated from the structure of D-erythrose (shown in the formulae for the four tetroses by turning the top part of the molecule (like a three bladed propeller) around the axis of free rotation that extends through C-2 and C-3.

Having got this far we can look at the D-aldoses. Their formulae (together with the names of the important ones) are shown in Fig. 3.1.

Before proceeding to discuss other types of sugars and the other "complication" in sugar chemistry let us be quite clear what the diagrams in the preceding table really mean. They are all conventional "flattened" pictures of the configurations of the molecules in the particular conformation already defined. This is not the "major" conformation of the sugars. To take the special case of glucose, which is the most familiar sugar in the table, you might guess that (taking into account the tetrahedral array of bonds round each carbon atom) the most stable conformation is the one in which the C–C backbone of the molecule forms a "linear" zig-zag. We shall see later that this guess is in fact erroneous; however, this conformation has the following structure:

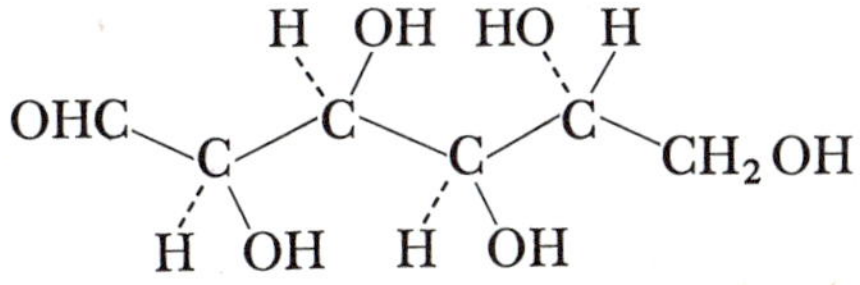

$$
\begin{array}{c}
\text{H} \quad \text{OH} \quad \text{HO} \quad \text{H} \\
\text{OHC}\diagdown\,\text{C}\diagup\,\text{C}\diagdown\,\text{C}\diagup\,\text{C}\diagdown\text{CH}_2\text{OH} \\
\text{H} \quad \text{OH} \quad \text{H} \quad \text{OH}
\end{array}
$$

Perhaps this fact is self-evident. If it is not, it is essential that the reader convinces himself by writing down the various structures, that this is the case so that the features of a molecule (such as glucose) which contains several asymmetric carbon atoms can be recognized. These substances are of vital importance in understanding the molecular achitecture of living cells.

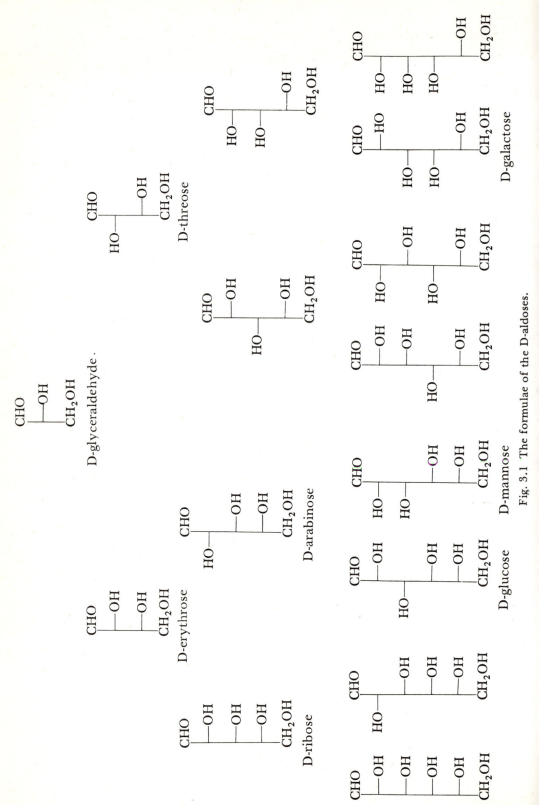

Fig. 3.1 The formulae of the D-aldoses.

3.4. Naturally Occurring Substances Simply Related to the Aldoses

In the following discussion, C-1 is the "top" carbon atom and C-*n* is the "bottom" carbon atom (C-6 for a hexose such as glucose). The examples that are used to illustrate the nomenclature of these compounds are derived from glucose (with one exception to be mentioned shortly).

Three groups of acids (related to the aldoses) occur. Those in which C-1 is present in an acidic function ($-CO_2H$) as opposed to the aldehyde ($-CHO$) of the aldoses are called aldonic acids. Those in which C-1 is $-CHO$ but C-*n* is $-CO_2H$ as opposed to the primary alcohol ($-CH_2OH$) of the aldoses are called alduronic acids. The dibasic acids (both C-1 and C-*n* are $-CO_2H$) are the aldaric acids. The suffixes (-onic, -uronic and -aric) apply to the nomenclature of the particular acids. Thus the five-carbon aldonic acids are called the pentonic acids; the six-carbon ones the hexonic acids. Similarly the individual acids are named after their parent aldose as seen below (for those related to glucose).

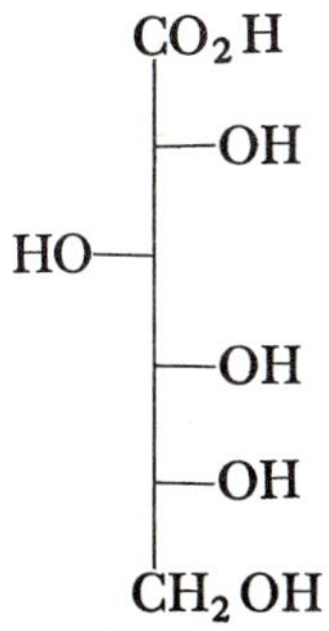

CO_2H	CHO	CO_2H
—OH	—OH	—OH
HO—	HO—	HO—
—OH	—OH	—OH
—OH	—OH	—OH
CH_2OH	CO_2H	CO_2H
Gluconic acid	Glucuronic acid	Glucaric acid

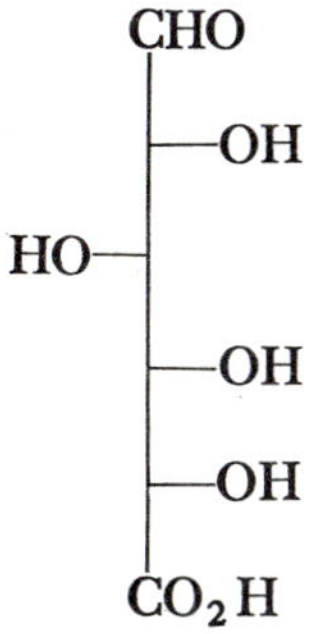

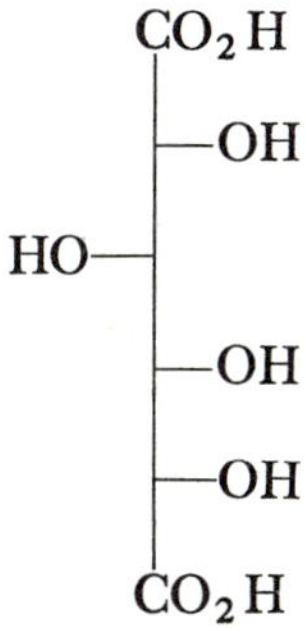

The acids can be regarded as oxidation products of the aldoses. Indeed gluconic acid is a product of the oxidation of glucose by familiar "aldehyde reagents" such as Fehling's solution or ammoniacal silver oxide. The other simple derivatives are the reduction products of the aldoses, the alditols. The alditols are polyhydric alcohols; they have the group $-CH_2OH$ at C-1. The hexitol derived from glucose is shown below. For reasons not discussed here it is not usually called "glucitol" but rather sorbitol.

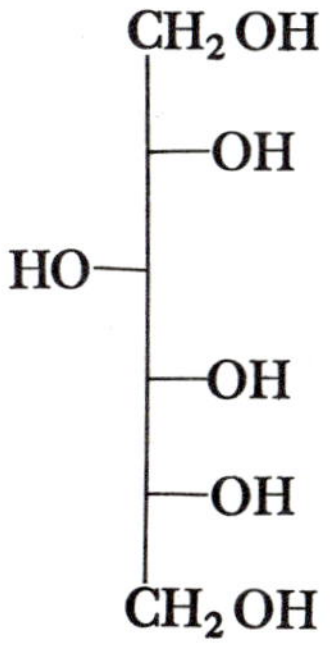

$$CH_2OH$$
$$-OH$$
$$HO-$$
$$-OH$$
$$-OH$$
$$CH_2OH$$

Sorbitol

Before proceeding to other aspects of sugar chemistry, it is worth just pausing to consider the isomerism of the alditols and aldaric acids. They differ from the aldoses in that the two ends of the molecules are the same. Consequently there are fewer optically active centres and fewer stereo-isomers. The familiar example is the aldaric acids of the tetroses (which are known by the trivial name of the "tartaric acids"). At first sight we might expect there to be a total of four tartaric acids corresponding respectively to D- and L-erythrose and D- and L-threose.

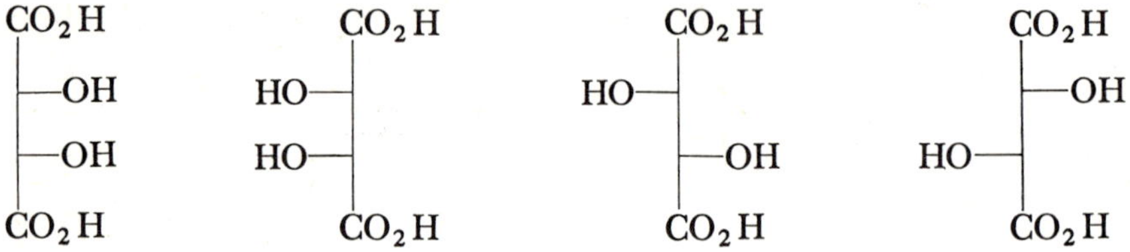

$$
\begin{array}{cccc}
CO_2H & CO_2H & CO_2H & CO_2H \\
-OH & HO- & HO- & -OH \\
-OH & HO- & -OH & HO- \\
CO_2H & CO_2H & CO_2H & CO_2H \\
I & II & III & IV
\end{array}
$$

Four possible structures for tartaric acids

Structures I and II are different views of the same molecule (the inactive meso tartaric acid). III and IV are non-superimposable and are D- and L-tartaric acid.

The other important simple natural "derivatives" of aldoses are the amino sugars. One of these compounds, glucosamine (strictly 2-deoxy 2-amino D-glucose), is of great importance in several structural polysaccharides. The structure of glucosamine is shown. The substituent at C-2 is $-NH_2$ (replacing the $-OH$ of glucose).

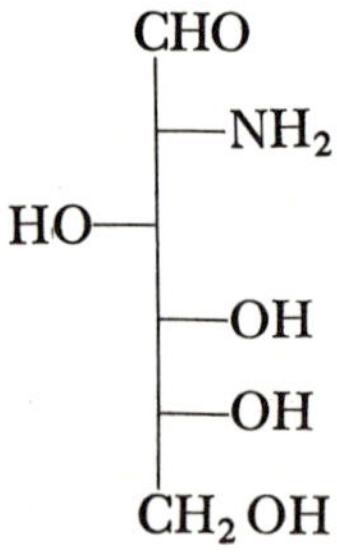

$$
\begin{array}{l}
CHO \\
-NH_2 \\
HO- \\
-OH \\
-OH \\
CH_2OH
\end{array}
$$

Glucosamine

3.5. Introduction to Ketoses

At first sight, the ketoses form a much more complicated pattern of possible structures than do the aldoses. The carbonyl group can occur in any one of a number of positions (C-2, C-3, C-4 or C-5 in ketohexoses). In practice, the only ketoses of metabolic importance are the 2-ketoses (with the C=O attached to the carbon atom next to the end of the chain). Moreover, there are only two of these whose structures require comment at this stage.

The simplest ketose (ketotriose) is dihydroxyacetone

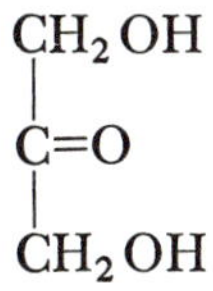

$$\begin{array}{l} CH_2OH \\ | \\ C=O \\ | \\ CH_2OH \end{array}$$

Dihydroxyacetone

Notice that when compared with the aldotriose, glyceraldehyde, there is one important difference; dihydroxyacetone has no asymmetric carbon atom. There is only one (enantiomorphic) stereoisomer of ketotetrose; only two ketopentoses and only four ketohexoses. One of these is our "other ketose", fructose. The structure of D-fructose is shown.

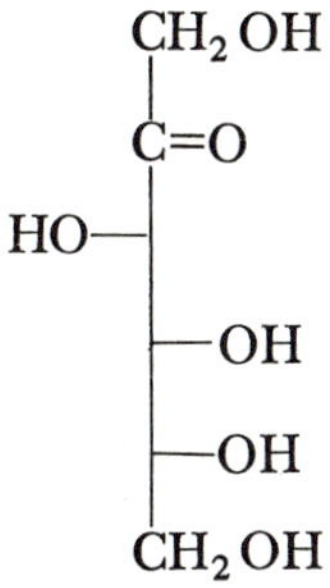

$$\begin{array}{l} CH_2OH \\ | \\ C=O \\ HO-| \\ -OH \\ -OH \\ CH_2OH \end{array}$$

D-fructose

Notice that this sugar (which is in fact laevorotatory) is named D-fructose for the same reason that D-glucose is so named; the configuration at C-5 is the same as that at the optically active centre in D-glyceraldehyde.

3.6. The Reactions of Aldoses and Ketoses

Aldoses and ketoses contain carbonyl groups and alcohol groups. We therefore expect them to have the chemical properties associated with these classes of compounds. Insofar as they are alcohols we expect them to have the reactions characteristic of alcohols, for example the formation of esters with acids. The esters of the monosaccharides are of enormous importance in metabolism. We shall leave them until the reactions of the carbonyl groups have been considered for a reason which becomes clear later.

The chemical properties of carbonyl compounds can be summarized as follows. By way of illustration, the simple compounds acetaldehyde (CH_3CHO) and acetone ($(CH_3)_2CO$) are used.

Condensation reactions are typical of both aldehydes and ketones.

They involve the elimination of water between a carbonyl group and certain reagents such as substituted hydrazines. Thus acetone reacts with phenylhydrazine to produce acetone phenylhydrazone.

$$
\begin{array}{c}
CH_3 \\
| \\
C=O + H_2N-NHC_6H_5 \\
| \\
CH_3
\end{array}
\longrightarrow
\begin{array}{c}
CH_3 \\
| \\
C=N-NHC_6H_5 + H_2O \\
| \\
CH_3
\end{array}
$$

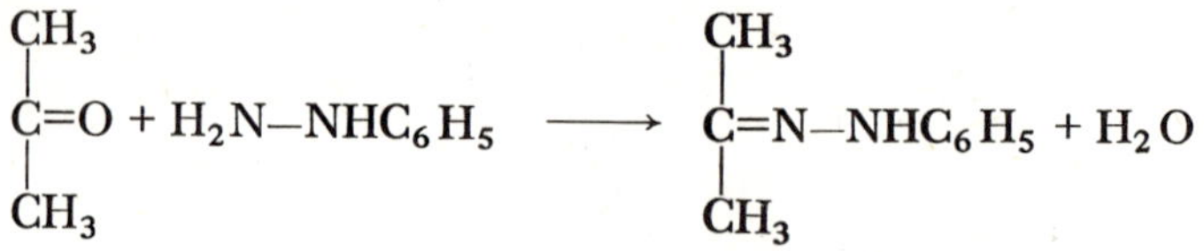

Formation of acetone phenylhydrazone

Addition reactions are also typical of both aldehydes and ketones. These reactions involve the reaction between the carbonyl compound and the reagent with the production of an addition product without the elimination of a small molecule. The reaction reflects the unsaturated character of the double bond between the atoms of the carbonyl group. As an example, acetaldehyde reacts with hydrogen cyanide to yield acetaldehyde cyanohydrin.

$$
CH_3-C{\Large\langle}{\substack{H \\ O}} + HCN \rightleftharpoons
\begin{array}{c}
H \\
| \\
CH_3-C-OH \\
| \\
CN
\end{array}
$$

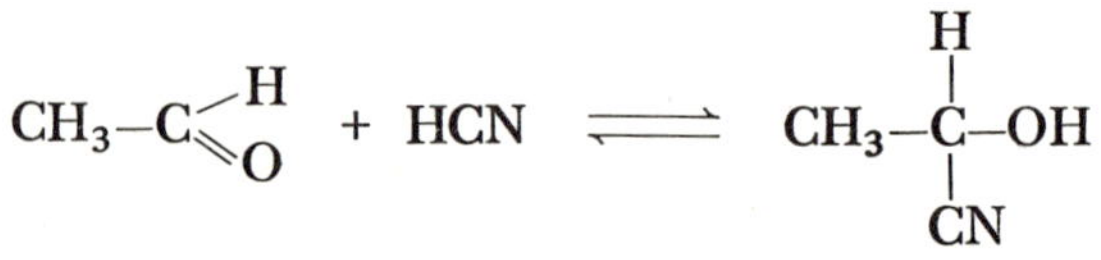

Formation of acetaldehyde cyanohydrin

We shall return to this reaction shortly as we shall see that it is analogous to one of the central reactions of monosaccharide chemistry. For the moment note that the product (acetaldehyde cyanohydrin) contains an asymmetric carbon atom. As this product has been formed from two inactive reagents, the racemic (+/−) mixture will be formed. The mixture cannot be resolved into its + and − components for (as shown) the addition is reversible and the pure cyanohydrin is difficult to isolate. However, if the product is hydrolysed (to convert the −CN group into $-CO_2H$) the product (racemate of lactic acid) can be resolved into the + and − enantiomers.

Aldehydes but not ketones are strong reducing agents. Thus they will reduce ammoniacal solutions of silver oxide to silver (the "silver mirror test"). The reason is that aldehydes (but not ketones) are readily oxidized to carboxylic acids. The reduction of silver oxide by acetaldehyde to produce acetic acid is illustrated below.

$$
Ag_2O + CH_3CHO \longrightarrow 2Ag + CH_3CO_2H
$$

Reduction of silver oxide by acetaldehyde

Some of the reactions of carbonyl compounds are only explicable in terms of the keto/enol equilibrium (illustrated for acetaldehyde):

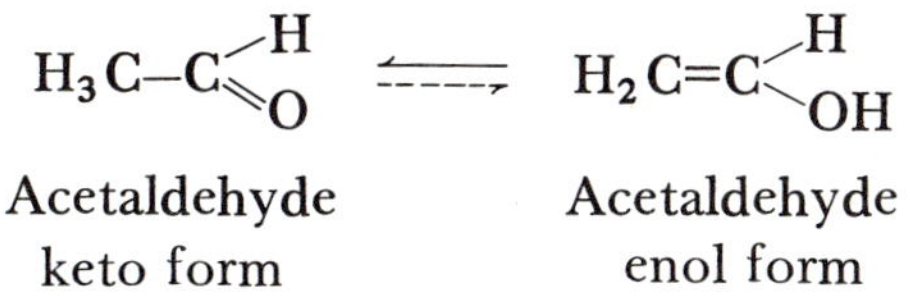

$$H_3C-C\diagup^H_{\diagdown O} \quad \rightleftharpoons \quad H_2C=C\diagup^H_{\diagdown OH}$$

Acetaldehyde	Acetaldehyde
keto form	enol form

This equilibrium implies that a solution of acetaldehyde in water consists of a mixture of molecules containing C=O groups ("keto form") and a very small number of molecules containing C=C–OH groups ("enol form"). The rate of equilibration is a function of pH. The rate of equilibration is minimal around pH 7. In alkali or acid the rate is faster. One particularly important consequence of the keto/enol equilibrium relates to the properties of hydroxy ketones. Fig.3.2 shows the keto/enol equilibrium applied to dihydroxyacetone and D- and L-glyceraldehyde.

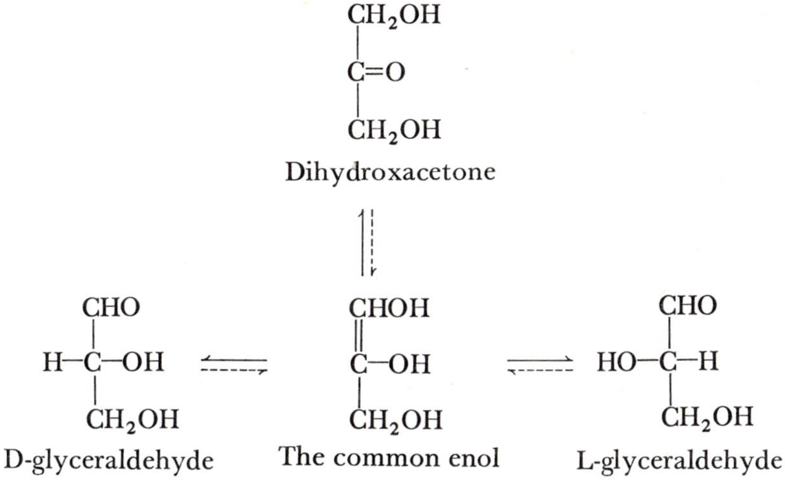

$$
\begin{array}{c}
CH_2OH \\
| \\
C=O \\
| \\
CH_2OH
\end{array}
$$

Dihydroxacetone

$$
\begin{array}{ccc}
CHO & CHOH & CHO \\
\| & \| & | \\
H-C-OH \rightleftharpoons & C-OH \rightleftharpoons & HO-C-H \\
| & | & | \\
CH_2OH & CH_2OH & CH_2OH
\end{array}
$$

D-glyceraldehyde	The common enol	L-glyceraldehyde

Fig. 3.2 The keto/enol equilibrium showing the enol common to dihydroxyacetone, D-glyceraldehyde and L-glyceraldehyde.

All these substances share the same enol. Thus a solution (in water) of either glyceraldehyde or dihydroxyacetone will slowly form an equilibrium mixture of the two sugars with the enol forming the intermediate. We have already said that at neutral pH, the equilibration is slow; the time taken to form the equilibrium mixture would be, therefore, extremely long (months or even years). However, in strong alkali, the rate of equilibration is faster (a day or so). Thus if we dissolved D-glyceraldehyde in strong alkali, we would achieve an equilibrium containing D-glyceraldehyde, L-glyceraldehyde and dihydroxyacetone. A more direct chemical consequence of this equilibration is that, unlike "normal ketones" (such as acetone), dihydroxyacetone and other hydroxyketones resemble acetaldehyde and other aldehydes in their ability to reduce reagents such as ammoniacal silver oxide:

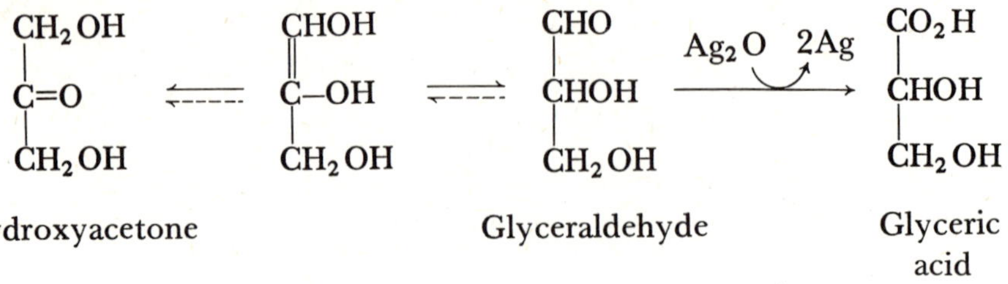

$$
\begin{array}{ccccc}
\text{CH}_2\text{OH} & & \text{CHOH} & & \text{CHO} \\
| & & \| & & | \\
\text{C=O} & \xleftarrow{} & \text{C--OH} & \xrightarrow{} & \text{CHOH} \\
| & & | & & | \\
\text{CH}_2\text{OH} & & \text{CH}_2\text{OH} & & \text{CH}_2\text{OH}
\end{array}
\quad \xrightarrow[\text{Ag}_2\text{O} \;\; 2\text{Ag}]{} \quad
\begin{array}{c}
\text{CO}_2\text{H} \\
| \\
\text{CHOH} \\
| \\
\text{CH}_2\text{OH}
\end{array}
$$

Dihydroxyacetone   Glyceraldehyde   Glyceric acid

3.7. Glucose, Mannose and Fructose

The facts described above are translated to the more complicated case of three hexoses in Fig. 3.3. The two consequences of this scheme are immediately seen. In strong alkali, a solution consisting initially of either glucose, or mannose or fructose will yield an equilibrium mixture of all

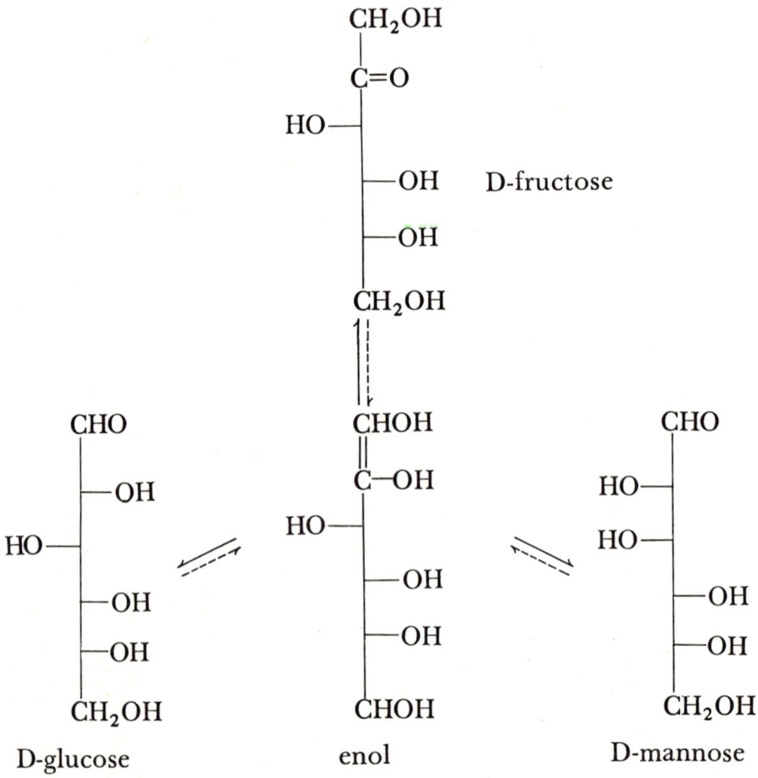

Fig. 3.3 The keto/enol equilibrium showing the enol common to D-fructose, D-glucose and D-mannose.

three sugars; fructose (like glucose and mannose) is a strong reducing agent and will give a positive silver mirror test (or other "aldehyde tests" such as the Fehling's reaction). The interconversion of sugars by the alteration of the configuration of one of the carbon atoms ("epimerization") is an important biological reaction.

There is another chemical process that "unites" these three sugars. This is the reaction between one of these sugars (for example glucose) with phenylhydrazine. From the reaction of this substance with acetone already described, the product should be glucose phenylhydrazone:

$$CH=N-NHC_6H_5$$

$$\overset{|}{-}OH$$

$$HO-$$

$$-OH$$

$$-OH$$

$$CH_2OH$$

Glucose phenylhydrazone

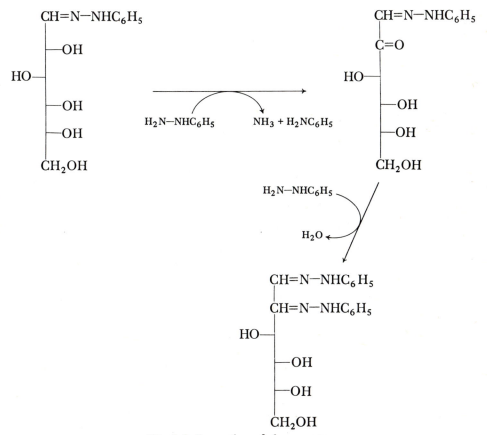

Fig. 3.4 Formation of glucosazone.

In practice, attempts to make this as a pure crystalline compound lead to the formation of gummy mixtures of products. The reason is that glucose phenylhydrazone is oxidized by phenylhydrazine. In order to make the reaction quantitative, the phenylhydrazine is used in excess and the solution is heated. Under these conditions, the glucose phenylhydrazone (Fig. 3.4) is converted to a ketone by a second molecule of phenyl-hydrazine; a third molecule of phenylhydrazine then reacts and forms a doubly substituted hydrazone known as an osazone. As the asymmetry about C-2 has been lost in this case, it is clear that the same product (which can equally be called "mannosazone" or "fructosazone") will be produced by mannose or fructose.

The osazones are employed for the characterization of sugars and the occurrence of three sugars with a common osazone is of great practical importance to chemists.

3.8. Addition Reactions of Carbonyl Groups in Monosaccharides

The addition reactions of the carbonyl groups in aldoses and ketoses give carbohydrate chemistry its characteristic flavour. It is a particular reaction of this class that results in sugars being fundamentally different from the simpler alcohols and carbonyl compounds. We have already commented on one example of an addition reaction (Section 3.6). A more relevant case is the addition of an alcohol (such as ethanol) to acetaldehyde:

$$CH_3-C{\overset{H}{\underset{O}{}}} \ + \ C_2H_5OH \ \rightleftharpoons \ CH_3-\overset{H}{\underset{OC_2H_5}{C}}-OH$$

Acetaldehyde   Ethanol   Hemi-acetal

The reaction is simple enough; it tells us that in alcoholic solution, acetaldehyde exists partly as a free aldehyde and partly as an addition product, called a hemi-acetal. However, it is important that you familiarize yourself with it before considering the more complex case of hemi-acetal formation by a sugar such as glucose. The O-atom originally present in the C=O of the aldehyde is present in an OH group in the hemi-acetal; the O-atom originally present in the OH of the ethanol is present in an ether linkage (C–O–C) in the hemi-acetal. Note also that (as with the cyanohydrin; Section 3.6) we have introduced an asymmetric centre into the molecule by forming the hemi-acetal.

Glucose is a flexible molecule containing a C=O group and several OH groups. Can the C=O group ever approach one of the OH groups sufficiently closely for an internal addition reaction (to yield a hemi-acetal) to occur? A consideration of models implies that the OH groups on C-4 and C-5 can approach the C=O closely enough for this to occur readily. As the actual products involve addition of the OH on C-5, let us confine ourselves to discussing that particular case.

The first thing we require is a convention for representing the conformations of glucose for which this addition is feasible. The one that is used here and elsewhere in this book is to draw the molecule as follows:

$^6CH_2OH$
—OH
OH   CHO
HO
OH

Glucose

The structure is twisted round so that C-1 is on the right and the bond between C-2 and C-3 is nearest to the reader. The lines (or bonds) joining C-4 to C-1 and C-5 to its OH are "seen" in planes behind C-2 to C-3.

The formation of the hemi-acetal in this case produces a new asymmetric centre. However, as the molecule already contains asymmetric centres, the two possible products (Fig. 3.4) will not be enantiomers as was the case in the reaction between acetaldehyde and ethanol; but rather two stereoisomers. These two isomers are referred to as α and β and the special type of isomerism that occurs in this case is called anomerism (so we may call the products the α- and β-anomers).

CH_2OH CH_2OH CH_2OH
—O H   —OH   —O OH
OH   OH   CHO   OH
HO   OH   ⇌   HO   ⇌   HO   H
OH   OH   OH

α-D-Glucopyranoside   Straight-chain form   β-D-Glucopyranoside

Hemi-acetal formation in glucose

Note two other points from this scheme. The convention for naming the anomers is based on the orientation of the hydroxyl on C-1 in the new "anomeric" form, if it is sticking up, the anomer is β; if it is sticking down it is α. Notice also that we have referred to the hemi-acetals as "pyranoses". This is a reference to the fact that, as a consequence of the addition of hydroxyl on C-5 to the C=O of C-1, a six-membered ring (five C-atoms and one O-atom) has been formed. If the hydroxyl of C-4 had added to the C=O, the products (α- and β-anomers) would have contained five-membered rings (four C-atoms and one O-atom) and would have been called furanoses. If the scheme is compared with the formation of acetaldehyde hemi-acetal, it will be seen that the anomeric hydroxyl contains the O-atom present in C=O in the straight-chain form and the ring oxygen ("ether link") is present as the —OH on C-5 in the straight chain form.

3.9. Cyclic Structures of Monosaccharides

A solution of glucose in water is essentially an equilibrium mixture of the α- and β-pyranoses. The straight-chain form is present in extremely small amounts but this form is important in two regards. It is the essential intermediate for the interconversion of the anomers (Section 3.8). Additionally it is the reactive species that results in glucose having the typical aldehydic properties.

Not all monosaccharides occur in solution entirely as pyranoses. Furanoses are found in certain cases and furanose structures are important in more complex sugars. These other cases are described as they arise later. It is worth considering the cyclic structures of some of the simple derivatives of glucose already mentioned. Clearly glucosamine and glucuronic acid can form pyranoses in the same way as glucose. The α-anomers of these structures are shown as follows:

CH_2OH — O H — OH — HO — OH — NH_2

$COOH$ — O H — OH — HO — OH — OH

Glucosamine   Glucuronic acid

Sorbitol is clearly incapable of forming cyclic structures; but what of gluconic and glucaric acids? We shall ignore glucaric acid as its role in medical biochemistry is negligible. Gluconic acid clearly cannot form a hemi-acetal as it is not an aldehyde. On the other hand alcohols and carboxylic acids do react. The reaction in this case is not an addition but a condensation and the product is an ester. Esterification is not such a spontaneous reaction as hemi-acetal formation. Nevertheless, gluconic acid does form internal esters readily and one of these esters is of great metabolic importance. In the reaction shown note that internal (cyclic) esters are given the trivial name "lactones".

CH_2OH — OH — OH — HO — CO_2H — OH $\longrightarrow$ CH_2OH — O — OH — HO — =O — OH + H_2O

Gluconic acid   Gluconolactone

Three pieces of evidence for the occurrence of the cyclic structures of glucose are presented here. They are not given in their historical order and are not the only (nor necessarily the most unequivocal) proofs for the structures; they have been chosen as they each illustrate important aspects of the structure and properties of sugars and other carbohydrates.

Glucose is an optically active substance and, therefore, a solution in water rotates the plane of polarized light passing through it. If you make up a solution of glucose in water (using solid glucose out of a bottle) and measure its rotation, a characteristic value (depending on the concentration of the solution) will be observed. If readings are taken (on the same solution) at later time intervals it will be found that the rotation is not constant but falls to a certain value and then remains constant. If the glucose is dissolved in dilute acid or dilute alkali the same phenomenon will be observed except that the time taken to reach the equilibrium value is much shorter. Indeed, in the case of dilute alkali, the equilibration will be essentially instantaneous. The explanation is that solid glucose (as crystallized from water) is pure α-anomer.* With time the α-anomer forms an α/β equilibrium mixture. The straight-chain form is the intermediate.† The specific rotations‡ of α- and β-glucose are different; that of the β-anomer is much lower so that as more and more α is converted into β the rotation falls until equilibrium is achieved.

The second reaction is "glycol cleavage". A glycol is a substance that contains hydroxyl groups on adjacent carbon atoms. The sugars discussed so far in this chapter are all glycols. Glycols have a number of characteristic reactions. Of particular importance to us is the oxidative cleavage of glycols by periodic acid (HIO_4). The reaction is shown below for the simplest of all glycols, ethylene glycol ($(CH_2OH)_2$).

$$\begin{array}{c} CH_2OH \\ | \\ CH_2OH \end{array} + HIO_4 \longrightarrow \begin{array}{c} CH_2{=}O \\ \\ CH_2{=}O \end{array} + H_2O + HIO_3$$

Glycol cleavage of ethylene glycol

The products are two molecules of carbonyl compound(s), both formaldehyde in this very simple case, water and iodic acid. The reaction occurs with more complex glycols, such as glycerol ($CH_2OH–CHOH–CH_2OH$). In this case two cleavages occur. The reason is not obvious at first sight. Following a cleavage, glycerol should yield formaldehyde and glycollic aldehyde;

$$\begin{array}{c} CH_2OH \\ | \\ CHOH \\ | \\ CH_2OH \end{array} + HIO_4 \longrightarrow \begin{array}{c} CH_2{=}O \\ \\ CH{=}O \\ | \\ CH_2OH \end{array} + H_2O + HIO_3$$

Glycol cleavage of glycerol

* And this is how commercially available solid glucose is made. If glucose is crystallized from pyridine and certain other organic solvents, pure solid β-anomer is obtained. Mutarotation in the case of a solution of such crystals in water produces an *increase* in rotation until the equilibrium value is achieved.

† This reaction is catalyzed by dilute alkali. It is not to be confused by the reaction catalyzed by strong alkali in which the enol is the intermediate (p. 54).

‡ Specific rotation is the rotation of a defined concentration (1% w/v) of a substance measured with a defined pathlength (10 cm) for the polarized light. It is thus a value characteristic of the substance and independent of the conditions of the measurement. Rotation is a simple linear function of both concentration and pathlength.

However, all hydroxyaldehydes (and hydroxyketones) which contain an —OH group on a carbon adjacent to a C=O behave as glycols in this reaction. The reason is that aldehydes and ketones exist in solution partly in a hydrated form. Consequently a —OH group is generated on the "carbonyl C-atom".*

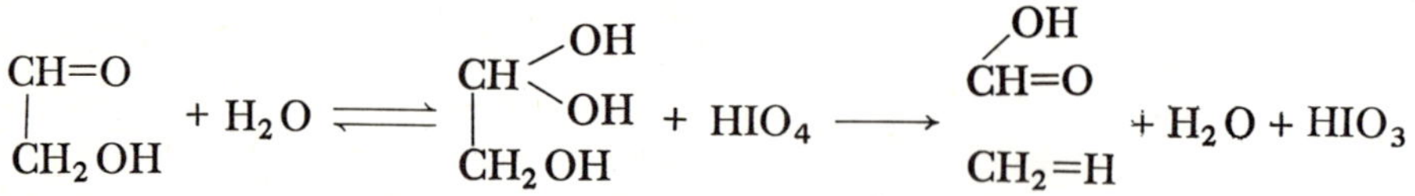

$$
\begin{array}{c}
\text{CH=O} \\
| \\
\text{CH}_2\text{OH}
\end{array}
+ \text{H}_2\text{O} \rightleftharpoons
\begin{array}{c}
\text{CH}{<}^{\text{OH}}_{\text{OH}} \\
| \\
\text{CH}_2\text{OH}
\end{array}
+ \text{HIO}_4 \longrightarrow
\begin{array}{c}
\text{CH=O}^{\text{OH}} \\
\text{CH}_2\text{=H}
\end{array}
+ \text{H}_2\text{O} + \text{HIO}_3
$$

If these reactions are put together, it is seen that one molecule of glycerol reacts with two molecules of periodic acid to produce two molecules of iodic acid, two molecules of formaldehyde and one molecule of formic acid.

As periodic and iodic acids are very easy to assay in solution by a volumetric method (iodometric titration), it follows that the reaction may be used to "titrate" the number of glycol groups in a sugar. The straight-chain structure of glucose (drawn below with the aldehyde group hydrated) implies that there are five positions for glycol cleavage (marked X) and one would, therefore, anticipate a rapid reduction of five moles of periodate (to iodate) by one mole of glucose.

$$
\begin{array}{c}
\text{CH}{<}^{\text{OH}}_{\text{OH}} \\
\overset{*}{|}\text{—OH} \\
\overset{*}{|} \\
\text{HO—}| \\
\overset{*}{|}\text{—OH} \\
\overset{*}{|}\text{—OH} \\
\overset{*}{|} \\
\text{CH}_2\text{OH}
\end{array}
$$

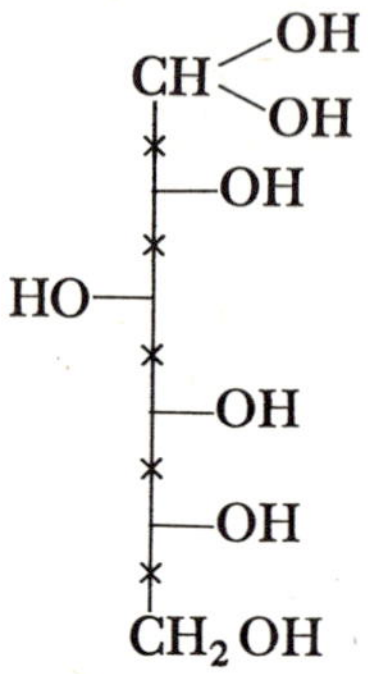

Glucose

The experimental finding is that there is a rapid reduction of three moles of periodate (by one mole of glucose) followed by a very slow reduction of a further two moles. This is a consequence of the fact that glucose occurs as a cyclic structure and there are three glycol groups in the molecule.

The initial products from this reaction will be two moles of formic acid (from C-2 and C-3) and a dialdehyde, which is slowly hydrolyzed in water to yield formic acid and a molecule (glyceraldehyde) that consumes two further moles of periodate.

---

* Hydration of aldehydes and ketones is entirely analogous to the reaction (in alcoholic solution) to form hemi-acetals. In the formation of a hydrate, water (HOH) adds across the double bond in the way that ethanol (HOC_2H_5) did in the reaction with acetaldehyde.

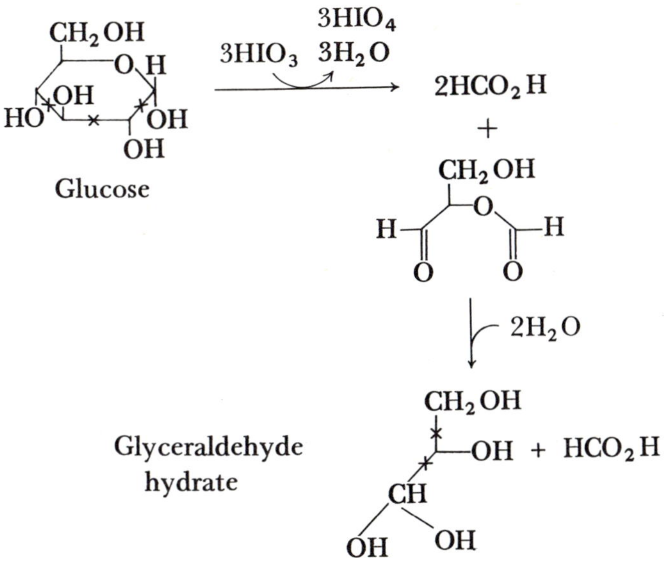

First stage in the oxidation of glucose by periodate. The glycol bonds in glyceraldehyde hydrate (marked X) consume two further moles of HIO_4.

For the third reaction characteristic of cyclic sugars, we go back to the reaction between acetaldehyde and an alcohol. The product (a hemi-acetal) results from a reversible addition. If acetaldehyde is warmed with an excess of ethanol and an acid catalyst, the hemi-acetal is formed as an intermediate and condenses irreversibly with a second molecule of ethanol:

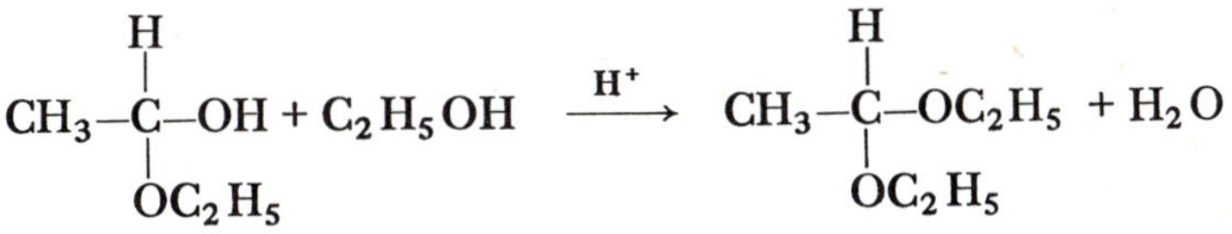

Formation of true acetal from hemiacetal

The product of this reaction is known as a "true acetal". This reaction is an extremely important one as it introduces an arrangement of bonds that links monosaccharide residues together in more complex carbohydrates. It is, therefore, important to recognize a true acetal structure in an organic molecule. A central carbon atom is surrounded by one –H group, two "ether" groups (C–O–) and one C–C bond. We have said that the formation of a true acetal is not readily reversible. Thus the compound shown above is not in equilibrium with acetaldehyde and ethanol. On the other hand it can be hydrolyzed to yield these two substances if it is boiled with dilute aqueous acid.

Now glucose is a hemi-acetal. Therefore, if it is warmed with an acidified alcohol it forms a true acetal. Fig. 3.5 is an illustration of the reaction of α- and β-glucose with methanol to yield the two (α and β) true acetals.

In this scheme α- and β-glucose are shown to be in equilibrium with one another. The two true acetals are not in equilibrium with one another. These two true acetals are, therefore, distinct substances (called α- and β-methyl glucopyranoside) and can be separated by fractional crystallization or chromatography. If α- (or β-) methyl glucopyranoside is heated with aqueous mineral acid it is hydrolyzed to form glucose and methanol. The separation of these two substances and their recognition as true acetals was historically the reaction that gave the vital clue in elucidating the hemi-acetal cyclic structure of glucose. As we shall see

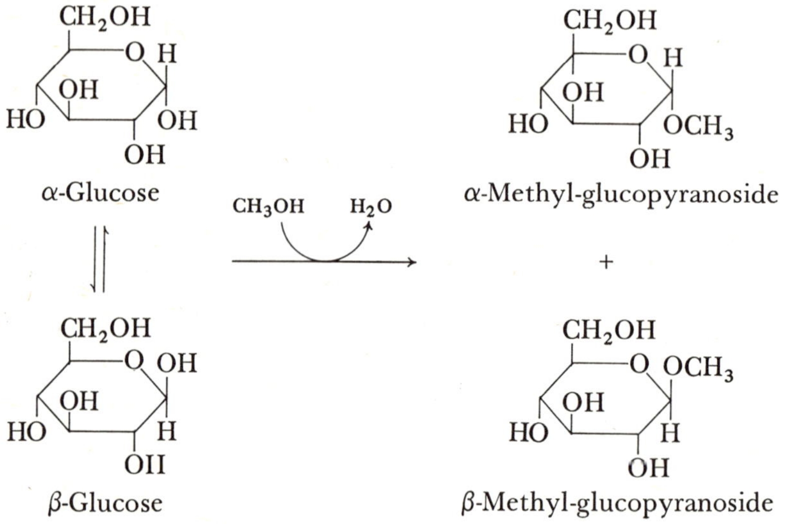

Fig. 3.5 Formation of α- and β-methyl-glucopyranosides.

these true acetals of monosaccharides are an important class of carbohydrates. The general name for them is glycoside. Those in which the sugar is glucose are called glucosides; similarly those in which the sugar moiety is galactose are called galactosides, those in which it is mannose are mannosides and so on.

3.10. Sugar Phosphates

Several types of sugar ester occur in cells but by far the most important are the phosphates. Sugar phosphates play central roles in metabolism in the structure of important enzyme co-factors and in nucleic acids. Here we shall briefly consider the general properties of sugar phosphates.

In the last chapter we saw that phosphoric acid (or orthophosphoric acid) is a covalent molecule in which phosphorus has a valency of five:

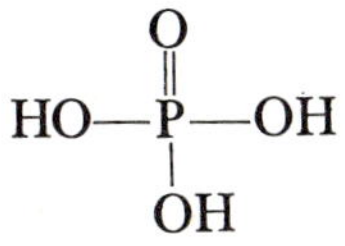

$$\underset{\underset{OH}{|}}{\overset{\overset{O}{\|}}{HO-P-OH}}$$

Phosphoric acid

The phosphorus atom is surrounded by four oxygen atoms which lie at the corners of a slightly irregular tetrahedron (compare the completely regular tetrahedron defined by the hydrogen atoms of methane). Phosphoric acid is tribasic and forms three classes of salts (for example it forms three potassium salts: KH_2PO_4, K_2HPO_4 and K_3PO_4). It also forms three classes of ester. If we use the ethyl esters as examples, the classes are:

Phosphomonoesters, e.g., ethyl phosphate $C_2H_5OPO_3H_2$,
Phosphodiesters, e.g., diethyl phosphate $(C_2H_5O)_2PO_2H$,
Phosphotriesters, e.g., triethyl phosphate $(C_2H_5O)_3PO$.

The phosphomonoesters are dibasic acids; the phosphodiesters are monobasic acids and the phosphotriesters (which are not of biochemical importance) are not acids at all. The pK-values for phosphomonoesters are roughly the same as the first two pK-values of phosphoric acids and the pK-values for phosphodiesters is roughly the same as the first pK-value for phosphoric acid. Thus the phosphate esters are strong acids.

Like all hydroxyacids, phosphoric acids form anhydrides. The simplest one (pyrophosphoric acid) is the tetrabasic acid produced by the elimination of one molecule of water between two molecules of phosphoric acid:

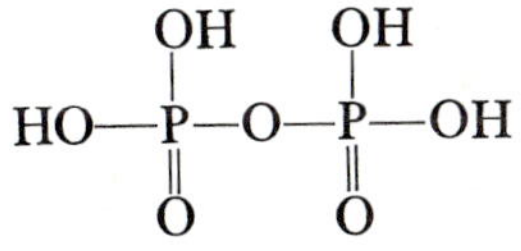

$$\underset{\underset{O}{\|}}{\overset{\overset{OH}{|}}{HO-P}}-O-\underset{\underset{O}{\|}}{\overset{\overset{OH}{|}}{P-OH}}$$

Pyrophosphoric acid

It will be seen later that the type of bond linking the phosphate groups in pyrophosphoric acid is extremely important in metabolism.

A straightforward (and metabolically important) phosphate ester of glucose is glucose 6-phosphate;

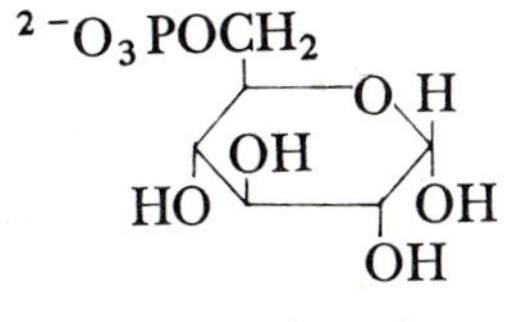

$$^{2-}O_3POCH_2$$

Glucose 6-phosphate

The molecule is shown as fully ionized as the two pK-values are lower than 7 and hence under most physiological circumstances it will exist in solution in this form. The substance is a typical phosphomonoester; it is not readily hydrolyzed. Other glucose phosphates (such as glucose 3-phosphate) have similar properties. However, there is one type of sugar phosphate, exemplified by glucose 1-phosphate, which is far from being a typical phosphomonoester.

$$
\begin{array}{c}
\text{CH}_2\text{OH} \\
\text{O}\,\text{H} \\
\text{OH} \\
\text{HO} \quad \text{OPO}_3^{-2} \\
\text{OH}
\end{array}
$$

Glucose 1-phosphate

The oxygen of the ester bond in this case is an anomeric oxygen—it is derived from the hydroxyl groups of a hemi-acetal. The unique properties of this substance derive from the special properties of such hydroxyl groups. Unlike glucose 6-phosphate, glucose 1-phosphate is more readily hydrolyzed and its hydrolysis is highly exergonic.

Glucose 1-phosphate is the first example discussed so far of a substance which is a "high energy compound"; that is to say, when its phosphate group is transferred to another molecule (such as to water during its hydrolysis) energy is released in such a way that a series of linked metabolic reactions can be "driven".

3.11. Disaccharides

Now we must return to some true acetals. We have seen that a hemi-acetal sugar can form a derivative in which it is linked to an alcohol moiety. This type of linkage, between the anomeric centre of a sugar and an alcohol occurs in naturally occurring substances in which two or more monosaccharide units are linked together.

Disaccharides are sugars containing two monosaccharide units linked together by a glycoside (true acetal) bond. The structures of three disaccharides containing glucopyranose residues are shown in Fig. 3.6. Maltose and cellobiose are related to one another in the same way as are α- and β-methyl glucoside (Fig. 3.5). In the case of these two sugars, the substituent is not methyl (as with the methyl glucosides) but rather 4-glucosyl. Notice that the artificial convention of "bending" the glycoside linkage is adopted so that we can represent the monosaccharide moieties from a conventional orientation. Isomaltose (like maltose) is also an α-glucoside. In this case, however, the substituent is 6-glucosyl.

All these three disaccharides are reducing agents (like glucose) as there is a free hemi-acetal group on the "right hand" glucose moiety. Equally, of course, their solutions consist of α/β anomeric mixtures in this position (only the α-anomers are illustrated). However, C-1 of the "left hand" glucose moiety is a centre of neither reducing properties nor of equilibration. Maltose and cellobiose are distinct substances (just as are the α- and β-methyl glucosides). If any of these three disaccharides is boiled with aqueous mineral acid, it is hydrolyzed. The product is glucose and the number of reducing equivalents in the solution is doubled.

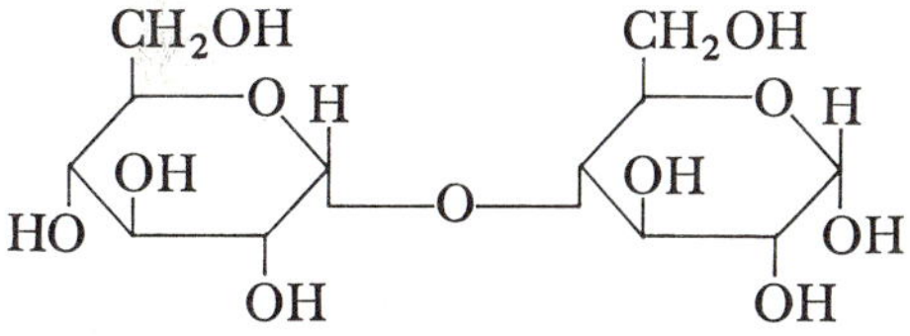

Maltose

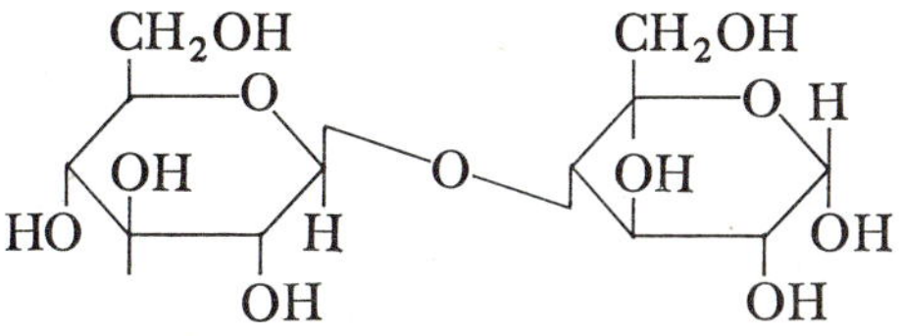

Cellobiose

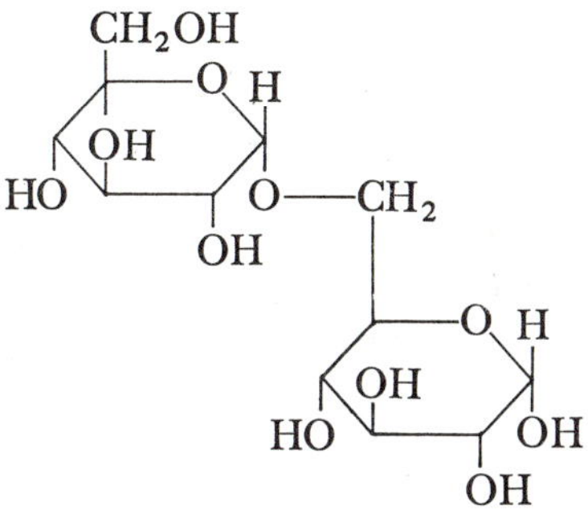

Isomaltose

Fig. 3.6 Three disaccharides.

Lactose is a β(1,4)-linked disaccharide like cellobiose. It differs from cellobiose in that it is a β-galactoside:

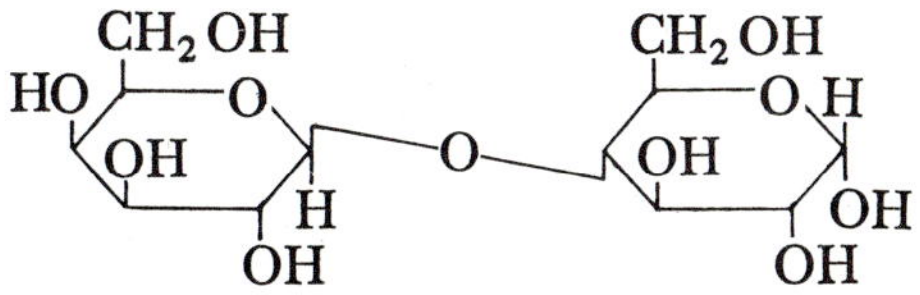

Lactose

The "left hand" monosaccharide moiety is the β-pyranose of galactose. Lactose has the same chemical properties as maltose, cellobiose and isomaltose except that following acid hydrolysis the product is an equimolar mixture of glucose and galactose.

Sucrose is a disaccharide of a different character to the foregoing. The structure of sucrose and the products of its acid hydrolysis are shown below.

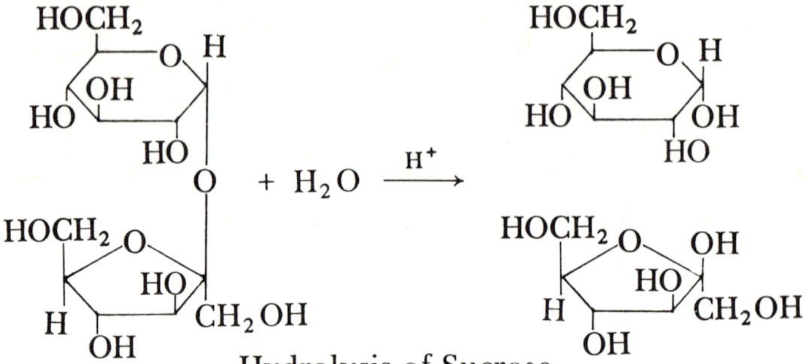

Hydrolysis of Sucrose

The glycoside bond (the "long vertical" one) is hydrolyzed to yield the two component monosaccharides; the upper one is clearly glucose. The lower furanose is one of the anomeric configurations of fructose. Just as glucose exists in solution as a mixture of hemi-acetals, so fructose exists as a mixture of "hemi-ketals". A hemi-ketal is the ketone equivalent of a hemi-acetal. The formation of two hemi-ketals (from acetone and ethanol and the formation of a cyclic form of fructose) are shown in Fig. 3.7. Note that convention "sticking down is α, sticking up is β" applies to the anomeric hydroxyl groups in ketoses as it does in aldoses.

$$(CH_3)_2C{=}O + C_2H_5OH \rightleftharpoons (CH_3)_2C\begin{matrix} OH \\ OC_2H_5 \end{matrix}$$

Ethyl hemi-ketal
of acetone

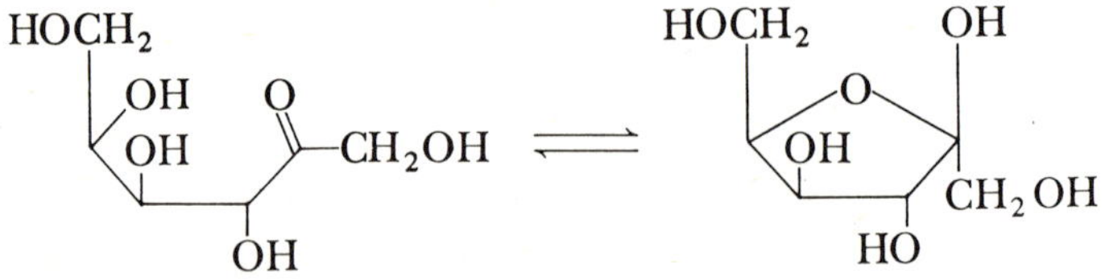

Fructose straight-chain form   Fructose β-furanoside form

Fig. 3.7 Hemi-ketals. [By way of illustration the Fig. illustrates the formation of one of the fructofuranoses as fructose occurs as a furanoside in its important biochemical derivatives. In solution, fructose exists predominantly as a mixture of its pyranoses (formed by the addition of the OH on C-6 to the carbonyl of C-2).]

Returning to sucrose, notice that the glycoside bond links the two anomeric centres (C-2 of fructose and C-1 of glucose). Sucrose, therefore, is not a reducing agent itself. Following acid hydrolysis it yields an equimolar mixture of two reducing sugars (fructose and glucose).

3.12. Polysaccharides

The glycoside bond can be used to link very large numbers of mono-saccharide moieties together to form polymeric chains. Three such naturally occurring polymers are illustrated in Fig. 3.8.

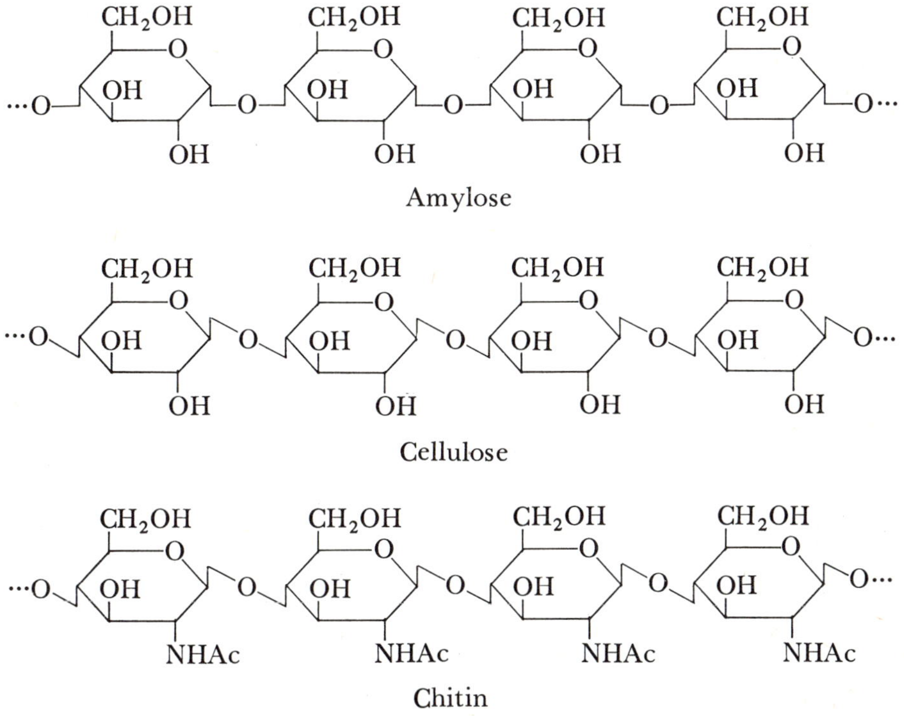

Fig. 3.8 Naturally occurring polysaccharides of glucose (Ac = CH_3CO-).

Amylose is one of the two components of starch (the major carbohydrate storage product of higher plants); cellulose is the major structural polysaccharide of higher plants; chitin is the stuff that forms the exoskeleton of insects, lobsters and crabs.

Amylose and cellulose are clearly related to maltose and cellobiose respectively. They contain essentially no reducing groups (unless there is one at one end of the chains) and on acid hydrolysis will yield glucose. It is a familiar observation that paper or cotton wool (which both consist largely of cellulose) disintegrate to yield a strongly reducing solution if treated with hot hydrochloric acid. Chitin has a similar structure to cellulose. Its monosaccharide moieties are N-acetyl glucosamines (Ac— = $CH_3-C\lessgtr^O$). If chitin is hydrolyzed with acid, the products are the glucosamine salt of the acid and acetic acid.

Several more complex linear polysaccharides are important structural polymers. Examples are hyaluronic acid, heparin and bacterial cell wall.

These substances differ from the foregoing polymers in that the mono-saccharide units are not all the same (typically two monosaccharides alternate along the chain) and are highly substituted with ester or ether groups. However, they all contain glycoside ("true acetal") bonds linking the residues together.

Glycogen is the major carbohydrate reserve in animal cells; amylopectin is the other major component in starch. Both of them are branched-chain glucose polysaccharides. The two polymers differ in their branching pattern, but have the same types of chemical bonding. The molecules contain chains of $\alpha(1,4)$ glycoside linked glucose moieties (just like amylose) joined at branching points by an $\alpha(1,6)$ bond just like that in isomaltose. The structure in the region of such a branch point is illustrated:

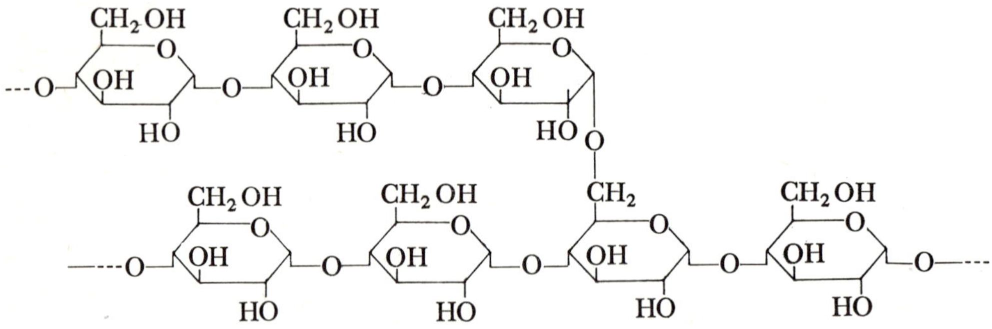

The $\alpha(1,6)$ bond in amylopectin and glycogen

3.13. Nitrogen Glycosides

We have discussed the structure of glycosides and recognized them to be true acetals (or true ketals). Let us just recall the analogy with the simpler derivatives of acetaldehyde and pursue the argument a little further. Four hypothetical reaction schemes are shown in Fig. 3.9.

Scheme (1) is the reversible formation of a hemi-acetal (addition reaction) followed by the reaction with excess alcohol to form a true acetal (condensation reaction). Scheme (2) shows the analogous processes applied to the formation of one of the cyclic structures of glucose (reversible addition) and the reaction with an alcohol (ROH) to form a glycoside. Scheme (3) is analogous to the formation of a true acetal except that the other reactant (apart from the hemi-acetal) is not excess alcohol, but diethylamine. Again water is eliminated and the other product is a nitrogen analogue of a true acetal. Scheme (4) shows a reaction such as that in (3) applied to the possible reaction between α-glucose and a secondary amine (R_2NH).

The product of scheme (4) is an example of an *N*-glycoside. The "*N*" stands for nitrogen; the other glycosides are sometimes referred to as

$$CH_3CH{=}O + C_2H_5OH \rightleftharpoons CH_3CH{\Large\langle}{\small\begin{array}{l}OH\\OC_2H_5\end{array}} \xrightarrow[C_2H_5OH]{excess}$$

$$CH_3CH{\Large\langle}{\small\begin{array}{l}OC_2H_5\\OC_2H_5\end{array}} + H_2O \qquad (1)$$

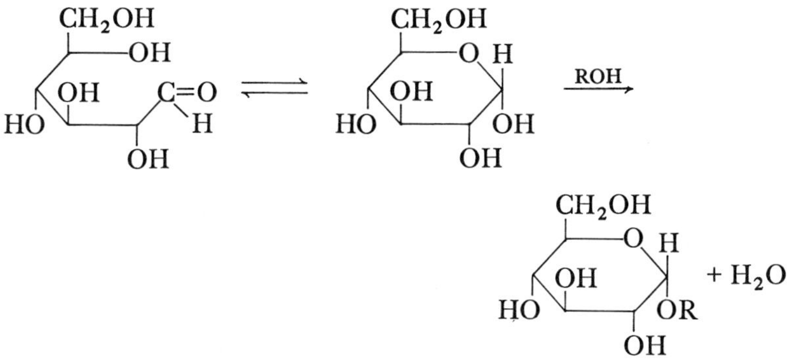

$$\text{(open chain form)} \rightleftharpoons \text{(ring form)} \xrightarrow{ROH}$$

$$\text{(glycoside)} + H_2O \qquad (2)$$

$$CH_3CH{\Large\langle}{\small\begin{array}{l}OH\\OC_2H_5\end{array}} + (C_2H_5)_2NH \longrightarrow CH_3CH{\Large\langle}{\small\begin{array}{l}N(C_2H_5)_2\\OC_2H_5\end{array}} \qquad (3)$$

$$\xrightarrow{R_2NH} \qquad + H_2O \qquad (4)$$

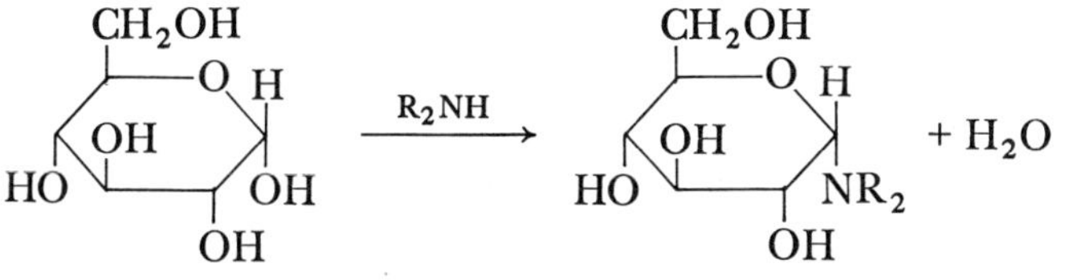

Fig. 3.9 Hypothetical reaction schemes illustrating the formation of *O*- and *N*-glucosides. [Scheme (3) of this Fig. is purely theoretical. Although one might expect the reaction to occur, as NH is isoelectronic with —OH, the hemi-acetal is always in equilibrium with ethanol and acetaldehyde. Aldehyde and amines react according to a different scheme not relevant to the present discussion.]

oxygen glycosides (*O*-glycosides). The *N*-glycosides share most of the properties of *O*-glycosides previously described; they are not reducing sugars, they do not equilibrate with other anomers (thus the α-*N*-glucoside of scheme (4) is not in equilibrium with the β-anomer). Also (like the *O*-glycosides) they can be hydrolyzed with hot mineral acids. If the α-*N*-glucoside of scheme (4) were boiled with aqueous hydrochloric acid, then the products would be glucose and the salt $R_2NH_2^+Cl^-$. This property can be used to distinguish between the two sorts of sugars that contain C—N bonds. These bonds are acid-labile in *N*-glycosides but are not acid-labile in amino sugars such as glucosamine.

The *N*-glucosides are of negligible biochemical interest. On the other hand, the *N*-glycosides of the pentose, D-ribose, are among the most important of all natural products. These substances are known by the trivial name ribonucleosides. The sugar is invariably present in the β-furanose configuration. The structure of a ribonucleoside and its hydrolysis is illustrated in Fig. 3.10.

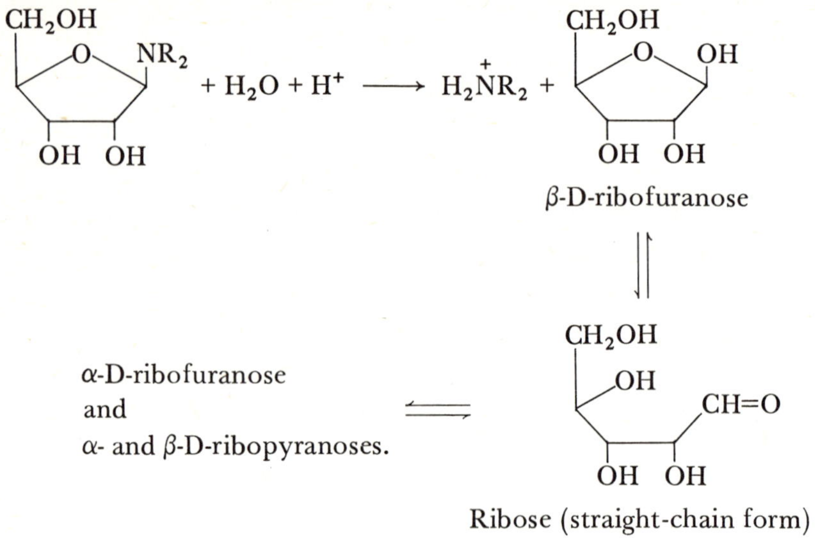

β-D-ribofuranose

α-D-ribofuranose
and
α- and β-D-ribopyranoses.

Ribose (straight-chain form)

Fig. 3.10 Hydrolysis of a ribonucleoside.

3.14 Pyrimidines and Purines

So far we have just represented the secondary amine as R_2NH. What are the secondary amines that contribute to the structure of naturally occurring ribonucleosides? There are four commonly occurring ones and they are usually referred to as "bases". They are all heterocyclic compounds. The ring system in two is that of pyrimidine, in two that of purine. The structures of pyrimidine and purine are shown with the conventional numbering of the atoms.

Pyrimidine   Purine

Pyrimidine can be regarded as being formally derived from the structure of benzene with two of the CH groups replaced by the isoelectronic N. Purine contains two heterocyclic systems, pyrimidine and imidazole, fused so that they share a common C—C bond.

One point will have struck the reader already. Although purine itself is a secondary amine and can be attached to ribose via N-9, pyrimidine is a tertiary amine. The pyrimidines that occur in ribonucleosides are invariably substituted with an oxygen at C-2. This generates an effective secondary amine group at N-1. The reason for this seemingly esoteric fact, that has enormous biological implication, is best seen by considering the simpler case of the heterocyclic substance pyridine:

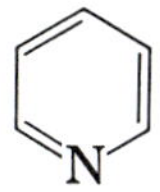

Pyridine

Pyridine (like pyrimidine) is a benzenoid compound. We might expect it to form "typical" aromatic derivatives, e.g., we might expect α-hydroxy-pyridine to be analogous to phenol.

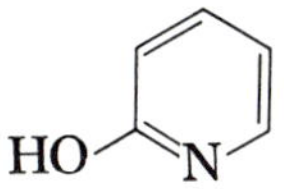

α-Hydroxypyridine

   Attempts to synthesize this compound yield a substance which is far from similar to phenol—rather it is the same as the internal cyclic amide produced by the removal of water from the unsaturated amino acid shown below.

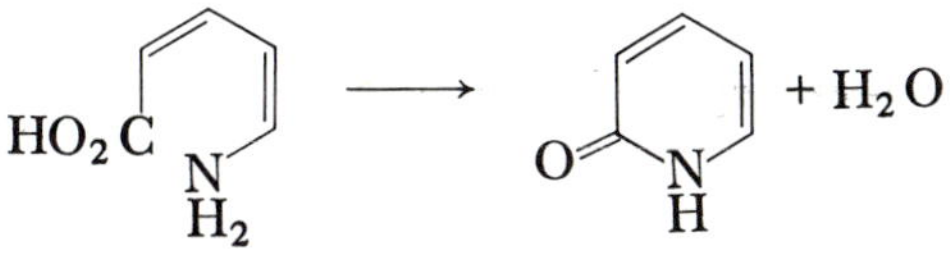

$$+ H_2O$$

Formation of α-pyridone

Internal cyclic amides of this type are called "lactams" (cf. lactones). This lactam is called α-pyridone. The explanation is that α-hydroxy pyridine and α-pyridone are in tautomeric equilibrium, with the equilibrium position strongly favouring the lactam (α-pyridone). *

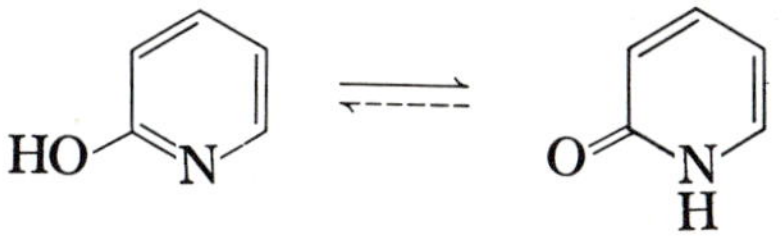

α-Hydroxy pyridine   α-pyridine

   This equilibrium occurs whenever an oxygen is substituted at a carbon next to a nitrogen atom in pyridine or pyrimidine. We always draw the favoured (lactam) structure. A similar equilibrium occurs if the carbon (next to a ring nitrogen) is substituted by $-NH_2$. In this case the

---

* This type of equilibrium is sometimes erroneously referred to as a keto/enol equilibrium. In no sense is α-pyridone a ketone. Rather should it be called a lactam/lactim equilibrium (α-hydroxy pyridine being the lactim).

equilibrium strongly favours the amino tautomer and this is the tautomer used in structural formulae. The equilibrium (illustrated for α-amino-pyridine) is as follows:

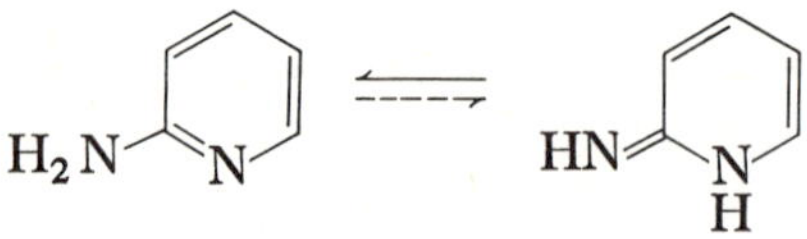

α-Amino pyridine

Now let us look at the four common bases that occur in ribonucleosides.

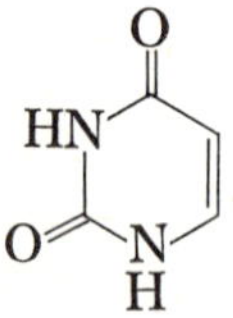

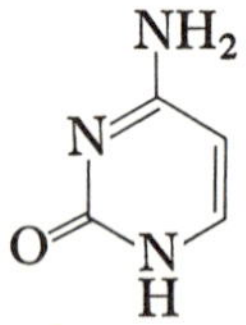

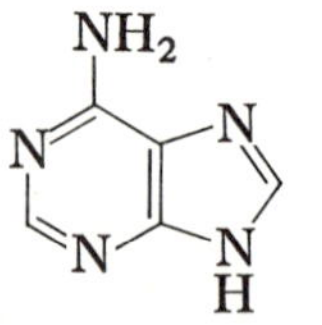

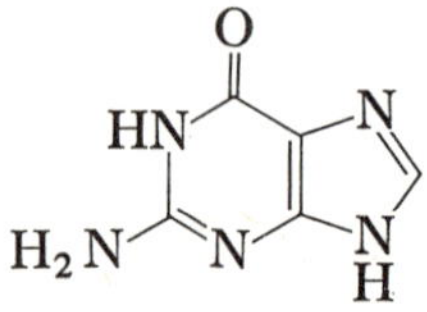

Uracil   Cytosine   Adenine   Guanine

The corresponding ribonucleosides are as follows:

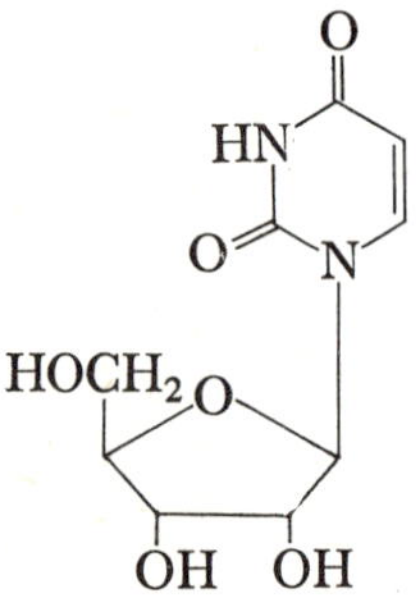

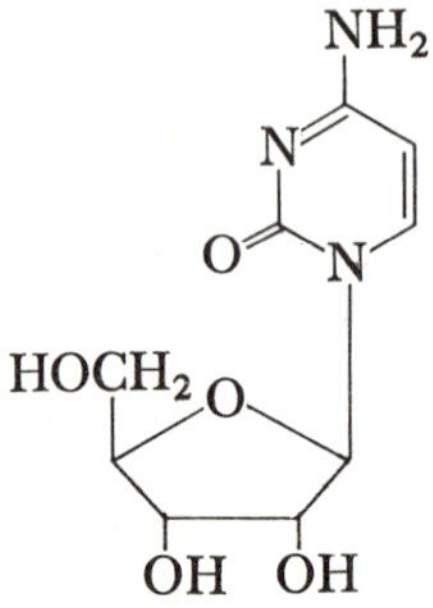

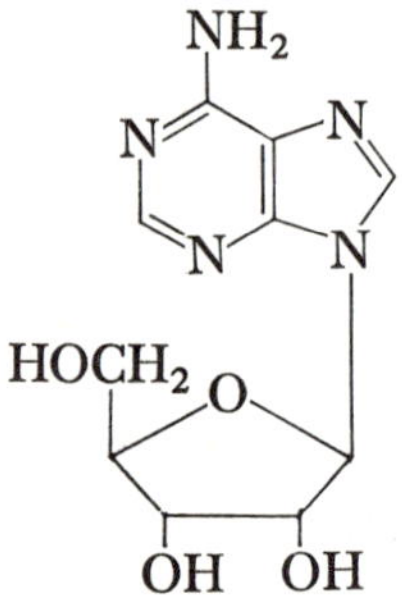

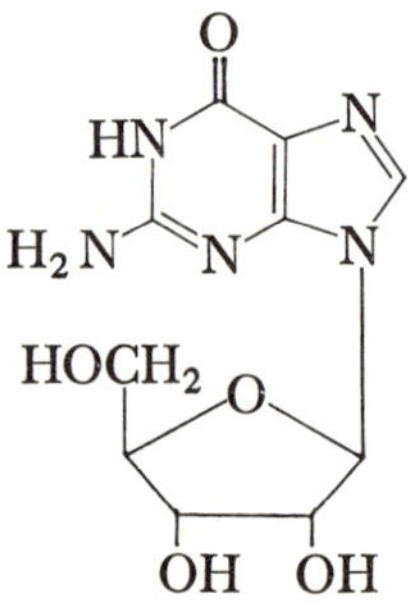

Uridine   Cytidine   Adenosine   Guanosine

We shall see shortly that it is necessary to refer to the atoms in both the bases and sugar moieties. To avoid confusion, the ring atoms in the bases are numbered normally and those in the ribose are primed thus, $1'$ $2'$, $3'$, $4'$, and $5'$. This means that in the pyrimidine ribonucleosides (uridine and cytidine) N-1 is attached to C-$1'$ via an *N*-glycoside bond and in the purine ribonucleosides (adenosine and guanosine) N-9 is attached to C-$1'$ (again by an *N*-glycoside bond).

The four ribonucleosides themselves are not major natural products— their importance derives from their phosphate-ester derivatives, the ribonucleotides.

3.15. Phosphomonoesters of Ribonucleosides

The phosphomonoesters of the four ribonucleosides are the "simple ribonucleotides". There are in all twelve such substances. For each ribonucleoside there are 2′-, 3′- and 5′-phosphates. Those derived from uridine are known as the uridylic acids or uridine monophosphates (UMP for short), those derived from cytidine are the cytidylic acids (CMP), from adenosine the adenylic acids (AMP) and from guanosine the guanylic acids (GMP).

The four classes of ribonucleotides are degradation products of ribonucleic acid. Additionally, the 5′-nucleotides have an important set of derivatives in which the phosphate group is linked via a phospho anhydride link of the type that occurs in pyrophosphate to another phosphate. Examples of derivatives containing this type of structure are 5′-adenosine diphosphate and triphosphate (ADP and ATP—see below) and uridine diphosphate glucose.

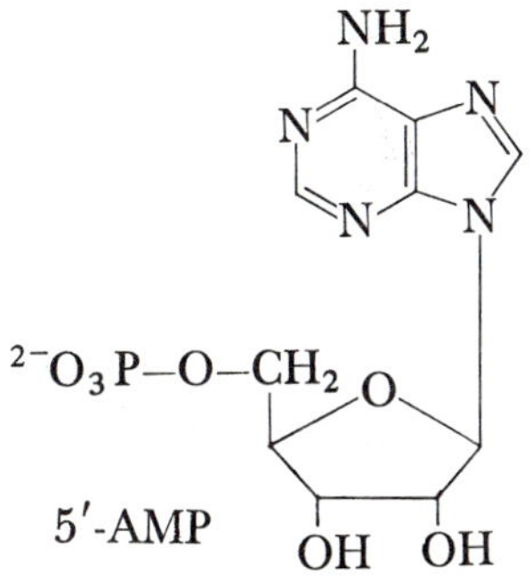
NH2
N
N
N
N
2−O3P–O–CH2
O
OH OH

5′-AMP

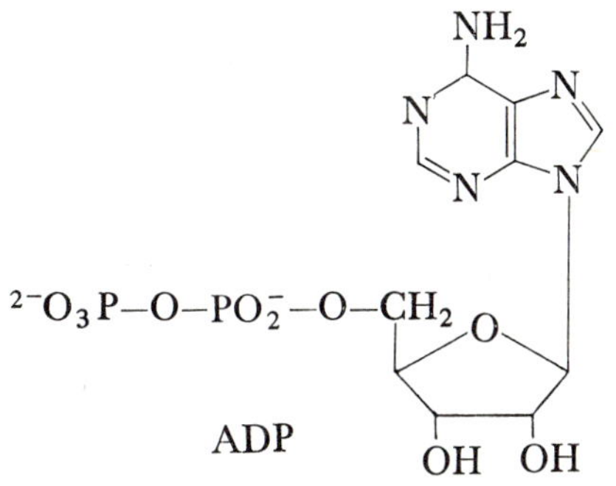
NH2
N
N
N
N
2−O3P–O–PO2−–O–CH2
O
OH OH

ADP

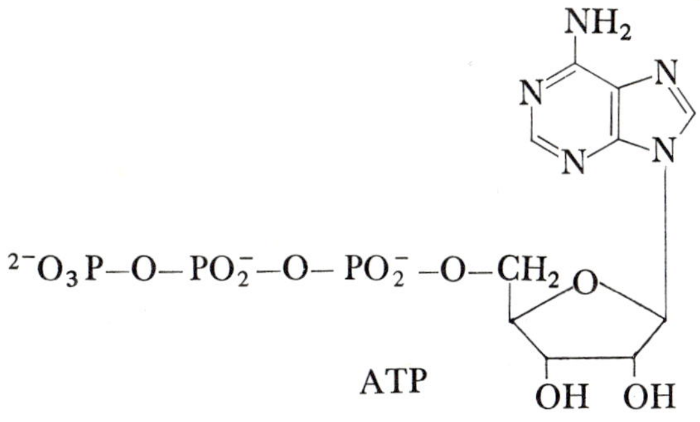
NH2
N
N
N
N
2−O3P–O–PO2−–O–PO2−–O–CH2
O
OH OH

ATP

Nucleotide structures are found in many co-factors including the oxidized and reduced forms of NAD, NADP and FAD and also coenzyme A. These substances and their metabolic roles are described in Chapters 6 and 9. At this point in the book, we briefly consider a novel feature about NAD and NADP. These substances (nicotinamide adenine dinucleotide and nicotinamide adenine dinucleotide phosphate) contain an *N*-glycoside bond between a ribose moiety and nicotinamide:

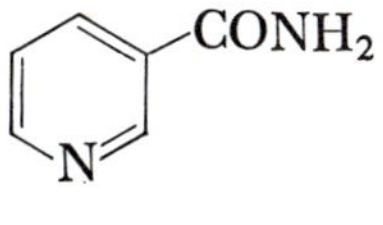
CONH2
N

Nicotinamide, unlike the purines, α-pyridone, uracil and cytosine, is a tertiary amine. Its riboside and ribotide (the latter is called nicotinamide mononucleotide, NMN) therefore contain positively charged quaternary ammonium ions:

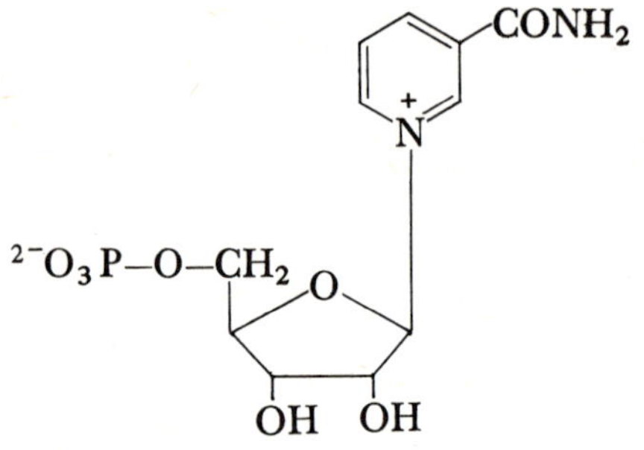

It is seen in Chapter 9, that the co-factors that contain the NMN groups (NAD and NADP) are important biological oxidizing agents. The reason for this is that their reduced forms are *N*-glycosides of a hypothetical *secondary* amine. In looking at the formulation of this, remember that acceptance of H-atoms by the co-factor is *reduction*, and the molecule that has donated the H-atoms (a secondary alcohol) has been *oxidized*, Similarly the reduced form of NMN is a *reducing* agent.

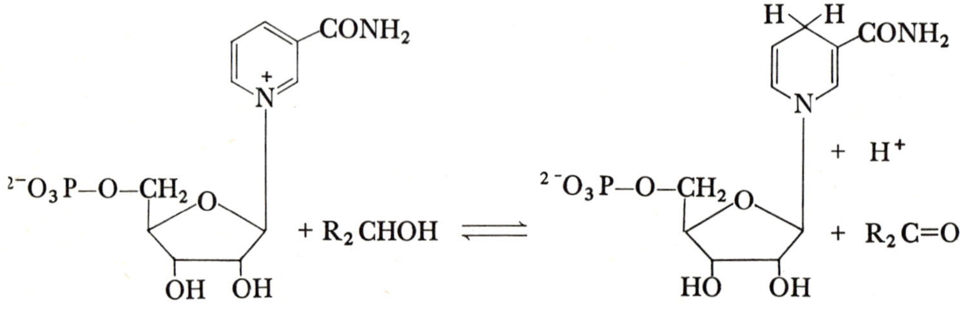

$$+ \; R_2CHOH \;\rightleftharpoons\; + \; H^+ \qquad + \; R_2C=O$$

| Oxidized form of NMN | Secondary alcohol | Reduced form of NMN | Ketone |

3.16. Phosphodiesters of Ribonucleosides

Two types of phosphodiester of ribonucleosides are important. In one group, the phosphate links two oxygens in the same nucleotide. In the other, the phosphate bridges two separate nucleoside units.

Steric considerations limit the number of phosphodiesters of the former type to two classes, those in which the oxygens on C-2' and C-3' are bridged (the 2',3'-cyclic nucleotides) and those in which the oxygens on C-5' and C-3' are the bridge (the 3',5'-cyclic nucleotides). The two adenosine nucleotides are shown to illustrate the difference between these groups.

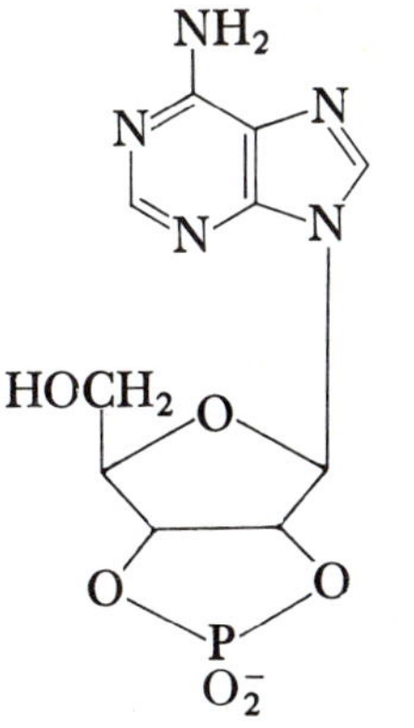

2',3'-cyclic AMP 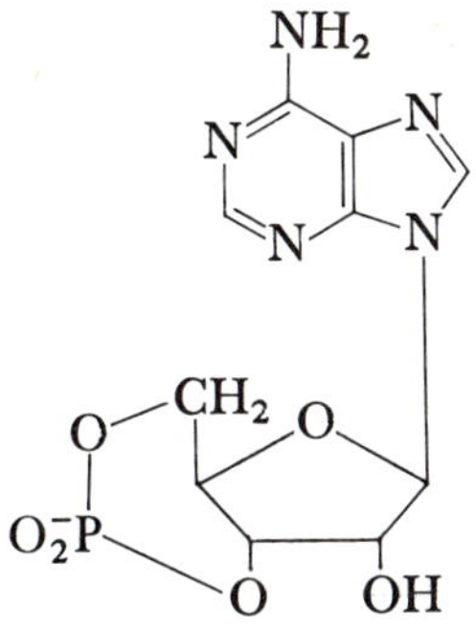   3',5'-cyclic AMP

The 2',3'-cyclic nucleotides are of significance in considering the degradation of certain other nucleotides (Section 3.14); 3',5'-cyclic AMP is a naturally occurring ribonucleotide of great biological importance (Chapter 13).

The phosphodiesters in which two different ribonucleosides are linked together are exemplified by the diribonucleoside monophosphate adenylyl uridine,

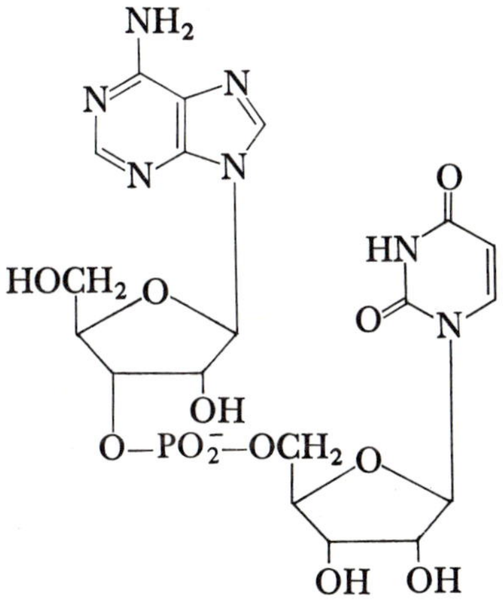

Adenylyl uridine

This is not a naturally occurring substance but it contains a type of linkage that occurs in nucleic acids (next section). The C-3' of adenosine is linked via a phosphodiester to C-5' of a cytidine residue. There exist several shorthand conventions for representing structures such as this. The simplest one uses a single capital to represent the nucleoside (U for uridine, C for cytidine, A for adenosine and G for guanosine) and a hyphen to represent the phosphate. A symbol to the left of a capital letter refers to something attached to C-5' of the ribonucleoside, one to the right to something attached to C-3'. Thus 5'-AMP would be represented as

-A, and 3'-AMP as A- (2'-AMP could not be represented at all). The diribonucleoside monophosphate above would be represented as A-U. Note that A-U is not the same substance as U-A. An alternative convention that conveys more information for certain purposes, represents each nucleoside as a little stick representing the "side view" of the furanose ring labelled with its initial letter (U, C, A or G). The substituents on the sugar ring (OH for hydroxyl and P for phosphate) are shown. Again the convention 5' on the left and 3' on the right is adopted. A-U is then represented thus:

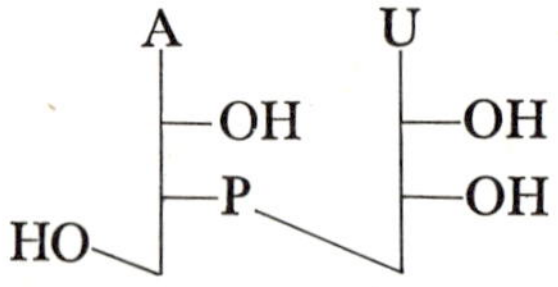

Adenylyl uridine

The two OH's closest to the capital A and U represent the 2'-hydroxyls.

Diribonucleoside monophosphates are atypical phosphodiesters in one respect: they are readily hydrolyzed by alkali. The bond between the phosphorus and the C-5' is hydrolyzed to yield (in the case of A-U) uridine and a mixture of 2'- and 3'-AMP. The explanation lies in the fact that the 2',3'-cyclic phosphate is an intermediate. The production of this intermediate (and the consequent yield of a 2'/3' ribonucleotide mixture), occurs both in the alkaline hydrolysis of a substance such as A-U and in the hydrolysis of such substances by enzymes (ribonucleases).

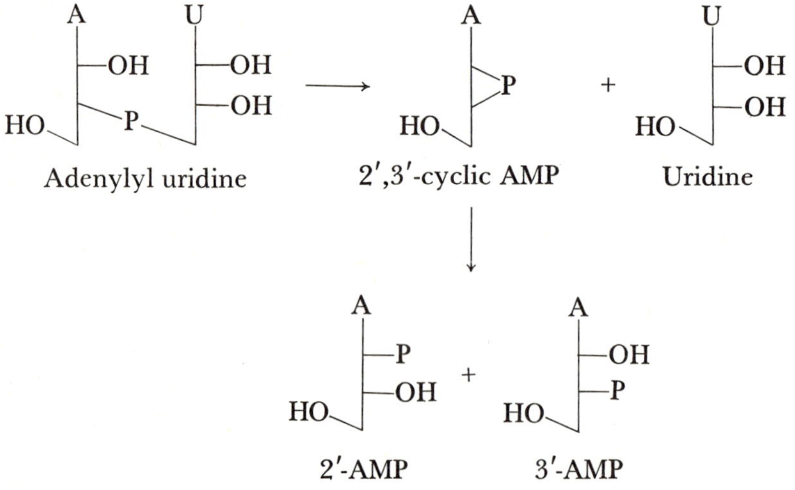

Adenylyl uridine   2',3'-cyclic AMP   Uridine

2'-AMP   3'-AMP

3.17. Primary Structure of Nucleic Acids

Nucleic acids are macromolecules with molecular weights in the range 2×10^4 to an unknown limit in excess of 10^9. They have a common basic

structure

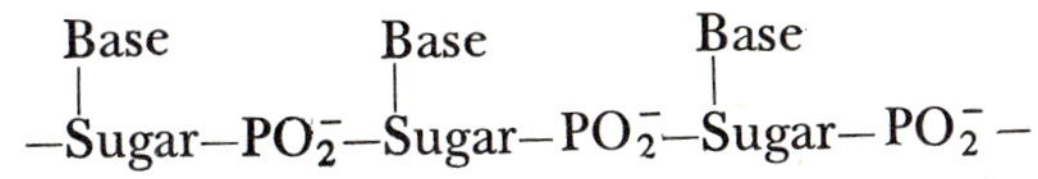

$$
\begin{array}{ccc}
\text{Base} & \text{Base} & \text{Base} \\
| & | & | \\
\end{array}
$$

$$-\text{Sugar}-\text{PO}_2^- -\text{Sugar}-\text{PO}_2^- -\text{Sugar}-\text{PO}_2^- -$$

In other words they have a backbone consisting of monosaccharide sugar moieties linked by phosphodiesters. A base is attached to every sugar via an *N*-glycoside bond. Only two sugars occur in the backbone of nucleic acids, D-ribose and 2-deoxy-D-ribose:

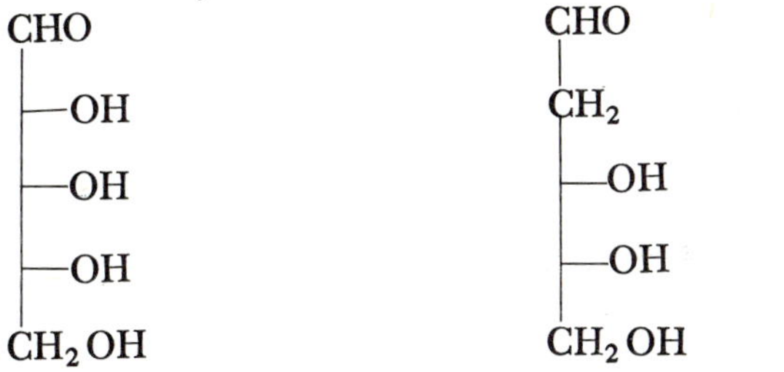

D-ribose   2-deoxy-D-ribose

Moreover any one nucleic acid molecule contains only one of the sugars, and so there are two types of nucleic acids: those that contain ribose, ribonucleic acid or RNA; and those that contain deoxyribose, deoxyribonucleic acid or DNA. Notice that although these substances are referred to universally as acids, the pK of the phosphodiester groups is so low (less than 1) that they are invariably studied as the nucleate polyanions.

An RNA molecule consists of ribonucleosides linked $3' \rightarrow 5'$ by phosphodiester bonds. A typical RNA molecule contains between 80 and several thousand nucleotide residues thus:

$$5'-\text{G}-\text{G}-\text{G}-\text{U}-\text{A}-\text{U}- \cdots \text{C}-\text{C}-\text{A} \ 3'$$

Notice that there is a free $5'$-phosphomonoester at one end of the molecule ("the $5'$-end") and no phosphate moiety at the $3'$-end, i.e., both $2'$- and $3'$-hydroxyls of the terminal adenosine are free. If such a molecule is hydrolyzed by alkali it will yield essentially a mixture of $2'/3'$-ribonucleotides together with the nucleoside diphosphate $-\text{G}-$ (guanosine containing a phosphate on $5'$ and another on $2'$ or $3'$) and free adenosine from the two ends.

DNA has an analogous structure to RNA; it contains deoxyribose in place of ribose and it contains four bases. Of these bases three (cytosine, guanine and adenine) are the same as those in RNA. There is no uracil in DNA, it is replaced by thymine (5-methyluracil)

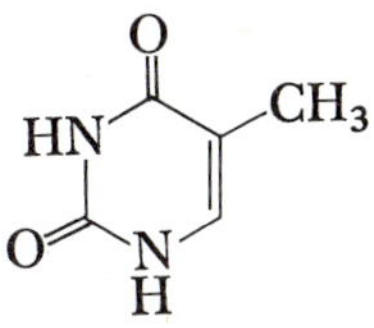

Thymine

As no drawings of deoxyribonucleosides or nucleotides have been shown so far, a fragment of DNA drawn in the three conventions is shown in Fig. 3.11.

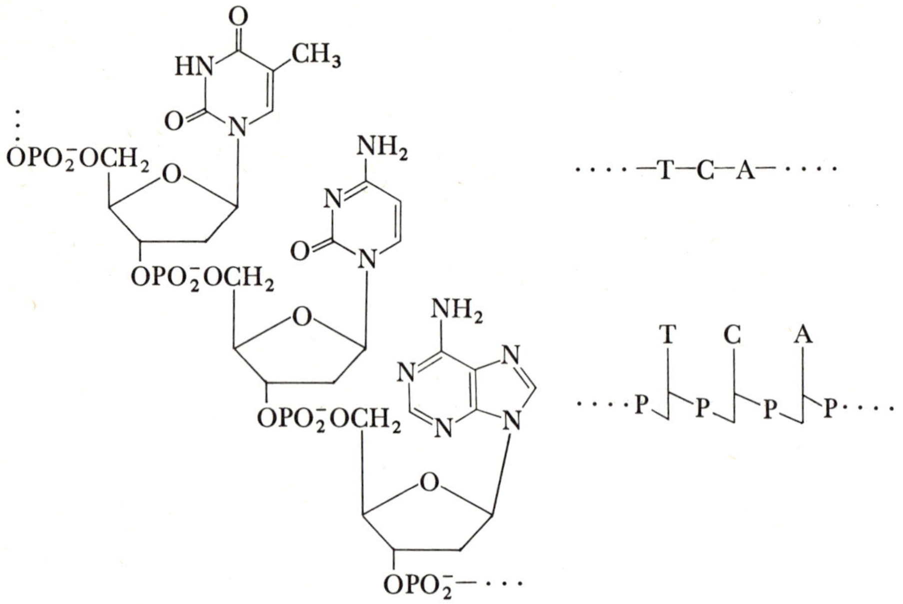

Fig. 3.11 Three ways of drawing a fragment of a DNA sequence.

Unlike RNA, DNA is not readily hydrolyzed by alkali, although alkali does affect the secondary structure of DNA.

Both RNA and DNA are hydrolyzed by acid (the *N*-glycoside bonds are acid-labile). In practice, nucleic acids (notably RNA) are surprisingly resistant to acid hydrolysis for a reason not discussed in this book. However, conditions are available for the quantitative conversion of nucleic acids into their constituent bases, together with degradation products of the sugar-phosphate backbone. The bases may be separated by chromatography and their relative amounts calculated by spectrophotometry. These relative amounts (corrected to molar percentages) are referred to as the base ratios or base composition of the nucleic acid.

3.18. The Double Helix

The double helix is the characteristic secondary structure of nucleic acids. A double helical region of a nucleic acid structure consists of two sugar-phosphate backbones oriented in an anti-parallel sense, that is, with the 5'-3' direction of one strand opposite the 3'-5' direction of the other strand. These two strands are wound round one another to form a right-hand helix. The bases lie in the centre of the helix paired in a characteristic fashion: and A of one strand opposite a T (or U) of the other and a G opposite a C. Two such strands (with Gs corresponding to Cs and As corresponding to Ts) are said to be complementary. The pairings of A with T and G with C are achieved by hydrogen-bonding. The structures of G : C and an A : T base pair are given in Fig. 3.12.

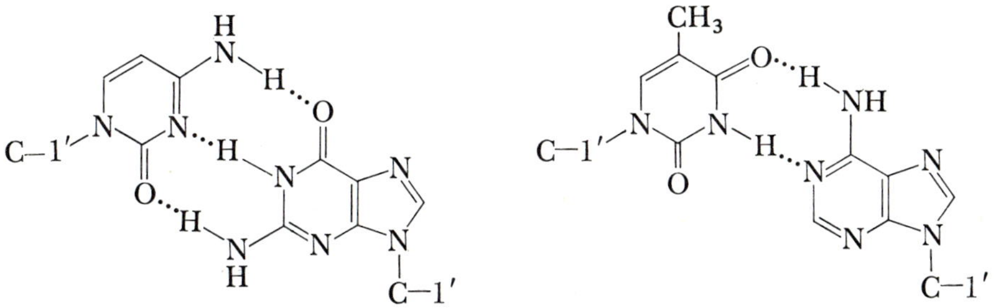

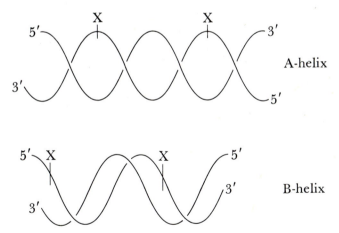

Fig. 3.12 Structure of base pairs in DNA (G : C on the left, A : T on the right).

These are not the only hydrogen-bonded structures that are possible involving the bases. They are not even the two most stable structures; their universal occurrence in nucleic acids (which has profound biological consequences) derives from the identity of the overall geometry of these two structures. They are both planar structures and the distance between the two nucleoside C-1' atoms are identical, 1·11 nm.

The two helix types that occur commonly in nucleic acids are shown in a highly simplified fashion in Fig. 3.13.

Fig. 3.13 Simplified form of A- and B-helices (see text for explanation). The two lines that form the helix in each case represent the sugar phosphate backbone. The base-pairs are not represented.

The distance between the two Xs in Fig. 3.13 is the repeat distance along the axis of the helix. In the B-structure, which is easier to envisage, this distance is 3·4 nm apart. The base pairs are perpendicular to the helix axis (i.e., they would appear as thin lines in this figure as the planes would be perpendicular to the paper). Notice that the two grooves lying between the black lines (sugar-phosphate back-bones) are of unequal size. In the A-form the two grooves are of equal size. In this case there are eleven base pairs between the two Xs and these pairs are inclined to the helix axis. If models of the helices are inspected, you can see light through the A-helix (there is a hole down the middle). In the B-helix, the perpendicular base pairs completely "fill" the projection seen by looking down the helix.

3.19. DNA

DNA isolated from cells consists of pairs of complementary molecules associated (by hydrogen bonding) in a double helix. An example of a fragment of such a double sequence is given below:

5′ ...–A–A–T–C–A–G–A–T–C–...3′

3′ ...–T–T–A–G–T–C–T–A–G–...5′

DNA can be precipitated from solution as a fibrous material. By the technique of X-ray diffraction it is possible to determine the geometry of the molecules. Both A- and B-helices are found and the transition from A-form to B-form can be observed. The equilibrium between the two is determined by the water-content of the fibres and other parameters. In solution, it is believed that DNA assumes the B-structure.

If solutions of DNA are heated, or subjected to extremes of pH, the two strands become dissociated to yield a solution of denatured, single-strand DNA molecules. The denaturation is readily followed spectrophotometrically as native DNA has a lower optical absorbance in the region of the maximum at 260 nm than does single-strand DNA.

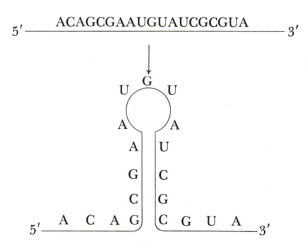

Fig. 3.14 Formation of a short double helical region in an RNA sequence with intramolecular complementarity.

3.20. RNA

RNA is only capable of assuming the A-helix conformation. In the types of RNA encountered in normal cells double stranded complementary molecules do not form. However, RNA contains short double helical sections (A-form) as a consequence of intramolecular complementarity as illustrated schematically in Fig. 3.14.

4

The Structure and Properties of Proteins and Polypeptides

We shall consider the properties of proteins in considerable detail because proteins are so ubiquitous in living tissues, and have such important functions. Proteins constitute about half of the dry weight of the body and a knowledge of their structure and properties holds the key to many important aspects of biochemistry. The word "protein" is derived from the Greek word, meaning first.

4.1. Functions of Proteins

We can summarize the functions of proteins in living tissues as follows:

(a) Enzymes
All the enzymes so far discovered are proteins and to date over 1000 different enzymes have been characterized.

(b) Nutrition
The so-called storage proteins provide a source of nutrient to the offspring, examples are egg white (ovalbumin), and the major protein of milk (casein). Many other proteins in the body also no doubt serve as a source of amino acids for metabolism.

(c) Transport

It is necessary for certain substances to be carried from one tissue to another and proteins serve as the carriers. Thus haemoglobin carries O_2, serum albumin fatty acids, and other serum proteins combine specifically with lipids, iron, copper, thyroxine, vitamin A etc.

(d) Protection

The main defence mechanism of the body against both bacteria and viruses is the immunological response which is manifested in the presence of antibodies in the blood. These antibodies are proteins and as we shall see they have the property of combining specifically with substances which are foreign to the body. A quite different mechanism ensures that a wound does not lead to serious loss of blood. This protection involves the formation of a clot and the plasma protein fibrinogen plays an important role in this process.

(e) Contraction

There must be many different proteins involved in the processes of contraction and relaxation that go on in many parts of the body. The best characterized are the proteins myosin and actin which are involved in muscular contractions.

(f) Bodily structure

These may take many different forms and in quantitative terms are the most abundant of all proteins in the body. Keratin exists in a variety of structures, hair, skin, nails; collagen is important in connective tissue, elastin in the walls of the blood vessels; proteins in combination with carbohydrate, known either as glycoproteins or mucoproteins, are important in many secretions, e.g., mucus.

(g) Maintenance of osmotic pressure and pH

The plasma proteins and especially serum albumin have an important role in the maintenance of the conditions necessary for the cells of the tissues to survive. Particularly important is osmotic pressure and pH.

(h) Hormones

The various metabolic processes of the body are controlled by many different mechanisms but an important one involves the secretion of hormones by the endocrine glands. In many cases the hormones consist of polypeptides or proteins, e.g., insulin, growth hormone.

We see, therefore, that proteins play an extraordinarily versatile role in the body. If we had to pick out a unique characteristic that was responsible for this versatility it would be the ability of proteins to combine in a specific manner with a broad range of different substances. This specificity permits enzymes to combine with their substrates, antibodies with antigens, transport proteins with their passengers etc. It also constitutes the molecular basis of the differences which exist between species and even between individuals.

4.2. General aspects of Protein Structure

(a) Simple and conjugated proteins

Proteins are macromolecules, i.e., they have a relatively high molecular weight. The polypeptide chains, which constitute the main structure of the proteins, are polymers of amino acids. Twenty-four different amino acids occur in the polypeptide chains of proteins but, of course, not all twenty-four necessarily occur in all proteins. Of these amino acids twenty are common and may already be familiar to you while four are rather special. Thus we can say that the polypeptide chains are polymers of 24 different monomers (see Fig. 4.1).

The proteins may be divided into simple proteins which consist solely of polypeptide chains and conjugated proteins which, in addition to the polypeptide chains, also have other organic or inorganic components. The non-peptide portion of the conjugated proteins is called the "Prosthetic" group. Among the conjugated proteins we can recognize the following important groups.

(i) Nucleoproteins   The nucleic acid is associated with protein. Thus the DNA in the nucleus is associated with the small proteins known as histones and protamines. The ribosomes which are the site of protein synthesis in the cytoplasm consist of RNA and protein. Viruses have a similar composition but the protein portion usually consists of similar repeating units and comprises the coat protein of the virus.

(ii) Lipoproteins   There are two main groups of lipoproteins. Those that constitute an important part of the membrane of the cell and those that transport the lipid in the blood. The latter can be divided into

(a) α-lipoproteins (high density)
(b) β-lipoproteins (low density). There has been much interest in recent years in the concentration of cholesterol in the blood and the relationship of this to the incidence of coronary heart disease. The concentration of cholesterol in serum is correlated with the concentration of β-lipoproteins.
(c) pre β-lipoproteins (very low density)
(d) Chylomicrons. These are the minute globules that are present in the lacteals after the ingestion of a fatty meal.

(iii) Glycoproteins   Carbohydrate is linked to an amino acid in the polypeptide chain (usually either asparagine, or serine, or threonine). Most

of the proteins that are exported from cells are glycoproteins and all the plasma proteins have a carbohydrate prosthetic group with the exception of serum albumin.

(iv) Phosphoproteins The phosphate group is esterified to the hydroxyl group of either serine or threonine. Several important enzymes exist in both a phosphorylated and non-phosphorylated form. The storage proteins casein and ovalbumin are also phosphoproteins.

(v) Haemoproteins The haem group which carries a metal, i.e., Fe, may either be loosely associated with the polypeptide chain as in myoglobin and haemoglobin, or may be covalently linked as in cytochrome *c*. Other examples are peroxidase and catalase.

(vi) Flavoproteins Here the prosthetic group contains a flavin of which riboflavin (vitamin B_2) is an example. The prosthetic group may be flavin mononucleotide (FMN), which is the phosphate of riboflavin, or flavin-adenine dinucleotide (FAD). These flavoproteins are very important in reactions involving biological oxidation.

(vii) Metalloproteins Some enzymes contain a metal as an integral part of their structure. Examples are, Zn: alcohol dehydrogenase, Mg: phosphotransferases, Mn: arginase, Cu: tyrosinase.

(b) Homogeneity

Proteins vary widely in their molecular weight. Some are very large indeed and may approach 1 million daltons. The lower limit is more difficult to define, but we usually think of insulin with a molecular weight of 5733 as being one of the smallest proteins. This represents the demarcation point in size between a polypeptide and a protein although it is worth emphasizing that this is strictly a definition of convenience and that there is no marked difference in property between a large polypeptide and a small protein.

Proteins possess a wide variety of physical properties and may be separated from one another on the basis of these properties. In Chapter 2 mention has already been made of methods that involve differences in solubility of proteins at various concentrations of salt and their mobility in an electric field. If one applies a variety of methods to the purification of a protein it is usually possible to obtain a homogeneous product, i.e., a polypeptide of a defined composition.

(c) Species specificity

As stated in Chapter 1 proteins are the molecular basis of the differences that exist between biological species. How can we demonstrate this?

Suppose we isolate a homogeneous sample of serum albumin from a number of different species of animals. We then inject each protein into a rabbit and subsequently examine the serum of the rabbits to see if it has changed in composition as a result of the injection of serum albumin. The way we do this is to mix a solution of some of the original albumin (the antigen) with some of the rabbit's serum (antiserum) and see if there is a precipitate. This is known as the precipitin-reaction. We can plot our results in the form of a table. We see that the effect of the different serum albumins is related to their sources. The antigenicity of the serum albumin depends on the extent of the phylogenetic difference between the two species involved. Since the phylogenetic difference between rabbits and chickens is greater than between rabbits and rat, human and cow, the chicken serum albumin was most antigenic. (See Table 4.1.)

TABLE 4.1. The antigenicity of serum albumin in the rabbit

Serum albumin is purified from the serum of various animals and each preparation is injected into a rabbit. After a period of about 1 month a sample of blood is taken from the rabbits and the serum incubated with a small sample of the serum albumin that had been injected. The formation of a precipitate may be observed. One plus indicates a slight precipitate and three plus a heavy precipitate.

Source of albumin injected	Formation of Precipitate
Rabbit	None
Rat	+
Human	++
Cow	++
Chicken	+++

The serum albumin serves the same function in each of the various species of animal and since the structure of the different albumins is broadly similar we can predict that genetically they had a common ancestor. The serum albumins are, therefore, said to be homologous. The fact that the different serum albumins can be differentiated by their immunological properties implies that in spite of their overall similarity they do possess some structural differences. We say, therefore, that the structure of homologous proteins exhibits species specificity.

We will later illustrate the difference in structure of homologous proteins by reference to insulin, but take this opportunity to make an interesting point. When a diabetic patient requires insulin it would be quite impractical to provide human insulin. Hence we inject the patient with bovine or pig insulin which are biologically just as active as human insulin. They do, however, differ in structure from human insulin and hence might have been expected to have been antigenic. Fortunately they are very poor antigens and can usually be injected over many years without the patient developing antibodies. If it were not for this piece of good fortune it would not be possible to treat diabetic patients with insulin.

The phenomena described above have proved to be a stimulating challenge to biochemists, wanting to discover the structural basis for the unique properties of proteins. Much progress has been made over the last ten years or so, and we feel that we have a reasonable understanding of many of the basic principles involved. We aim, therefore, to describe these concepts to you using certain specific proteins which are of particular importance in medicine as our examples.

4.3. Nature of the Amino Acids in Proteins

Figure 4.1 shows the 20 amino acids that are found in virtually all proteins. You have already been introduced to them in Chapter 2 but this time the formulae have been written so as to emphasize the differences in the R groups of the amino acids. In addition, the following four amino acids are found in certain proteins,

| hydroxy-proline | phospho-threonine |
| hydroxy-lysine | phospho-serine |

All the amino acids listed are of the L-configuration and we do not find D-amino acids in proteins.

All the protein amino acids contain an α-amino group except for proline which is an imino acid. This is very important in protein structure for in this case the charged group present is NH_2^+ and not NH_3^+.

There is one rather special type of amino acid that has not been mentioned, namely the iodotyrosines, which are of particular interest clinically. The thyroid gland contains a protein, thyroglobulin, in which some of the tyrosine residues in the polypeptide chain are iodinated and some are present in a more complex structure, thyroxine (see Fig. 4.1). If a person is fed a diet lacking in iodine then the thyroid gland may become enlarged (goitre) and is deficient in these iodinated amino acid residues. The condition may be corrected by administration of iodine.

4.4. The Structural Organization of Proteins

In considering the structure of the polypeptide chain component of proteins, it is useful to distinguish four levels of organization. We should remember that the definitions that follow are merely to help our better understanding of protein structure, and it should not be surprising that there is some overlap between the different levels of organization. Before going into detail we may summarize as follows.

(a) Primary

The amino acid residues are covalently linked together through peptide bonds. The primary structure defines the number of amino acids in a single polypeptide chain and the order in which the different amino acids are present.

(b) Secondary

This involves interactions between the CO and NH groups of the peptide bonds. In fact the links are hydrogen bonds. These may be between amino

(1) Nonpolar or hydrophobic R groups

L-Alanine (ala)

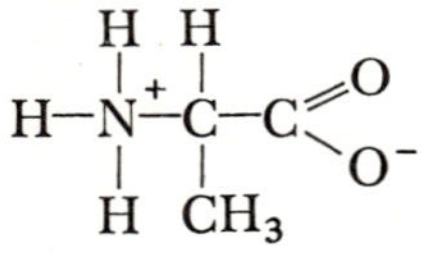

L-Valine (val)

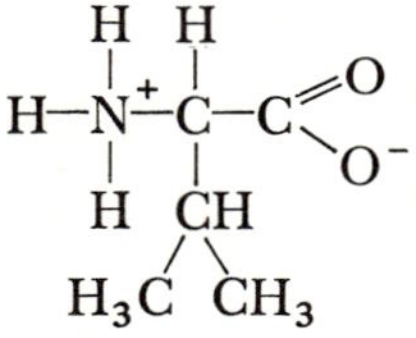

L-Leucine (leu)

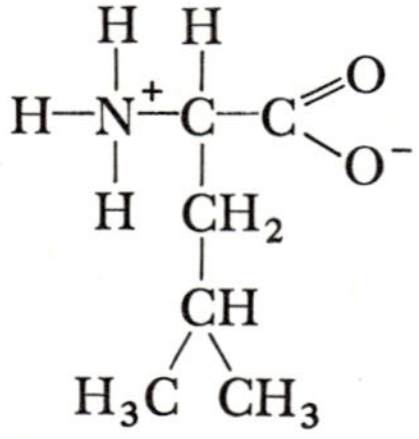

L-Isoleucine (ile)

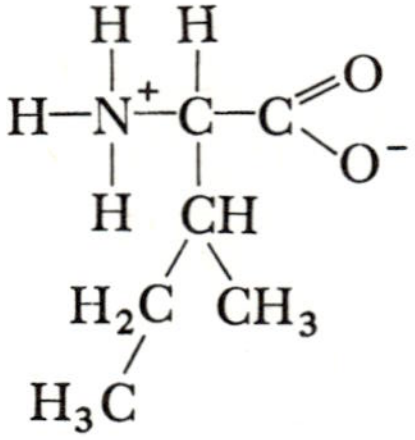

L-Methionine (met)

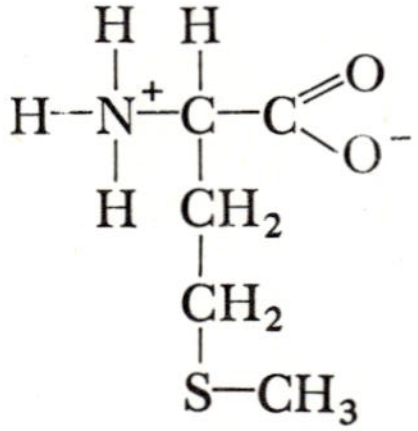

L-Phenylalanine (phe)

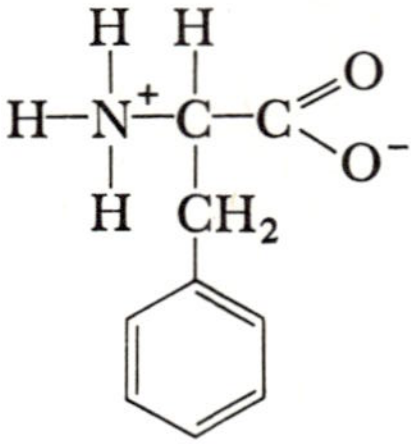

L-Tryptophan (trp)

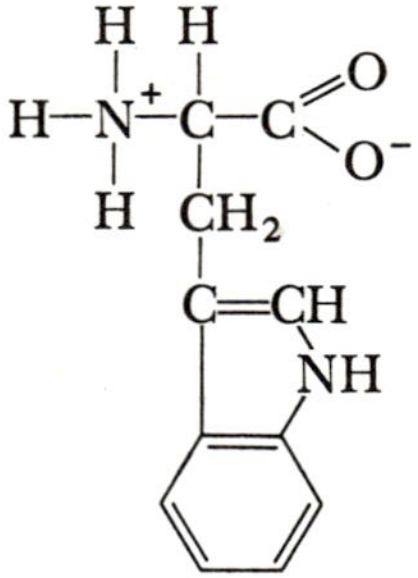

L-Proline (pro)

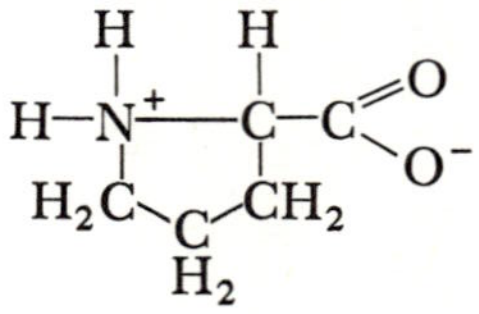

(2) Uncharged polar or hydrophilic R groups

Glycine (gly)

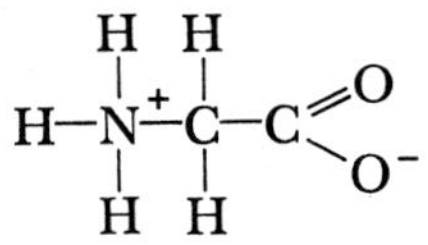

L-Serine (ser)

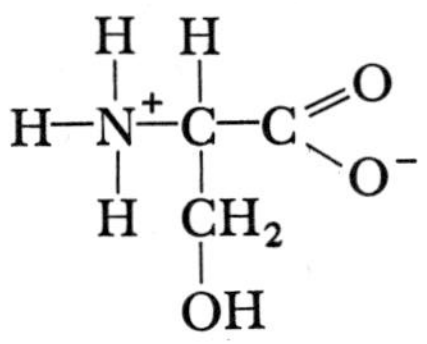

L-Threonine (thr)

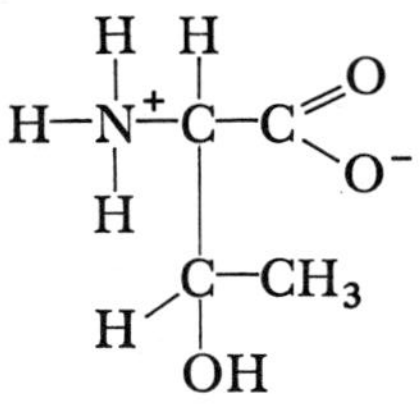

L-Tyrosine (tyr)

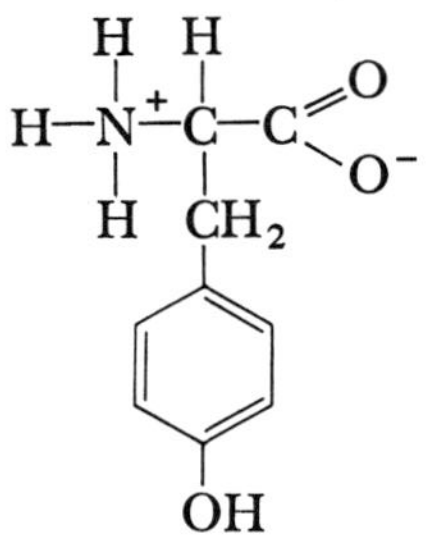

L-Cysteine (cys)

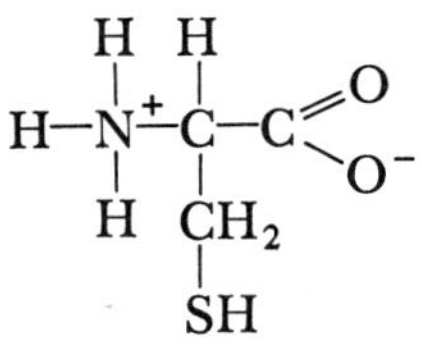

L-Asparagine (asn)

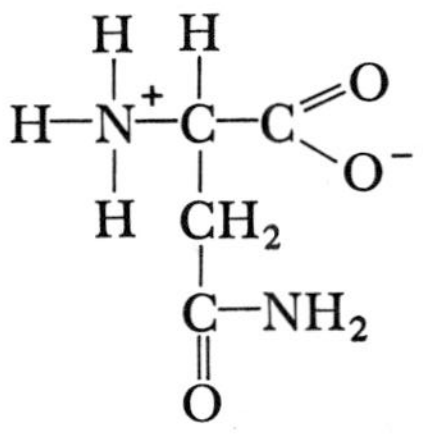

L-Glutamine (gln)

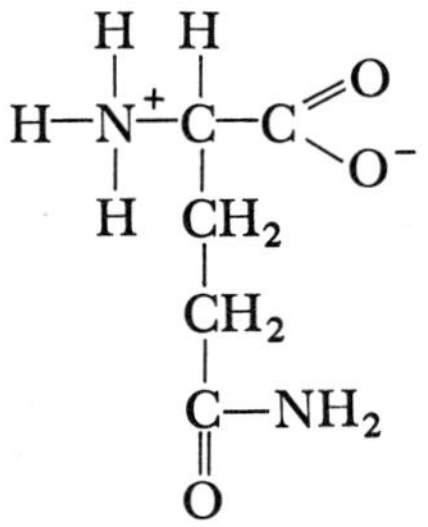

(3) Negatively charged R groups at pH 6-7

L-Aspartic acid (asp)

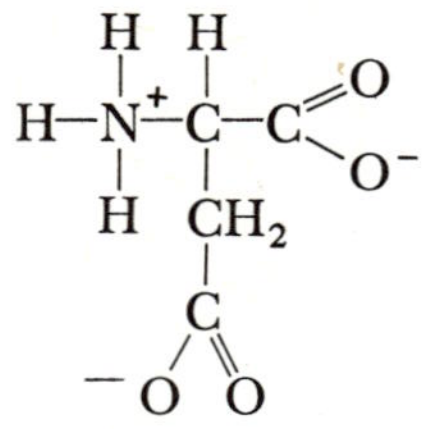

L-Glutamic acid (glu)

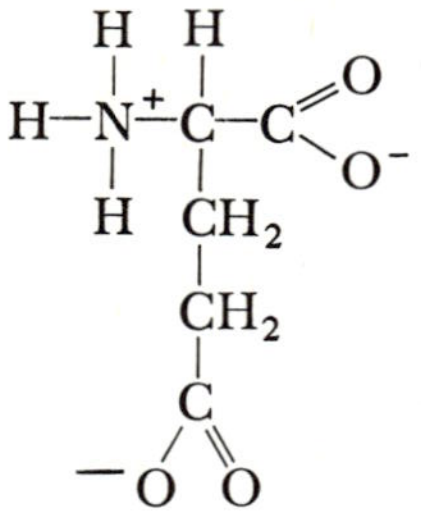

(4) Positively charged R groups at pH 6-7

L-Lysine (lys)

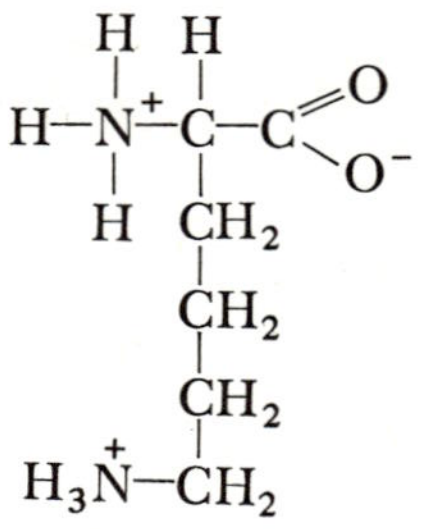

L-Arginine (arg)

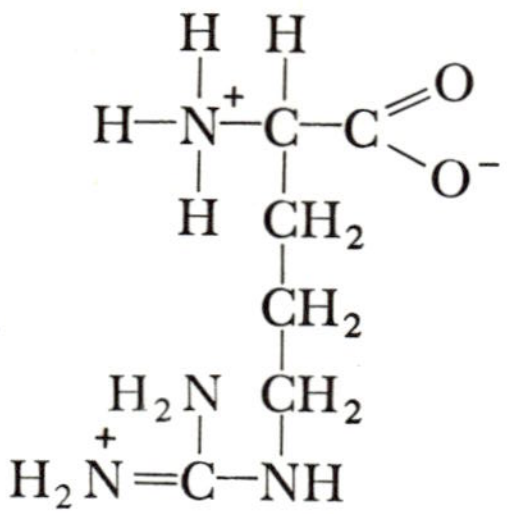

L-Histidine (his)

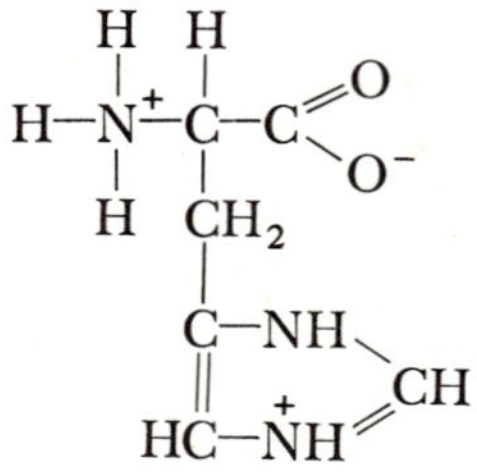

(5) Iodinated amino acids derived from Tyrosine

Diiodotyrosine

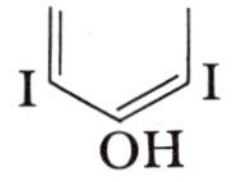

Triiodothyronine

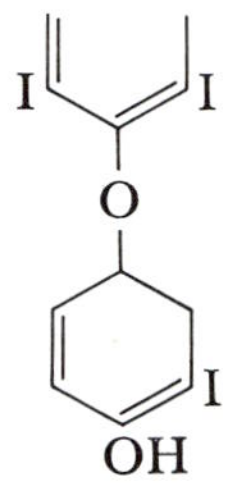

Thyroxine

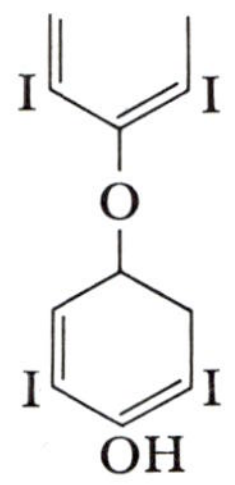

Fig. 4.1 The formulae of the twenty amino acids that occur in proteins arranged according to the properties of their R groups (four additional amino acids are also found in proteins, namely hydroxy-proline, hydroxy-lysine, phospho-threonine and phospho-serine). Also included are the iodinated derivatives of tyrosine which are present in thyroglobulin.

acid residues in the same chain or between amino acid residues in two parallel chains.

(c) Tertiary

This involves interactions between the R groups of the amino acid residues. The R groups can interact by various types of bond, e.g., H bond, salt linkage, hydrophobic interactions, S—S link. The tertiary structure is, therefore, concerned with the way in which a polypeptide chain, defined by its sequence of amino acids (primary structure) and "back-bone" interactions (secondary structure) may be further folded by side group interactions.

(d) Quaternary

Separate polypeptide chains each constituting a protein may sometimes interact one with another to provide a protein which, therefore, consists of several subunits. This interaction is known as quaternary structure.

The way in which the polypeptide chain folds upon itself to produce the protein with the correct biological properties is known as the conformation of a protein.

We will now consider each level of organization in turn.

4.5. Primary Structure

(a) The peptide bond

A peptide bond is formed by the interaction of the COOH group of one amino acid and the NH_2 group of another with the elimination of the elements of water.

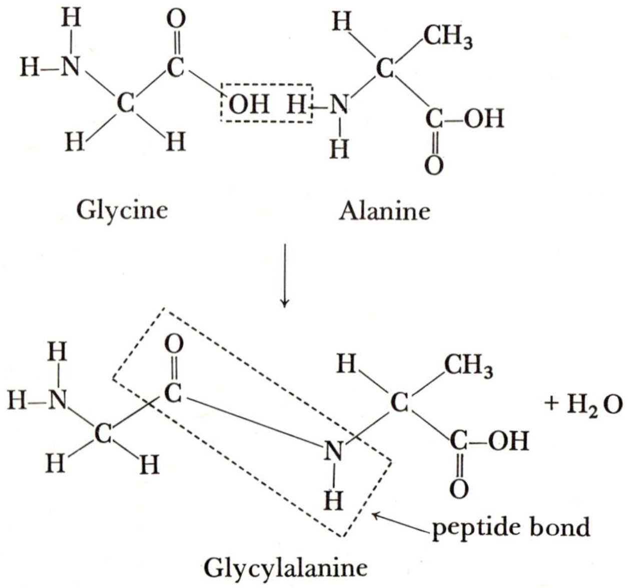

Glycine   Alanine

Glycylalanine

The nomenclature of a particular peptide always starts from the N terminus as shown in the diagram. Proline can also participate in a peptide bond but in this case we have

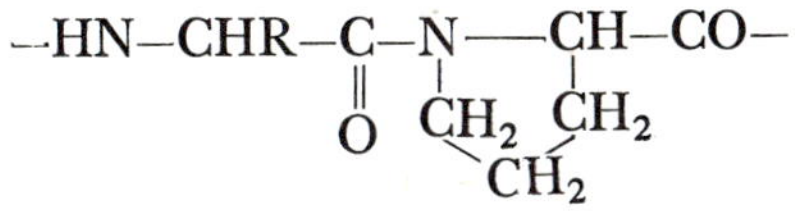

$$-HN-CHR-\underset{\underset{O}{\|}}{C}-N-\underset{\underset{CH_2}{|}}{CH}-CO-$$
$$CH_2\ CH_2$$

and there is no H available on the $\underset{\underset{O}{\|}}{C}-N-$ for hydrogen bonding.

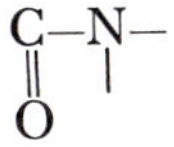

(b) Evidence of the polypeptide structure of proteins and the properties of polypeptides

(i) Titratable groups   Since most of the α-carboxyl and α-amino groups of the amino acids are engaged in the formation of peptide bonds the only groups of this kind available for titration are those at the N-terminus and C-terminus (see this Chapter 4.10 and Chapter 2.7).

When a polypeptide is hydrolyzed and the peptide bond broken the number of titratable groups should increase and this is found to be so.

(ii) Incomplete hydrolysis   Proteins may be hydrolyzed by heating with either acids or alkalis and it is possible to analyze the products of hydrolysis and show the presence of small peptides. The structure of such fragments has been checked by organic synthesis.

Another method of hydrolysis is to use proteolytic enzymes. Again it has been shown that peptides were produced as a result of the treatment of proteins with proteolytic enzymes.

(iii) The Biuret reaction   Biuret has the formula $NH_2CONHCONH_2$ and, therefore, is a simple substance containing a peptide bond. When Biuret is treated with $CuSO_4$ in alkaline solution a purple colour is produced. This is known as the Biuret Reaction and proteins give a strong reaction. It is concluded, therefore, that proteins contain peptide bonds.

(iv) Chemical synthesis   It has been possible to chemically synthesize certain proteins in the laboratory. Examples are insulin (51 amino acids), and the enzyme ribonuclease (123 amino acids). Since the products could be shown to possess the biological properties of the naturally occurring proteins we can be certain of their structure.

(v) X-ray crystallography   The examination of the diffraction pattern produced by passing X-rays through protein crystals supports the polypeptide structure of proteins. As we shall see such studies also tell us much about the conformation of proteins.

(c) Amino acid analysis of proteins

The first step in the understanding of the structure of a protein is the determination of the proportion of each of the 24 different amino acids in a sample of the homogeneous protein. The protein is first hydrolyzed by heating in 6 N hydrochloric acid *in vacuo* for 24 to 48 hours. This will satisfactorily hydrolyze all the peptide bonds between the amino acids but it also causes the destruction of some specific amino acids to a varying extent. In particular tryptophan is destroyed by acid so that other procedures must be used for the estimation of the amount of this amino acid in a protein.

Having obtained a hydrolysate containing only amino acids, the relative amount of each must be determined. This could be done by separating the amino acids by chromatography on paper or silica gel, but in practice ion-exchange chromatography on a column is now the routine method. This method was mentioned in Chapter 2.8. The absorbant is the cross-linked sulphonated polystyrene shown in Table 2.9.

The pH of the protein hydrolysate is adjusted to 3·0 at which pH all the amino acids will possess a NH_3^+ group. This will interact with the SO_3^- group of the resin and hence all the amino acids will be retained on the column. A buffer is now passed through the column, the pH and the ionic strength of the buffer gradually increasing. The increasing pH reduces the ionisation of the NH_3^+ group and so the ions of the buffer compete for the SO_3^- groups. As the pH rises so the acidic amino acids are eluted first followed by the neutral amino acids and finally the basic amino acids. The amount of each amino acid can be determined after reaction with ninhydrin which produces a purple colour except in the case of proline when the colour is yellow. A typical elution pattern is shown in Fig. 4.2. The estimation of the amino acid content of a protein hydrolysate by the

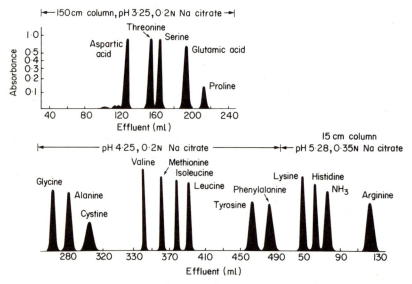

Fig. 4.2 Chromatographic analysis of amino acids separated on columns of an ion-exchange resin. The pH and ionic strength of the eluent containing sodium citrate is gradually increased. The three basic amino acids and NH_3 are separated on a second column. The shaded area under each peak is proportional to the amount of each amino acid in the mixture. (Modified from *Biochemistry* by A. L. Lehninger, Worth Publication.)

above method has now been automated thanks to the pioneer work of two American biochemists working at the Rockefeller Institute in New York, namely Moore and Stein. The apparatus used is known as an amino acid analyzer.

One use of such an apparatus is to check the phenylalanine content of the blood of children who are suffering from the inherited metabolic disease called phenylketonuria. These children tend to have a mental defect but can be reared normally if their blood phenylalanine concentration is maintained at near normal values by severely reducing the dietary intake of this amino acid.

(d) Amino acid sequence

The first problem in determining the sequence of amino acids in a polypeptide chain is to discover the nature of the amino acids at the N-terminus and C-terminus.

There are several methods of detecting the N-terminus amino acid but perhaps the best illustration is the method originally devised by Sanger for his work on the structure of insulin. The polypeptide chain is treated with the reagent dinitrofluorobenzene which reacts according to the following reaction

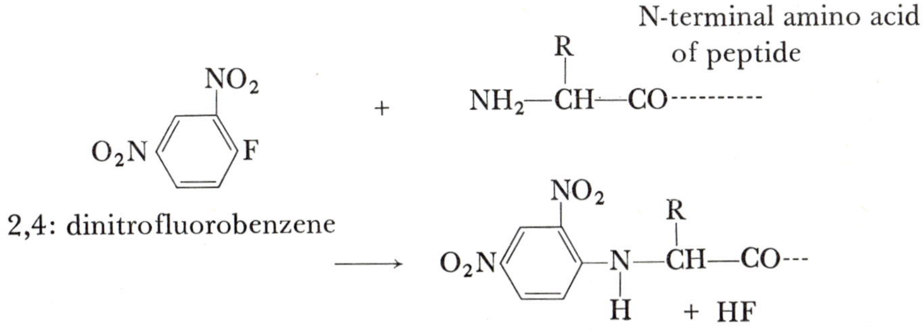

$$O_2N-C_6H_3(NO_2)-F \quad + \quad NH_2-\underset{\underset{R}{|}}{CH}-CO\cdots\cdots \quad \text{(N-terminal amino acid of peptide)}$$

2,4: dinitrofluorobenzene

$$\longrightarrow \quad O_2N-C_6H_3(NO_2)-\underset{\underset{H}{|}}{N}-\underset{\underset{R}{|}}{CH}-CO--- \quad + \quad HF$$

2,4: dinitrophenyl peptide

The product is a dinitrophenyl peptide and is bright yellow in colour. This is then hydrolyzed by acid. The dinitrophenyl amino acid is stable to hydrolysis and may be detected in the hydrolysate. It is possible to determine the precise identity of the dinitrophenyl amino acid and hence the amino acid at the N-terminus.

The easiest way to determine the C-terminal amino acid is to use the enzyme carboxypeptidase. Treatment of the polypeptide with this enzyme causes the C-terminal amino acid to be released first and its nature can then be determined by chromatography.

(e) Determination of the sequence of the amino acids in the interior of the chain

The aim must be to break the polypeptide into short peptides that are amenable to the determination of their precise structure. One wants also to produce peptide sequences that overlap. The methods of achieving this can be divided into (a) chemical and (b) enzymic.

(a) Chemical   The original method was to partially hydrolyze the polypeptide with acid but this procedure is not now used since it was not very reproducible and one could not predict which bonds were more likely to be susceptible to hydrolysis. A newer method is to use the reagent cyanogen bromide. This causes the conversion of all the methionine residues to homoserine and hydrolyzes the chain at that point leaving homoserine as the C-terminal amino acid. Since methionine is a comparatively rare amino acid in proteins the process often leads to the production of rather large peptides which can then be subjected to enzymic hydrolysis.

(b) Enzymic methods   Several of the proteolytic enzymes hydrolyze peptides at particular points so that one can predict the nature of the amino acids at the C-terminus of the resulting peptides. Trypsin is a particular favourite for it always produces peptides with C-terminal lysine or arginine. Chymotrypsin, in the main, breaks at aromatic amino acids so producing C-terminal tryptophan, tyrosine and phenylalanine. Other enzymes which are particularly useful are pronase, subtilisin and papain.

An example of the analysis of a portion of a polypeptide from milk α-lactalbumin is shown in Fig. 4.3. This shows the production of large peptides by cyanogen bromide and then the production of overlapping peptides by treatment with three different proteolytic enzymes. By piecing together the evidence from the analysis of the small peptides the precise sequence of the original polypeptide chain can be deduced with certainty.

(f) Proteins with more than one polypeptide chain

Sometimes a protein contains more than one polypeptide chain, the two chains being held by a covalent S—S bond involving the cysteine residues in the two chains.

$$\text{Cys-SH} + \text{HS-Cys} \xrightarrow{-2\text{H}} \text{Cys-S} - \text{S-Cys}$$

A typical example is found in insulin where there are two chains A and B

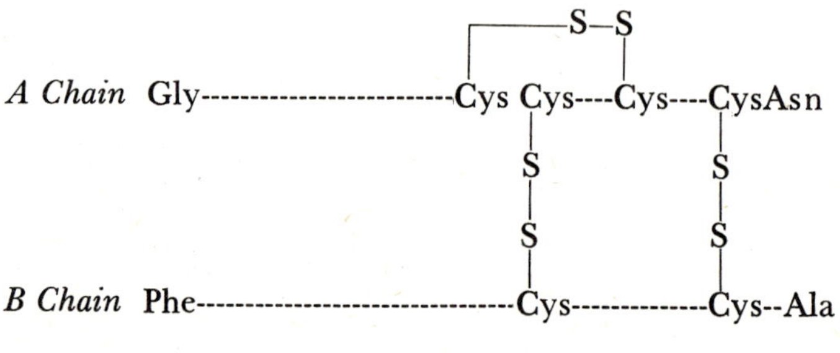

Cleavage Method

(a) Chemical e.g., CNBr.

N–I–C–D–I–S–C–D–K–F–L–N–D–N–I–T–N–N–I–M–C–A–K–K–I–L–D–I–K–G–I–N–O–W–L–A–H–K

(b) Enzymic

(i) Trypsin

N–I–C–D–I–S–C–D–K–F–L–N–D–N–I–T–N–N–I–M ‖ C–A–K–K–I–L–D–I–K–G–I–N–O–W–L–A–H–K

*(ii) Thermolysin

N–I–C–D | I–S–C–D–K | F–L–N–D–N–I–T–N–N | I–M ‖ C–A–K–K | I–L–D | I–K–G | I–N–O–W | L–A–H–K

*(iii) Chymotrypsin

N–I–C–D–I–S–C–D–K–F | L–N–D–N–I–T–N–N–I–M ‖ C–A–K–K–I–L | D–I–K–G–I–N–O | W | L–A–H–K

*Provides overlapping peptides

H = His	R = Arg	C = Cys	O = Tyr	N = Asn	S = Ser	D = Asp	I = Ile	A = ALA
P = Pro	K = Lys	M = Met	W = Trp	G = Gln	T = Thr	E = Glu	L = Leu	G = Gly
		F = Phe						V = Val

Fig. 4.3 The cleavage of a large peptide from the protein, α-lactalbumin, which occurs in human milk. The peptide containing 38 amino acid residues is cleaved first by cyanogen bromide and then into a number of overlapping peptides by three different proteolytic enzymes. (From the work of Findlay and Brew.)

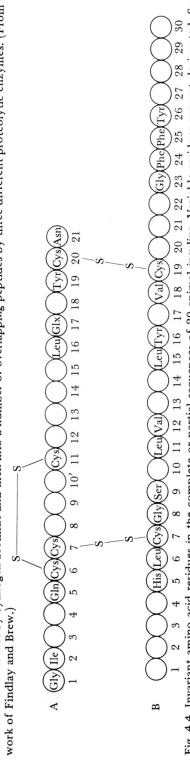

Fig. 4.4 Invariant amino acid residues in the complete or partial sequences of 20 animal insulins. Variable residues are not designated. Some insulins from fish have an additional N-terminal residue on the B (bottom) chain and the residue at B_{30} is deleted. (From P. T. Grant and T. L. Coombs: *Essays in Biochemistry*, vol. 6.)

Thus there are two inter chain S—S links and one intra chain link. Before the determination of the amino acid sequence of the chains it is necessary to separate the chains by oxidation. This is usually done with performic acid and so the cystine residues are oxidized to cysteic acid. Similarly S—S bonds between regions of the same chain are broken before the sequence determination is begun.

$$CysS\text{-}SCys \rightarrow 2Cys\ SO_3H$$

(g) Results of the determination of the sequence of amino acids

The primary structure of a large number of proteins has now been determined. The first was achieved by Sanger with insulin (51 amino acid residues) and the largest is by Edelman with immunoglobulin (1300). What general conclusions can be drawn from this work concerning the primary structure of proteins?

(i) That all possible combinations of amino acid exist. Nearly all the possible dipeptides have actually been found. One seldom finds repeating sequences.

(ii) That a particular homogeneous protein has a unique primary structure.

(iii) Enzymes which have rather similar properties often have rather similar structures particularly at the site where the substrate binds to the proteins. Examples are trypsin and chymotrypsin.

We mentioned before our interest in the structure of homologous proteins from different species. The primary structure of a number of such proteins has been determined and it might be helpful to consider a few of these.

(h) Examples of variations in the primary structure of homologous proteins

(i) Insulin We have previously mentioned that the insulin obtained from a wide variety of different species of animal has the same biological activity. The results of the determination of the primary structure of insulin shows that rather few of the amino acids are not subject to change, i.e., are invariant. This is shown in Fig. 4.4.

Since all the insulins are biologically active, it is clear that only a small proportion of the total amino acids are involved in the hormonal activity.

(ii) Haemoglobin As was mentioned in Chapter 2.7 haemoglobin has four polypeptide chains. Normal adult haemoglobin has two α chains and two β chains. Since the two chains serve the same function and are of similar size (α has 141 residues and β has 146 residues) they may be considered as homologous. Other types of chains also occur in haemoglobin, thus foetal haemoglobin has γ chains instead of β chains. In muscle, the function of haemoglobin is performed by myoglobin which only consists of one chain. One can, therefore, compare the primary structure of

haemoglobin and myoglobin chains from a wide variety of animal species. This has been done, and it is again found that only a small proportion of the amino acids are invariant so that we can conclude that not all the amino acids participate in the O_2 carrying function of haemoglobin. This point will be referred to again later.

Sometimes in comparing the primary structure of very similar proteins one can take a short cut and prepare what is known as a "finger-print" of the protein. This was done first by Ingram who wished to compare the structure of the so-called sickle-cell haemoglobin with normal adult haemoglobin.

In West Africa as many as 23% of the population have an unusual haemoglobin in their blood. In children bearing this haemoglobin, anaemia tends to develop, but the adults are normal. The effect of the presence of the abnormal haemoglobin in the red cells is to give them an abnormal shape because the abnormal haemoglobin is more insoluble than is normal haemoglobin. The abnormal red cells are crescent shape and, therefore, are called sickle cells and the anaemia that may result is called "sickle cell" anaemia. The abnormal haemoglobin is denoted as Hb S instead of Hb A for the normal adult haemoglobin.

Ingram, then working at Cambridge, set out to pinpoint the structural difference between Hb A and Hb S. He treated the proteins with trypsin, thus producing a number of tryptic peptides each of which will have a basic amino acid at the C-terminus. (This does not necessarily apply to the peptide containing the C-terminus end of the original protein.) The tryptic peptides are then separated on paper. The separation is achieved first by electrophoresis and then by partition chromatography. The peptides are detected by treating the paper with ninhydrin, and a map (referred to as the finger-print) of the tryptic peptides is obtained. When Ingram compared the finger-prints he obtained from Hb A and Hb S he found the results shown in Fig. 4.5. Just one peptide changed position. When he examined the composition of this peptide he found that the glutamic acid in Hb A had been changed to valine in Hb S. This result is fully in accord with the finger-print. The electrophoresis was performed at pH 6·4 at which pH glutamic acid would have a —ve charge. When this charge is removed on conversion of the glutamic acid to valine the peptide would move more rapidly towards the —ve electrode. The peptide concerned was in fact the N-terminal peptide of the β-chain.

Thus one can say that the phenomenon of sickle cell anaemia is due to the change of just one amino acid in one of the polypeptide chains of haemoglobin. Since sickle-cell anaemia is due to a mutational change in amino acid sequence within a species it may be called a "molecular disease". The sickle-cells and anaemia are only present in homozygotes. In heterozygotes there is no sickling or anaemia, but the persons concerned have a greater chance of survival from malaria. It is for this reason that the

Haemoglobin A

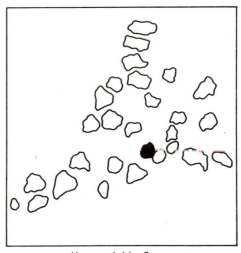

Haemoglobin S

Fig. 4.5 Finger-prints of the tryptic peptides of Hb A and Hb S. The only peptide that occurs in a different position in the two finger-prints is shaded. The peptides are applied to the paper in the lower right-hand corner and are separated from right to left by electrophoresis and in the upward direction by partition chromatography. (Modified from *Biochemistry*, by A. L. Lehninger, Worth Publication.)

disease is found in areas where malaria is still prevalent. This phenomenon is known as "heterozygote advantage".

There are many other cases of mutant haemoglobins, and so far, about 150 different haemoglobins have been examined. In nearly all cases the difference is due to the replacement of one amino acid in one of the two chains by another amino acid. Some of the haemoglobin mutations are harmless but some limit the physiological function of the haemoglobin as with Hb S.

(iii) Immunoglobulins   We have already described what happens when a rabbit is injected with a "foreign" protein such as chick serum albumin. The serum of the rabbit contains an antibody which precipitates when incubated with the antigen. The reason for this is shown in Fig. 4.6. The challenge to biochemists was the reason for the antigen and antibody interacting specifically with each other. We summarize now the evidence that has been obtained to answer this question.

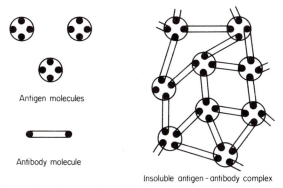

Antigen molecules

Antibody molecule

Insoluble antigen - antibody complex

Fig. 4.6 Interaction of antigen and antibody molecules which leads to the formation of an insoluble product.

Most antibodies are proteins with a molecular weight of 150,000 daltons. The basic structure of the protein has been elucidated through the work of Porter now in Oxford and Edelman in New York. The results are summarized in Fig. 4.7. Each molecule consists of two light chains and two heavy chains. The light chains have a molecular weight of 20,000 daltons and the heavy chains 50,000 daltons. The sequence of all 1300 amino acids in a typical immunoglobulin has now been determined and comparisons have been made between antibodies which are specific to different antigens. It is clear that the sequence of most of the amino acids is invariant, irrespective of the specificity of the antibody. It is the portions of the light and heavy chains towards the N-terminus that vary and which are responsible for the ability of the antibody to specifically interact with the antigens. We must presume that the differences in amino acid sequence cause subtle differences in shape so that the shapes of the antigens and the combining site on the antibody are complementary. This is a very good example of the structural complementarity mentioned in Chapter 1.

Another aspect of the structure of antibody has a special interest in respect to the aetiology of certain forms of myeloma (cancer of the blood). It was in 1847 that Henry Bence Jones, a physician at St. George's Hospital, London, described the presence of a peculiar protein in the urine of a patient of a General Practitioner named Dr. Watson. The protein was called subsequently "Bence Jones Protein" and is diagnostic of a certain form of myeloma. In this disease the patient has present abnormal amounts of an antibody—the "myeloma protein"—which is specific to a particular antigen. The Bence Jones protein is composed of the light chains of this antibody. The light chains pass through the glomerulus and are excreted.

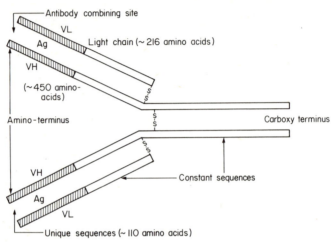

Fig. 4.7 The structure of immunoglobulin (IgG) shows two kinds of chain and regions in each. There are two heavy (H) and two light (L) chains. Towards the NH_2-terminus of H and L chains the amino acid sequence varies (dark strip) as between antibodies for different antigens whereas the rest of the chains have a constant amino acid sequence (white strip). (From A. L. Williamson, *Endeavour*, 1972, 31.) The variable regions of the L chains are indicated as VL and of the H chains VH. The antigen Ag interacts with the variable regions.

4.6. Secondary Structure

(a) Fibrous and globular proteins

From the known properties of proteins it is obvious that they cannot merely consist of long polypeptide chains but must be more sophisticated. So far as the polypeptide moiety of a protein is concerned they may be divided into two main groups; fibrous and globular proteins. The properties of these two groups may be summarized as follows.

Fibrous proteins are physically tough and are insoluble in water or dilute salt solution. They have polypeptide chains arranged along a long axis to give fibres or sheets. The fibrous proteins are the basic structural elements of connective tissue. The collagen of the tendons and bone matrix is a particularly important component. Elastin is another protein which is important in the connective tissue. Keratin is also a fibrous protein and may exist as hair, horn, feather or leather.

Globular proteins have the polypeptide chains tightly folded into compact spherical or globular shapes. Most globular proteins are soluble in aqueous solutions and they diffuse readily. The globular proteins usually have a dynamic or mobile function in the cell and are represented by the enzymes, antibodies, hormones, haemoglobin and serum albumin.

Some proteins fall into a category between fibrous and globular. Like the fibrous proteins, they are rod-like but like the globular proteins, they are soluble in aqueous salt solution. Examples of this category are myosin and fibrinogen.

We will now consider the structure of a few fibrous proteins. These examples are chosen because they illustrate the different types of structure possible and in the case of keratin and collagen because they are medically important.

(b) Silk fibroin

The fibrous protein with the simplest structure is fibroin which is the main component of silk.

A most important technique for the examination of protein structure has been X-ray diffraction. It was Astbury at Leeds who pioneered this work in his studies on fibrous proteins. The wavelength of X-rays is of the same order as the distance between the atoms in a solid. Therefore, when atoms are set in a systematic array as in crystals they act for X-rays in the same manner as an optical grating and produce a diffraction pattern which can be used to discover the systematic arrangement of the atoms. From such X-ray studies it was obvious that silk consisted of a series of polypeptide chains linked together. Moreover, it became apparent that the links were through H bonds, the nature of which have already been described in Chapter 2. Thus we have zig-zag chains linked together by H bonds as shown in Fig. 4.8.

These are known as pleated sheets because the side view of the linked sheet shows that they are rippled as shown in the Fig. 4.9.

The H bonds are individually very much weaker than covalent bonds, but if there are enough of them they make a very great contribution to the stability of the molecule. The R groups of the amino acids are in a different plane from the paper and do not participate in the interactions between the chains.

The chains in the pleated sheet may be either linked so that each is oriented in the same direction N → C (parallel) or in the opposite direction (antiparallel).

(c) Keratin

Before considering the structure of keratin we must mention certain essential properties of the peptide bond which contribute to protein structure. These are shown in Fig. 4.10.

(a) The six atoms making up the bond $C\alpha$—CO—NH—$C\alpha$ are co-planar.

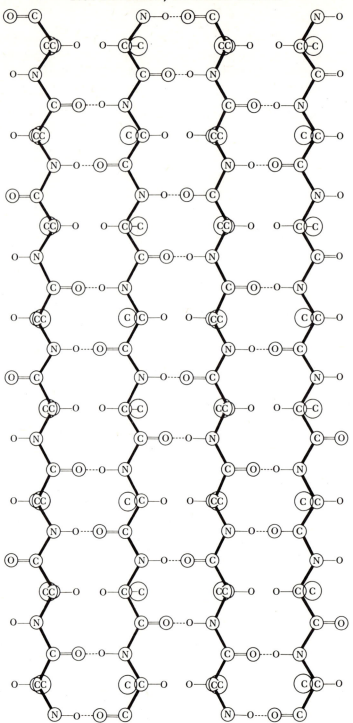

Fig. 4.8 Anti-parallel β-pleated sheet. The drawing represents the pleated sheet in the crystal structure of poly L-alanine as drawn by a computer. (This particular polymer may exist either as a pleated sheet or an α-helix.) Carbon, nitrogen and oxygen atoms are labelled C, N and O. The large balls labelled C represent methyl (CH₃) groups. The small unlabelled balls represent hydrogen atoms. Hydrogen bonds are indicated by dashed lines. (By kind permission of A. C. T. North.)

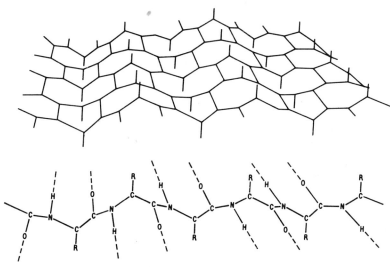

Fig. 4.9 A representation to show that the β sheet is pleated and rippled, with alternate side chains, extending above or below the plane of the H bonds. Dashed lines in the lower section indicate hydrogen bonds.

(Cα is the α-carbon of the adjacent amino acids) and in Fig. 4.10, these are shown with R', R" and R''' attached.

(b) The C=O and N—H are trans to one another.

(c) Each peptide bond can rotate about the Cα—C=O axis so that the chain can be twisted to produce a helix.

(d) The C—N bond of the link has some double bond characteristics and so cannot rotate freely.

The keratin proteins can be separated into two parts. The fibres or filaments which are fibrous proteins and have a comparatively small amount of sulphur and the amorphous proteins which are globular and which are rich in sulphur. The arrangement is like reinforcing steel rods (filaments) in a concrete foundation (amorphous proteins). The principal differences between the various mammalian keratins are due to the mode of packing of the filaments and the amount and constitution of the high sulphur-containing matrix proteins.

Part of the strength of keratins comes from the S—S links between the filaments and the amorphous protein but more especially from the S—S links in the amorphous protein itself. In the process of permanent waving the S—S bonds of the hair keratin are broken and then reformed after the fibres have been reshaped.

Astbury found that when the keratin fibres were soaked in water they could be stretched to about twice their original length. When he examined the stretched and unstretched forms by X-ray diffraction he found they were different. He called the unstretched form α and the stretched form β. The X-ray pattern of the β form was very similar to silk fibroin. Thus he argued that in some way the α chains must be more compact. It was

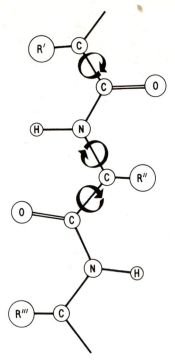

Fig. 4.10 The arrangement of the atoms in a polypeptide chain.

Pauling and Corey in California who concluded that the α-structure would have to provide the maximum possibility for H bonds within the single polypeptide chain. For this to happen the chain has to be twisted and this produces a helix which they called the "α-helix". This structure is shown in Fig. 4.11 and has been amply justified. It has the following characteristics.

(i) In contrast to the pleated sheets the H bonds in the α-helix are nearly parallel to the long axis of the helix. In the pleated sheets they are perpendicular.

(ii) All the C=O and N—H are H bonded. The links are between amino acid residues such that there are 3·6 residues per turn.

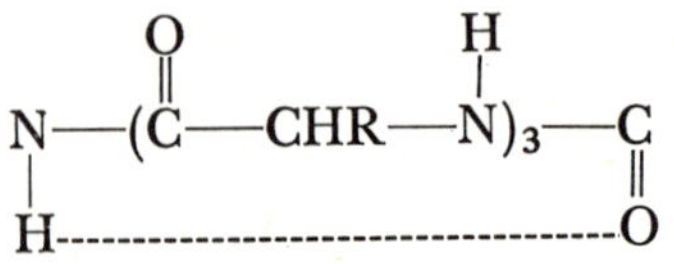

$$N-(\overset{\overset{\textstyle O}{\|}}{C}-CHR-\overset{\overset{\textstyle H}{|}}{N})_3-\overset{\underset{\textstyle O}{\|}}{C}$$
$$\overset{|}{H}\text{-----------------------------------}$$

(iii) The repeating unit is 5 complete turns of the helix which means 5 x 3·6 = 18 residues.

Fig. 4.11 The right handed α-helix. The drawing represents the helix as in the crystal structure of poly L-alanine as drawn by a computer. Symbols as in Fig. 4.8. (By kind permission of A. C. T. North.)

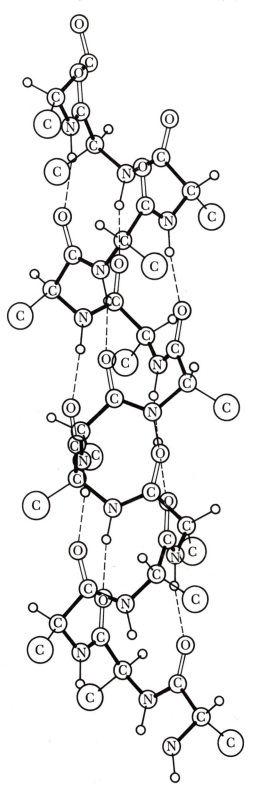

(iv) The helix can be either right-handed or left-handed but in fact virtually all known α-helices are right-handed.

(v) The R groups point outwards away from each other and do not interact.

(vi) The α-helix of keratin can be stretched to the β-form by breaking the H bonds.

(vii) We have previously mentioned that the imino acid proline differs from the α-amino acids in that in the peptide bond there is no H on the N which can participate in H bonding. It follows that the presence of proline in the chain will interrupt the α-helix.

(viii) The α-helix itself is not formed on a straight axis but is a little twisted. In hair keratin one finds three right-handed α-helices arranged in a left-handed coil.

(d) Collagen

Collagen is a fibrous protein which is an essential component of connective tissue, which in a variety of forms is found everywhere in the body. It is the material of skin, tendons, bones, cartilage and teeth.

The structure of all connective tissue is based on the same principle. Thus it contains relatively few cells and a preponderance of intercellular substance. In addition to collagen the intercellular space contains mucopolysaccharides, chiefly as protein complexes, and non-collagen proteins which are little understood. Collagen is synthesized in the cell in a soluble form known as tropocollagen and then after export from the cell changes its structure and insoluble fibrils are formed.

The proper maintenance of the connective tissue, and therefore collagen, is especially important for the well-being of the body. There are a number of diseases known collectively as the "collagen diseases". These diseases; rheumatic fever, rheumatoid arthritis, systemic lupus erythematosus and polyarteritis nodosa, are thought to be of immunological origin. Thus in some cases one can detect bacterial antibodies which cross-react with tissue antigens causing damage to connective tissue cells.

Metabolically collagen is interesting for it is in the main synthesized in the young. Collagen synthesis falls to a low level after the age of 18 although in the case of pregnancy collagen is synthesized in the uterus. Some believe that the ageing process depends on changes in the structure of collagen and hence much of the work in Gerontology is devoted to a better understanding of collagen structure.

Collagen naturally exists in a number of forms. Thus in the tendons it is built into long fibres but when these are heated it is converted into random coils and is then called gelatin. Collagen has an open sponge-like structure, absorbs water readily and shrinks on heating, e.g., drying of leather gloves.

The amino acid composition of collagen varies according to its source but all collagen is characterized by the presence of hydroxy-proline and hydroxy-lysine which together usually account for about 22% of the total amino acids, and the high proportion of glycine (33%). The content of tyrosine is low and there is no tryptophan or cysteine.

The fundamental structure of collagen is a triple polypeptide chain shown in Fig. 4.12. Since there is such a high content of proline and

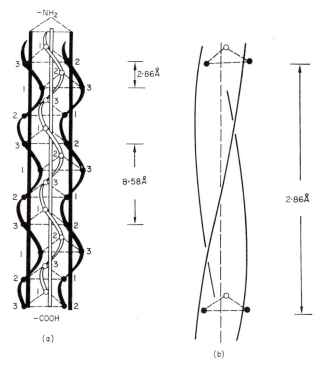

Fig. 4.12 Representation of model of collagen. The amino acids are indicated by numbers. Every third amino acid in a chain is glycine (1). The other amino acids (2) and (3) are variable but include proline and hydroxyproline. Fig. (a) shows the association of the amino acids in the three chains to allow H bonding. Fig. (b) shows the helix formed by the three chains. (From K. Kühn: *Essays in Biochemistry*, vol. 5.)

hydroxy-proline there is no α-helix in the chains. Every third amino acid is glycine so that the chains have the following sequence of amino acids: NH$_2$···Gly-X-Y-Gly-X-Y-Gly···COOH. H bonds are formed between the chains such that there is one H bond for each triplet of amino acids. This is between NH of glycine in one chain and C=O of proline or another amino acid in other chains. The three spirals are so arranged that the glycine residues that have no side chains lie inside the triple helix, while the bulky R groups of proline and hydroxyproline and the side chains of the polar amino acids are on the outside. There are other kinds of cross links between the chains of the triple helix of collagen apart from H bonds. Towards the N-terminus, these involve aldehyde groups and lysyl residues.

The chains in the collagen helix may be identical or different in amino acid composition. At either end, the chains are not organized in the helix, and it is at these points that they are more susceptible to proteolytic enzymes, particularly collagenase, which is specific for the linkage of hydroxy-proline and glycine.

The hydroxylation of the proline and lysine residues, takes place after their incorporation into the polypeptide chains in tropocollagen. The process of hydroxylation involves ascorbic acid (vitamin C) and there is no doubt that this is an important site of action of ascorbic acid. A deficiency of ascorbic acid causes scurvy, a condition involving connective tissue which leads to multiple tissue haemorrhages, for example, of the gums and periosteum.

4.7. Tertiary Structure

Most globular proteins exist in their native state as rather tight compact structures. While the secondary structure will account for the reduction of the overall size of the polypeptide chain it does not in itself account for the properties of the globular proteins. So far we have only mentioned in passing the interaction of the R groups of the amino acid residues in the polypeptide chains. We will see now that these interactions ensure that the chains fold one on another in a very specific way and do not merely collapse into a ball as with a length of knitting wool.

From X-ray studies on crystalline globular proteins, we now have a fairly precise idea of the way in which the chains of some proteins are folded.

(a) The characteristics of the R groups

The amino acids have already been written and listed in such a way that the various R groups predominate, see Fig. 4.1. If we consider these then we can classify them as follows.

(i) Amino acids with uncharged nonpolar	Alanine
	Isoleucine
	Valine
	Proline
or hydrophobic (water-hating) R groups.	Leucine
	Phenylalanine
	Tryptophan
	Methionine

(ii) Amino acids with uncharged polar or hydrophilic (water-loving) R groups. These are more soluble in water than those listed in (i) above.	Glycine Serine Threonine Cysteine Tyrosine
The polarity of serine, threonine and tyrosine is contributed to by their OH groups, that of asparagine and glutamine by their amide groups and that of cysteine by SH. Glycine is difficult to classify.	Asparagine Glutamine
(iii) Amino acids with negatively charged R groups at pH 6·0 to 7·0.	Aspartic Glutamic
(iv) Amino acids with positively charged R groups at pH 6·0. At pH 6·0 in the case of histidine about 50% of the molecules possess a +ve charge, but at pH 7·0 less than 10% have a +ve charge	Lysine Arginine Histidine

(b) Tertiary structure of myoglobin

The first convincing work in which X-ray crystallography contributed to our understanding of the tertiary structure of a globular protein, was done by Kendrew working at Cambridge on myoglobin. This is the O_2-carrying protein of mammalian muscle. It only has one chain of 153 amino acids and one Fe-containing haem group. It has a molecular weight of 17,500 daltons and does not contain S—S links.

The results of the X-ray work showed that myoglobin has the structure shown in the Fig. 4.13.

Some of the polypeptide chain is in the α-helical configuration but other parts are non-helical. In all 77% of the chain is α-helical. It was of particular interest to see which amino acids were at the points in the chain

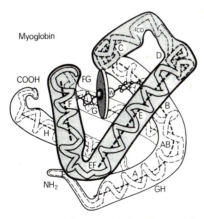

Fig. 4.13 The tertiary structure of myoglobin. The molecule is built up from 8 stretches of α-helix that form a box for the haem group. Histidines interact with the haem to the left and the right and the oxygen molecule sits at W. Helices E and F form the walls of a box for the haem, B,G_2 and H are the floor and the CD corner closes the open end. COOH indicates the carboxyl terminus and NH_2 the amino terminus. (Modified from R. E. Dickerson and I. Geis, *The Structure and Action of Proteins*, Harper and Row.)

where there is a change in direction. Not surprisingly proline was much in evidence at such points. Not only can proline not participate in the α-helix but the angle of the CO and N is different from that of the CO and NH of the α-amino acids.

The ferrous Fe atom of haem is coordinated both with the nitrogen atoms of the porphyrin ring and also with the N of histidine as shown in Fig. 4.14. The O_2 associates with the Fe but the precise manner of the interaction is unknown.

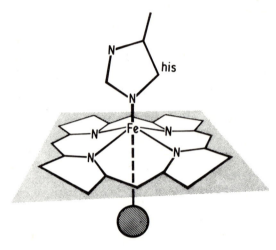

Fig. 4.14 The location of the haem group in haemoglobin. Haem is shown with its iron linked to the nitrogen of histidine above the plane of the porphyrin ring. Either water or oxygen can interact wth iron below the plane of the ring.

Apart from the link between Fe and histidine, the haem which is itself hydrophobic, is held in a hydrophobic milieu consisting in the main of non-polar amino acids (group 1).

What emerged from the studies on myoglobin was that its tertiary structure exactly fitted it for the interaction with O_2. This is a model for the way in which enzymes combine with their substrates. The general features to emerge were:

(i) The molecule is compact and only has room for a few molecules of water inside.

(ii) Nearly all the polar groups, including the charged ones, i.e., lys, arg, glu, asp, his, ser, thr are on the surface of the molecule and hence are exposed to the solvent water. This means that ionic forces play little part in holding the structure together, but explains why a change in pH of the environment has a marked affect on the solubility of the protein.

(iii) The interior of the molecule has nonpolar residues such as val, leu, ile and phe. This is why proteins in general are rather insoluble in organic solvents. A rather crude description of a globular protein is that it is an oily drop with polar coatings.

(c) Other globular proteins

As we have already stated haemoglobin has 4 chains, each of which has a structure which is rather similar to that of the single chain of myoglobin. The work on this protein was done by Perutz working at Cambridge. The four chains of haemoglobin fit together to give a compact globular structure as shown in Fig. 4.15. When O_2 combines with the Fe of the haem of one of the β chains, the distance between the two β chains decreases. This movement is important in that it assists the binding of subsequent molecules of O_2. This explains the very unusual O_2 saturation curve of haemoglobin mentioned in Chapter 2, and will be referred to in detail in Chapter 5.11 on enzymes.

The Fe in both haemoglobin and myoglobin is in the ferrous form, and does not undergo valency change as O_2 is bound and lost. The Fe can be oxidized to ferric by ferricyanide and the colour of the solution of haemoglobin changes from red to brown. This is due to the formation of methaemoglobin which cannot bind O_2. In another group of haemo-proteins, the cytochromes, the Fe undergoes reversible changes from ferrous to ferric, but they act as electron rather than O_2 carriers.

Carbon monoxide competes with O_2 for binding to haemoglobin. In human haemoglobin the CO affinity to haemoglobin is 200 times that of O_2 which accounts for its toxic effect. Since these triumphs with myoglobin and haemoglobin many other globular proteins have been studied. The first true enzyme to be worked on successfully was lysozyme by Phillips then in London. This was followed by ribonuclease, carboxy-peptidase, elastase and chymotrypsinogen.

The general ideas about the structure of globular proteins, deduced from the early work, have held good, except that the proportion of α-helical regions in the later proteins was much less than with myoglobin. Instead it is common to find the β-pleated sheet structure we described for silk fibroin. In other words the chains are often folded so that they run parallel or antiparallel and H bonds are formed between them.

(d) Nature of the interactions between the R groups of the amino acid residues

The interactions which play so important a part in the maintenance of the tertiary structure of proteins are shown diagrammatically in Fig. 4.16. The only interaction which is covalent is that between the two halves of the cystine residues. We have already seen this bond in insulin, and it plays an important part in many other proteins, particularly enzymes that have to retain their activity in an unfavourable milieu such as gastric juice. The S—S links in egg-white lysozyme and the whey protein α-lactalbumin are shown in the Fig. 4.17. The tendency of hydrocarbons to associate together in aqueous solutions leads to the formation of apolar or

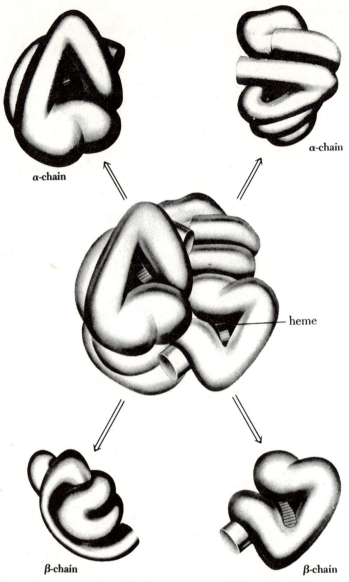

α-chain

α-chain

heme

β-chain

β-chain

Fig. 4.15 The arrangement of the four chains of haemoglobin showing how they interact to give a compact structure. (From *Biochemistry* by R. W. McGilvery, W. B. Saunders.)

hydrophobic bonds. These are certainly very weak bonds, but if there are enough of them, which there will be in the interior of a globular protein, they add to the stability of the molecule. The amino acids that are particularly important here are val, leu, ile and phe. These all have a common lack of affinity for water and are thereby pushed together out of the network of water molecules that are bound together by internal H bonds.

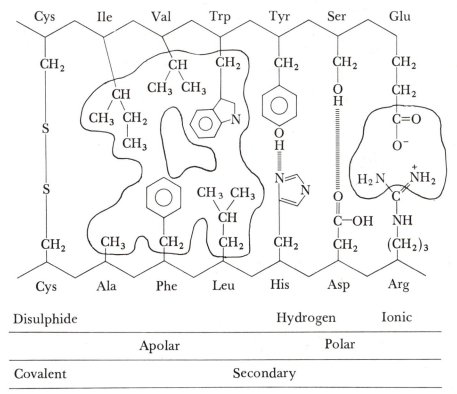

Fig. 4.16 The variety of bonds or interactions stabilizing the tertiary structure of protein molecules. (Modified from Loewy and Siekevitz, *Cell Structure and Function*, Holt, Rinehart and Winston Inc. N.Y.)

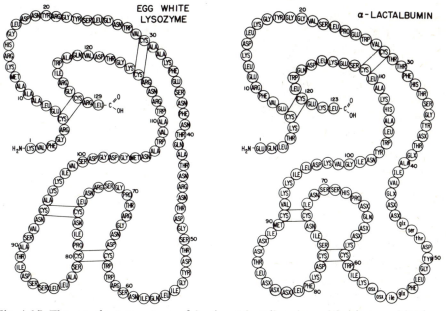

Fig. 4.17 The covalent structure of bovine α-lactalbumin and hen's egg-white lysozyme. (From Hill, Brew, Vanaman, Trayer and Mattock: *Brookhaven Symposium in Biology*, 1968, 21, p. 139.)

Apart from the H bonds formed in the α-helix and in pleated sheets, H bonds are formed between other groups. As shown in Fig. 4.16 the imidazole ring of his, the OH of tyr, ser and thr and the β-COOH of asp and the γ-COOH of glu participate. Also involved are the amide groups of gln and asn.

Finally there are some ionic interactions between the charged amino acids. We have seen that most of these groups occur on the outside of the molecule but there will be some interactions between lys and arg on the one hand and asp and glu on the other.

4.8. Quaternary Structure

Many functional proteins are composed of several separate polypeptide chains. The chains are not covalently linked but are held together by the same forces that are involved in the tertiary structure. An example of quaternary structure is the four chains of haemoglobin. Here the chains are easily dissociated and reassociated in solution. In the case of many enzymes, which have been shown to consist of subunits, it is more difficult to separate the subunits. The types of interaction responsible for the tertiary structure of a protein are also important for holding together the subunits. Since H bonds are important in this respect, one way of separating the subunits is to suspend the protein in a strong solution of urea which competes for the H bonds. The result is the separation of the subunits which may be detected by electrophoresis.

A good example of an enzyme with a subunit structure is lactate dehydrogenase. Here there are four chains. If the enzyme is isolated from a variety of different tissues of man it is found to consist of two slightly different chains A and B in a variety of combinations. Thus one can find an enzyme that consists of AAAA, BBBB, ABBB, AABB and AAAB. The actual properties of the resulting enzyme differ a little and satisfy the requirements of the particular tissue. Further reference to this enzyme will be made in Chapter 5.11

4.9. Denaturation

From the description of the various levels of organization of proteins, we reach the conclusion that each protein has a particular conformation which is peculiarly fitted to the biological functions of that protein. On the other hand, it would be quite wrong to think of a globular protein as having a fixed structure, as might be deduced from the work on crystals. A protein in solution has great mobility and will be changing in structure over quite a wide range. Thus it is to be likened to an accordion. Any damage to the structure of a globular protein may cause it to change markedly with loss of biological activity. Proteins which are likely to be exposed in an unfriendly environment often have S—S bonds to increase their rigidity. One finds no S—S bonds among the intracellular proteins of a bacterium such as *E. coli* but they are very common in proteins exported from animal cells such as ribonuclease and among the structural proteins like keratin.

The act of changing the conformation of a native protein is known as denaturation, and the resulting protein is said to be denatured. Denaturation may be accomplished by heat, high concentration of urea, extremes of pH, or organic solvents. Even shaking to cause a foam, can denature a protein.

During denaturation there is not normally any cleavage of peptide bonds. There are many ways of testing for the phenomenon, such as increase in viscosity, decrease in solubility, or changes in the sedimentation and diffusion coefficients. Denaturation will also result in the loss of activity of any enzyme. Thus, if ribonuclease is treated with a strong solution of urea and also a reducing agent such as mercaptoethanol to convert the S—S bonds to SH, it loses all its enzyme activity. We have mentioned before that the urea effectively breaks the H bonds of the protein. The result is that the protein is now merely a polypeptide chain with no secondary or tertiary structure and this is why it then has no enzymic activity. As a result of the removal of the urea and mercaptoethanol by dialysis the H bonds and S—S links are reformed and the enzymic activity is regained. This is the phenomenon of reversible denaturation. This kind of result has been achieved with several enzymes besides ribonuclease, viz., trypsin, amylase and lysozyme.

That it is possible to reverse the process of denaturation, indicates that the primary structure is responsible for the tertiary structure and that the polypeptide chain folds into the most thermodynamically stable structure which possesses the desired biological activity. That it is possible to chemically synthesize ribonuclease in the laboratory, is confirmatory evidence for this conclusion.

We have previously mentioned the studies on the human mutant haemoglobins, and said that the differences usually only involve one amino acid in one of the two chains. If one considers the total number of amino acid changes that have now been found in each chain of about 150 amino acids, one finds that *in toto* a surprisingly large number of amino acid residues are subject to change. In fact only about 80% of the residues in a chain are identical in all cases. How is it then, that these haemoglobins have the same physiological properties, and the α and β chains have a very similar tertiary structure? The answer must be that there are only certain positions in the amino acid sequence that determine the tertiary structure, and that some changes are possible in the other positions without affecting the tertiary structure.

Denaturation may also be irreversible. This is the more usual form of

denaturation especially where heat, extremes of pH, or organic solvents
are the agents. Proteolytic enzymes are slow to catalyze the breakdown of
native proteins but after the destruction of the tertiary structure through
denaturation the polypeptide chain is much more readily degraded. This is
one of the reasons why cooked protein is more digestible than native
protein. After denaturation proteins tend to aggregate. The reason for this
is that in solution the molecules of native proteins repel each other but
after their structure is destroyed this property is lost. This is what happens
when the ovalbumin of an egg is denatured by heating.

4.10. Protein as Electrolytes

In some ways a solution of a protein behaves like a mixture of amino acids
from which it is composed. Thus it is dipolar and the proteins act as
buffers against the change of pH. The pK of the ionizable groups of amino
acid residues within the polypeptide chains are shown in Fig. 4.18. We can
summarize for ribonuclease which we may regard as a typical globular
protein as shown in Table 4.2.

TABLE 4.2. The titratable groups of ribonuclease

Group	Amino acid involved	No. of residues	pK
α-NH$_2$	N-Terminal	1	7-8
Side-chain COOH	asp, glu	10	4-5
Side-chain NH$_2$	lys	10	10-11
Guanidyl	arg	4	12-13
Imidazole	his	4	6-7
α-COOH	C-Terminal	1	3-4

We may note that in a protein the proportion of α-NH$_2$ and α-COOH
groups is very small so that the buffering capacity must come from the
ionisable side chains. In fact it is the imidazole groups of histidine which
contribute most to the buffering capacity under physiological conditions
for it is only they that change their state of ionisation in the range 6-7.
Haemoglobin is exceptionally rich in histidine having 8% and this plays an
important role in the intracellular buffering capacity of the red cell.

The physical properties of a protein are very different from those of a
mixture of the amino acids comprising it. Thus protein structure is
extremely sensitive to pH changes and temperature. This is because the
molecule depends on the existence of many ionic interactions and weak
bonds between different parts of the chain.

At a certain pH value the net charge on a protein will be zero, and this
is known as the isoelectric point (I.E.P.), (see Chapter 2 and Table 2.8).

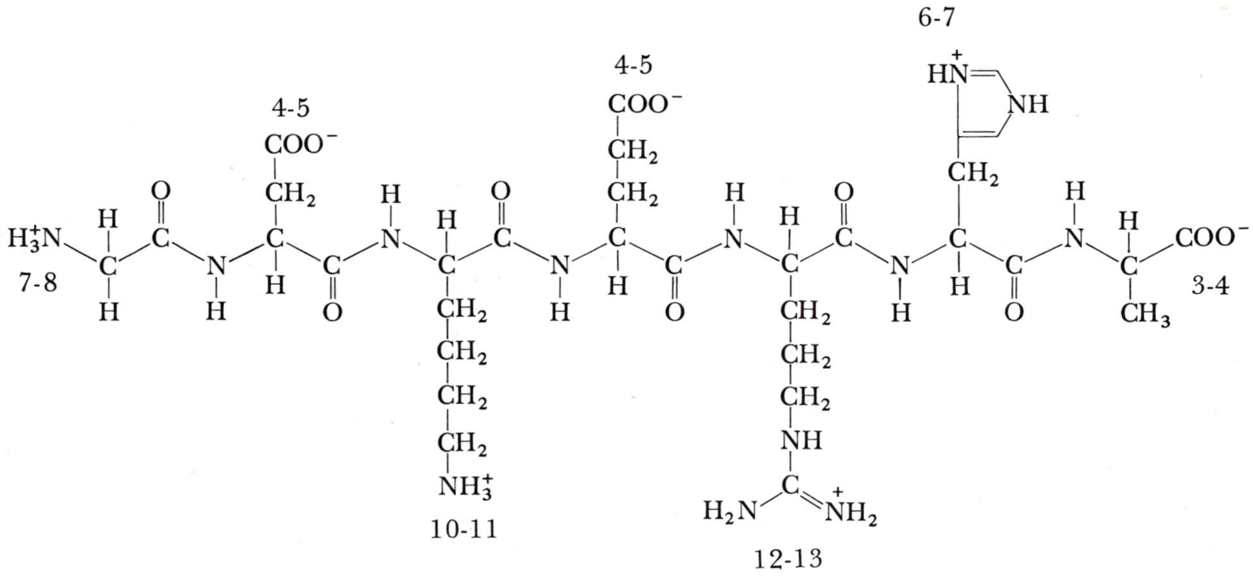

glycyl—aspartyl—lysyl—glutamyl—arginyl—histidyl—alanine

Fig. 4.18 A hypothetical polypeptide containing all the groups that normally contribute charges to proteins. The numbers represent the pK range of each dissociating group. (From Loewy and Siekevitz, see Fig. 4.16.)

At this pH the protein does not move under the influence of an electric field, i.e., in electrophoresis. At the I.E.P. the protein has its lowest solubility. This is because with charged molecules there is a repulsion which keeps them apart. If this charge is removed the molecules aggregate. As shown in Table 2.8, for many proteins the I.E.P. is between pH 4·5 and 6·5, but in the case of pepsin, it is 1 and for lysozyme it is 11. At any pH above the I.E.P., the protein carries a net negative charge and at any pH below the I.E.P., the protein carries a net positive charge.

4.11. Ion-binding

An important property of proteins is to bind ions. Since proteins are amphoteric they bind inorganic ions. The binding of cations such as Ca^{2+} and K^+, is of considerable physiological importance for 30-50% of the inorganic Ca^{2+} in the blood is bound to protein, and most of the K^+ in the red blood cell is bound to haemoglobin. In general the inorganic cations are bound to the carboxylate ion of proteins and in the case of phosphoproteins also by the phosphate groups attached to the OH group of serine.

Proteins will also combine with the anions of organic acids. Thus picric acid, salicylsulphonic acid and trichloroacetic acid are protein precipitants. For binding to take place the protein must have a positive charge, i.e., must be at a pH which is below the I.E.P. A similar phenomenon is responsible for the binding of dyes to proteins after they have been denatured on paper or cellulose acetate following electrophoresis.

Apart from the simple binding of anions and cations there is also the phenomenon of chelation. This involves the binding of metal ions at two or more sites and is especially important for the proteins with prosthetic groups containing metals. Many different metals are involved, such as, Cu,

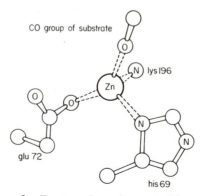

CO group of substrate

lys 196

Zn

glu 72

his 69

Fig. 4.19 The interaction of a Zn atom in carboxypeptidase A. The Zn chelates with glu 72, his 69 and lys 196. It also interacts with the carboxyl oxygen of the bond to be cleaved in the substrate.

Co, Mn and Zn. For the activity of the proteolytic enzyme carboxy-peptidase A, already referred to in section 4.5(d), a Zn atom is required. The interaction of the Zn in this enzyme with three amino acids and the substrate is shown in Fig. 4.19.

In addition to ions, proteins also bind a wide variety of other substances such as steroids, hydrocarbons, long-chain alcohols and fatty acids. Many drugs circulate in the blood with a substantial proportion bound to plasma proteins.

4.12. Ionic Strength and Solubility

In a medium of low ionic strength, the proteins become more soluble as the strength is increased. This is known as salting in. The salt stabilizes the

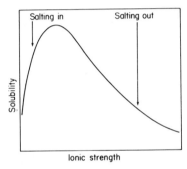

Fig. 4.20 A representation of the relationship between the solubility of a globular protein and the ionic strength of the aqueous medium.

charged groups in the protein. As the ionic strength of the solution is further increased, e.g., with ammonium sulphate, then the protein becomes less soluble. This is known as salting out, (Fig. 4.20). This is thought to be because of the competition between the protein and the salt for the available water, so that instead of protein-water interactions, one gets protein-protein interactions and the protein precipitates.

4.13. Determination of Molecular Weight

The large size and lability of proteins means that rather special methods had to be devised for determining their molecular weight. Amongst the most important methods available are:

(a) Ultracentrifugation

The rate at which a macromolecule moves in a centrifugal field will depend on, among other factors, the molecular weight. Although proteins are macromolecules they are not so large that they move rapidly in a centrifugal field and hence acceleration of the order of 200,000 g must be applied for the measurement of their sedimentation coefficients. This is a function of both the weight and shape of the particle. A value of 1×10^{-13} sec is called a Svedberg unit, abbreviated S. Provided one can determine the shape, density and hydration of a protein by other methods, then the molecular weight can be determined from its S value.

(b) Gel filtration

A very convenient method derives from the principle of gel filtration. In this procedure the substances in solution are passed through a column containing a dextran which has a structure rather similar to glycogen. The dextran forms a molecular sieve so that when a mixture of substances of varying molecular weight are passed through a column of the dextran (known as Sephadex) the smaller substances are retained in the molecular mesh whereas the larger molecules pass through with varying degrees of retention depending on their size. By measuring the degree of retention, the molecular weight of the proteins may be determined. This procedure is also useful for removing small molecular weight substances, e.g., salts, from a protein solution.

A further development of this procedure is to carry out electrophoresis of a protein solution on polyacrylamide gel in the presence of sodium dodecyl sulphate (SDS), which binds to the surface of the proteins by hydrophobic interaction. Under these conditions the rate of migration of the protein will be proportional to the molecular weight since the proteins will virtually carry the same charge due to the presence of sulphonic acid groups.

4.14. Protein Purification

Owing to the delicate nature of proteins, special methods must be used in their purification in order to avoid denaturation. It is also not easy to define a state of purity in very precise terms. In this case crystallization in itself is not adequate, for the crystals may retain small molecules and may indeed be mixed crystals of more than one protein, e.g., pro-insulin and glucagon may be present in crystalline insulin. It is usual, therefore, to attempt to purify a protein by a number of different methods and to analyze the product in as many different ways as possible. If, by all these methods, the evidence is that there is only one component, then we say that the protein is homogeneous. We mean by this, that according to all the criteria used, all the molecules present in our preparations are identical. We prefer the term homogeneity to purity.

The methods most commonly used in purification are:

(a) Salting out with either ammonium sulphate or sodium sulphate as already described.

(b) Organic solvents. Most of the fractions of serum proteins that are used in hospitals are purified by the use of either ethanol or ether in the presence of salts at various pH values. In order to prevent denaturation by the organic solvent the fractionation is done at a low temperature. In this case three variables are employed all of which have an effect on the solubility of protein.

(c) Isoelectric precipitation as already described.

(d) Gel filtration as already described.

(e) Ion-exchange chromatography. We have mentioned the use of ion-exchange chromatography on resins of sulphonated polystyrene for the separation of amino acids. These resins would cause denaturation of proteins. Instead weak acid and basic groups are attached to cellulose and these modified celluloses are used for the ion-exchange chromatography of proteins. Examples of modified celluloses have been given in Table 2.9. As stated there, the two most common materials are DEAE-cellulose (anion exchange) and CM-cellulose (cation exchange).

(f) Preparative electrophoresis. Since electrophoresis is a very discriminating method for the separation of proteins it is in principle a good method for their purification. There are, however, difficulties in the recovery of the native protein from the supporting medium such as polyacrylamide gel and the quantity of proteins that can be treated in this way is very limited.

At present the best single criterion of homogeneity, is that the protein gives a single-band when subjected to electrophoresis in gels of polyacrylamide. Sometimes even a homogeneous protein will give two bands on electrophoresis because the monomer aggregates to a dimer. A good check on whether the two bands are due to different polypeptides is to treat each band with trypsin and examine the tryptic finger-print as was done for Hb A and Hb S. Finally a homogeneous protein should have only one N-terminal amino acid per polypeptide chain.

4.15. Plasma Proteins

It was traditional to consider in detail the classification of proteins into various groups. Nowadays this is unnecessary since our better understanding of both structure and function of individual proteins has superseded the traditional, and often, arbitrary, system of classification. However, there is one group—the plasma proteins—that a medical student needs to study in detail. The plasma contains about 7% protein of which more than half is albumin and the rest globulin.

(a) Albumin and globulin fractions

These two terms need a word of explanation. Albumins and globulins are two classes of globular proteins distinguishable in their solubility. Table 4.3 shows the solubility criteria for albumins and globulins.

The classical definition of albumins and globulins refers to their solubility in distilled water. Some proteins which, according to this

TABLE 4.3. Some properties of albumins and globulins

	Albumins	Globulins
Dist. water	soluble	insoluble
Dilute salt	soluble	soluble
$\frac{1}{2}$ sat. $(NH_4)_2SO_4$	soluble	insoluble
Fully sat. $(NH_4)_2SO_4$	insoluble	insoluble

definition, are globulins are soluble at $\frac{1}{2}$ sat. $(NH_4)_2SO_4$ (e.g., β-lactoglobulin) so that it will be appreciated that the above definitions are not rigorous but are convenient.

The composition of the serum proteins is best examined by electro-phoresis on either filter paper or cellulose acetate (see Fig. 2.15). The pattern shown in Fig. 4.21 is typical of that obtained and indicates that in addition to serum albumin four bands of glubulins are detected.

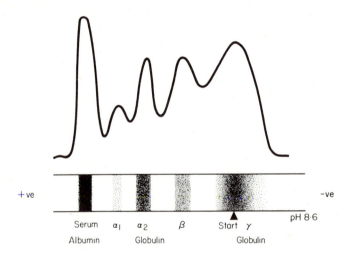

Fig. 4.21 The electrophoretic separation of the serum proteins of a patient suffering from a bacterial infection. The proteins have been separated by electrophoresis on a strip of cellulose acetate, the protein bands stained and the strip scanned. The area under the curves is approximately proportional to the amount of protein on the strip. Because of the infection the amount of γ-globulin in the patient's serum is greater than normal.

For fractionation the serum proteins are often separated into albumin and globulin fractions by ammonium sulphate. If the components of the fractions are examined by electrophoresis, one finds that while most of the serum albumin is present in the albumin fraction, some is also present in the globulin fraction. Similarly most of the α_1-globulin will be in the globulin fraction but some is in the albumin fraction. This indicates that salt fractionation while useful is not very precise.

A cursory examination of the electrophoresis pattern in Fig. 4.21 might suggest that the serum proteins only consisted of 5 proteins, but this would be a quite erroneous conclusion. The only band which represents a single homogeneous protein is that due to serum albumin. The α_1, α_2, and β globulin bands are due to a very complex mixture of proteins and the γ-globulin band, as we have mentioned before, contains a broad range of antibodies each of which differ in primary structure.

(b) Serum albumin

This protein has a molecular weight of 69,000 and consists of a single polypeptide chain of about 575 amino acids. Although it is the smallest of the serum proteins, it is a comparatively long polypeptide chain for a globular protein. Serum albumin is the most abundant of the serum proteins; it has a rather low I.E.P. at pH 4·7. This and its high net charge explains why it migrates most rapidly when electrophoresis is performed at pH 8·6. At a physiological pH of 7·4 serum albumin has a net negative charge of 18. This also explains why it is very soluble in water so that it is possible to get a solution containing 40% protein at pH 7·4.

The shape of serum albumin is that of an ellipsoid. Thus, while it is much more symmetrical than the fibrous proteins, it is not as compact as haemoglobin. The shape accounts for the fact that solutions of serum albumin have a low viscosity and this is important since the amount of work to be done by the heart depends, amongst other things, on the viscosity of the blood.

The osmotic effect of albumin is very important. While quantitatively it accounts for 50-60% of all serum proteins it is responsible for 75-80% of the osmotic pressure of the serum proteins. However, the main physiological function of serum albumin is probably in transport. Many substances bind to it, but in particular fatty acids, so that it is the major carrier of plasma fatty acids. The precise function of this protein is not completely clear for patients are known who have virtually no serum albumin and they appear to lead a normal life.

(c) Gamma-globulin

The structure of the immunoglobulins, which are found in the γ-globulin band on electrophoresis, has been mentioned earlier. The commonest antibody (about three quarters of the total circulating) is known as IgG and has an S value of 7. It is particularly effective in binding to the surfaces of bacteria and in neutralizing bacterial toxins which act as antigens. IgM with an S value of 19 is a pentamer of the basic four chain structure of IgG the chains being linked by S-S bonds near their carboxy-terminus. Other immunoglobulins are IgA, which although present in serum occurs in higher concentration in external secretions such as colostrum and saliva, and IgE concerned with allergic responses. There are differences in the primary structure of the chains of each class.

(d) Other globulins

The α_1-globulins include fractions that combine with bilirubin and the α_1-lipoprotein. Also prominent is a glycoprotein which increases in concentration when the body contains rapidly growing tissue such as a foetus or a tumour. This protein has, therefore, been much studied in connection with cancer.

The α_2-globulins include another glycoprotein and also haptoglobin. The latter binds haemoglobin derived from the breakdown of red cells and the combined protein is then taken up by the reticuloendothelial systems.

The β-globulins include the β-lipoproteins and the globulin responsible for the transport of iron, namely transferrin. This fraction also includes caeruloplasmin which transports copper. The absence of this protein is the basis of Wilson's disease in which the toxic effects of free copper are observed. Prothrombin, the precursor of thrombin, is also present as a β-globulin.

(e) Fibrinogen

This protein is, of course, not present in serum but is present in plasma. This is because of the reaction that occurs when plasma clots and the fibrin is lost in the insoluble clot. Fibrinogen is an elongated molecule,

$$\text{Fibrinogen} \xrightarrow[\text{Proteolysis}]{\text{Thrombin}} \text{fibrin} + \text{fibrino} - \text{peptides A and B}$$

twenty times as long as its diameter. During the conversion of fibrinogen to fibrin there is a very specific cleavage of peptides.

4.16. Naturally Occurring Peptides

(a) Glutathione

For a long time research was directed at the discovery of naturally occurring peptides in the body. At one time it was thought that they would play an intermediate role in the synthesis of proteins from amino acids, but this is not so. The predominant peptide which occurs in the tissues is glutathione. This was discovered in 1930 by Gowland Hopkins who is generally considered to be the father of biochemistry in the U.K. and earlier had played a leading role in the discovery of the vitamins.

Glutathione is a tripeptide γ-glutamylcysteinylglycine, as shown below.

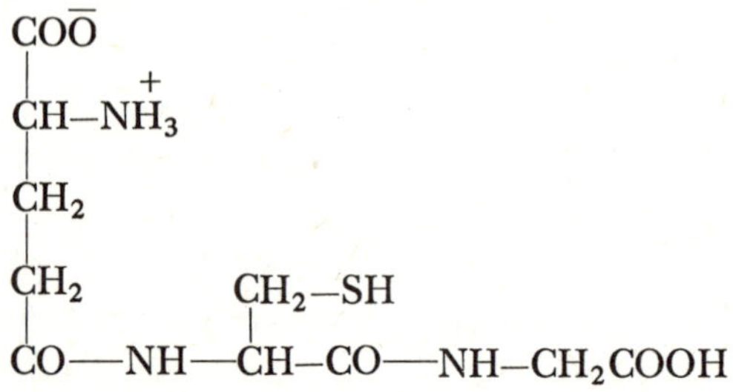

$$
\begin{array}{l}
\overline{\text{COO}} \\
| \\
\overset{+}{\text{CH}}-\text{NH}_3 \\
| \\
\text{CH}_2 \\
| \\
\text{CH}_2 \qquad\quad \text{CH}_2-\text{SH} \\
| \qquad\qquad\qquad | \\
\text{CO}-\text{NH}-\text{CH}-\text{CO}-\text{NH}-\text{CH}_2\text{COOH}
\end{array}
$$

The first peptide bond is through the γ-COOH of glutamic whereas this never occurs in proteins. The cysteine imparts to glutathione an SH group and so we can have oxidized and reduced glutathione. The biological role of glutathione will be discussed later in Chapter 7 under nitrogen metabolism.

(b) Physiologically active peptides

We have seen that the precise conformation of a protein is very specific for its biological function, but the presence in the body of small peptides reminds us that even such small molecules can also be very active and possess great specificity. There are three groups of such peptides. (a) Those that are hormones such as oxytocin, vasopressin, adreno-corticotrophic hormone (ACTH), melanocyte stimulating hormone (MSH), glucagon, calcitonin and parathyroid hormone. (b) Those active in digestion such as gastrin and secretin. (c) Those that are derived from an α_2-globulin in the serum, such as angiotensin and kallidin.

As an example of a peptide possessing great specificity, thyrotropic releasing hormone (TRH) may be mentioned. This is one of the factors released by the hypothalamus on receiving a neural stimulus. Each factor is relatively specific for an individual pituitary hormone so that TRH causes the release of thyrotropic-stimulating hormone (TSH). It has now been shown by chemical synthesis that TRH in pig and sheep is a tripeptide pyroglutamyl-histidyl-proline amide. It seems likely that the human hormone has the same structure. (In pyroglutamic acid a peptide

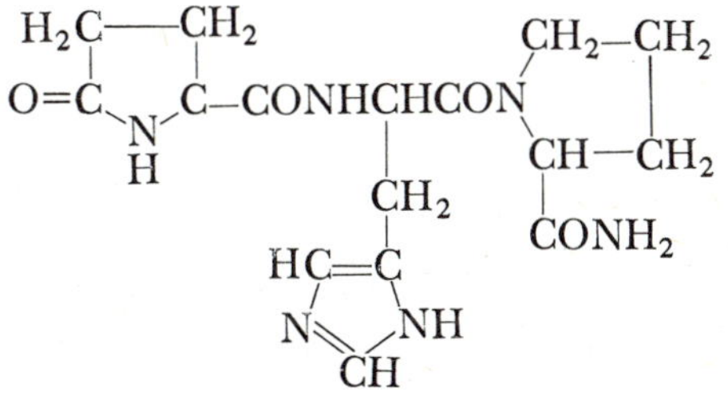

Thyrotropic releasing hormone   *TRH*

bond is formed between the free αNH$_2$ group and the γ-COOH.) There is great structural specificity as shown by the preparation and testing of a large number of structural analogues. Even the substitution at the N-terminus of glutamine renders the tripeptide almost inactive. Only a few nanograms (10^{-9} of a g) of TRH injected intravenously into a mouse causes a measurable increase in the amount of TSH released from the pituitary.

5

Enzymes

5.1 The Importance of Enzymes

A clear grasp of the nature and properties of enzymes is essential to the fundamental understanding of biochemistry, since enzymes are the highly selective tools used by the living organism to carry out the many thousands of interrelated chemical changes in the cell, involving the formation or destruction or interconversion of an extremely wide range of chemical compounds. A knowledge of enzymology is also important to medical students because enzymes are involved in many aspects of health and disease, diagnosis and treatment, as a few examples may illustrate.

(a) Inborn errors of metabolism

Some clinical conditions are now known to be due to an hereditary defect in a particular enzyme. Thus in *galactosaemia* there is an abnormally high level of the sugar galactose in the blood, caused by the absence of a key enzyme required to convert this sugar into the more easily metabolized glucose. *Phenylketonuria*, which can lead to severe mental retardation if untreated, is caused by the patient's inability to carry out the enzymic conversion of one amino acid into another (phenylalanine into tyrosine).

(b) Toxicity

Many chemicals owe their poisonous action to an interference with essential enzymes. For instance, human deaths have been caused by organophosphorus insecticides since these are powerful inhibitors of acetylcholinesterase, a key enzyme involved in nerve transmission. Cyanide kills by blocking cytochrome oxidase, a vital enzyme for cellular respiration.

(c) Chemotherapy

Some anti-bacterial drugs may owe their effectiveness to an ability to interfere with bacterial enzyme systems without seriously affecting the metabolism of the host, e.g., penicillin may act by upsetting the smooth functioning of enzymes involved in building bacterial cell walls.

(d) Vitamins

Enzymology throws light on the mode of action of these since the majority form essential parts of enzyme systems. If the supply of vitamins in the diet is inadequate, defects develop in the enzymic mechanisms, the chemistry of the body is disorganized and can lead to serious consequences as seen in deficiency diseases such as *beri-beri* and *pellagra*.

(e) Diagnosis

Assay of particular enzymes can sometimes provide useful information. For instance, *acute hepatitis* can be recognized by a distinct rise in the activity of a transaminating enzyme in the serum well before the more characteristic symptoms of jaundice appear. *Myocardial infarction* (a "coronary heart attack") causes damage to heart muscle cells and this leads to the release of the cellular enzymes into the blood.

(f) Treatment

A number of applications of enzymes in treatment have been suggested but much of the work is still in the experimental stage. There is, for example, some interesting work in progress on the use of the enzyme asparaginase in the control of some forms of cancer since there is the possibility that this could starve tumour cells of the asparagine, which they require to be supplied to them ready made, unlike normal cells which can make it for themselves.

5.2. Nomenclature of Enzymes

As the existence of catalytic agents in living systems began to be increasingly recognized in the early nineteenth century, names were conferred on them which were not systematic except that the suffix *-in* was usual, e.g.,

Emulsin—which liberated benzaldehyde from oil of bitter almonds
Ptyalin—the enzyme in saliva which hydrolyzed starch

Pepsin and Trypsin—enzymes from the stomach and pancreatic juice which digested proteins.

The practice of allocating such trivial names (as if to racehorses) naturally caused confusion as the number of known enzymes increased, and in 1883, a more systematic nomenclature was proposed by Duclaux which has been widely adopted. This involves the addition of the suffix "*-ase*" to the name (often slightly modified) of the first compound—the substrate—on which the enzyme was observed to act, e.g.,

substrate	*enzyme*
urea	urease
arginine	arginase
cellulose	cellulase

This idea was extended to the addition of the enzyme suffix *-ase* to the name of the chemical reaction carried out by the enzyme, so that it became possible to talk of classes of enzymes, e.g.,

decarboxylases—which split CO_2 from carboxylic acids

transaminases—which transfer an amino group from one compound to another

isomerases—which catalyze isomerizations.

However, when it became apparent that a single compound could act as a substrate for a number of different enzymes and so undergo various chemical changes, the names given to the enzymes often specified both the substrate and the reaction; e.g., pyruvic carboxylase, pyruvic dehydrogenase, pyruvic phosphokinase. As the properties of an enzyme may depend on the biological species in which it occurs, or on the organ of a particular species or even its location in the cell, a further qualification is sometimes necessary, e.g., yeast lactic dehydrogenase, intestinal phosphomonoesterase and mitochondrial malic dehydrogenase.

An International Enzyme Commission has now devised a classification for enzymes and has allocated an identifying classification number and a systematic name to each, showing the reaction catalyzed, e.g., EC2.7.1.2. ATP:D-glucose 6-phosphotransferase, which clearly indicates that the enzyme will facilitate the transfer of a phosphate group from ATP to the 6-position in glucose, which must belong to the D-series. As some systematic names are complex and unwieldy, shorter trivial names have received the Commission's blessing for everyday use. The enzyme EC2.7.1.2. above is more conveniently called glucokinase.

5.3. Some Key Events in Enzyme History

(a) From oils and ferments to proteins

Oil of bitter almonds played an interesting part in the early history of enzymes, as it has long been known that the odourless oil obtained from almond kernels could give rise to a fragrant substance in an unpredictable manner. The mystery was solved in the eighteen-thirties by the discovery that the oil contained amygdalin, a compound built up from benzaldehyde, hydrogen cyanide and sugar, which could be hydrolyzed to its components, and had an almond-like smell. This process was greatly facilitated by an agent, also derived from almonds, which if present acted as a catalyst for the hydrolysis. The agent was named emulsin by Liebig and Wöhler and was one of the first enzymes to be discovered. During the following decades, other enzymes were described, such as amylase from fermenting barley which converted starch to sugar, or pepsin and trypsin which aided the hydrolysis of proteins.

The phenomenon of chemical catalysis was becoming familiar and since these enzymic reactions could be carried out in a test-tube, a purely chemical theory of enzyme action was proposed by Liebig about 1870. Around this time, however, Pasteur in France was investigating the fermentation of sugar by yeast to alcohol or by acid-producing bacteria to lactic acid, and he was firmly convinced that living organisms were essential for this type of fermentation to occur. He drew a distinction between "unorganized ferments" such as emulsin and amylase, and living "organized ferments" such as yeast and bacteria.

A lively controversy between Liebig and Pasteur concerning the nature of fermentation was only settled in favour of Liebig twenty years after his death and three years after Pasteur had died. The evidence was supplied by a fortuitous discovery in 1897 by Büchner who was working on a theory that yeast had a therapeutic value. In the course of his work, he made a juice by grinding yeast cells with sand and squeezing the residue under high pressure. The clear cell-free fluid obtained was readily decomposed by bacteria and so Büchner tried adding sugar to act as a preservative, a function it performs well in jams. To his surprise, he found that this cell-free liquid was able to ferment sugar to alcohol and he thus showed that the presence of living organisms was not essential for fermentation. This juice provided a valuable material for research into sugar breakdown and eventually the whole pathway of glycolysis was elucidated.

Many would say that modern biochemistry really dates from Büchner's accidental discovery. Writing in 1932, Arthur Harden, a pioneer worker on glycolysis, commented: "As in the case of so many discoveries, the new phenomenon was brought to light, apparently by chance, as a result of an investigation directed to quite other ends, but fortunately fell under the eye of an observer possessed of the genius which enabled him to realize its importance and to give to it the true interpretation."

An ever-increasing number of enzymic reactions were described during the following decades and attempts often were made to purify the enzymes themselves. The process usually resulted in preparations showing enhanced activity and containing protein. This led to a belief that enzymes might be proteins although occasionally this was doubted by workers who obtained an enzymically active solution which gave no positive test for protein. We now know that this was due to the very high specific activity of some enzymes so that the concentration of the enzyme protein was below the limits of detection of the tests used.

Most chemical catalysts are small molecules such as platinum black and these active materials are often dispersed for experimental convenience on an inert carrier such as asbestos. Perhaps, by analogy, Willstätter around 1920 suggested that the protein in enzyme preparations was an inert colloidal carrier for the active groups in which the catalytic properties resided. It was clear that these doubts and theories would only be resolved if chemically pure enzymes could be examined. The complete purification of an enzyme was no easy matter since activity could be lost during the separation processes and at that time methods had not been highly developed for the separation of the complex mixtures of proteins which occur in most crude enzyme preparations. However, in 1926, Sumner isolated urease as very small crystals of protein and later in 1930-36, Northrop and his co-workers obtained pepsin, trypsin and chymotrypsin as crystalline proteins and the idea became generally accepted that all enzymes are distinctive proteins.

(b) Crystalline enzymes

Up to the present time, about two hundred crystalline enzymes have been obtained and some have been selected for more detailed structural studies. As was described in Chapter 4, the modern techniques of protein chemistry enable the molecular weight of an enzyme and the number, nature, and order of its constituent amino acid residues to be determined. Today the primary structure of several enzymes is known. This was first achieved in 1963 with ribonuclease which contains 124 residues. The elucidation of the sequence of the 333 residues in glyceraldehyde 3-phosphate dehydrogenase is one of the most impressive examples to date of this type of work (see Chapter 4.5).

For a substrate to react, it must first bind to some particular region of the enzyme molecule (the active centre) where the molecular environment provides structural complementarity. Moreover, the arrangement of chemical groupings at the active centre must confer catalytic activity. There are many examples in the literature of proposed mechanisms of reaction at active centres, derived mainly from studies with enzyme inhibitors. Real progress was hindered by a lack of definite knowledge about the conformation of the active centre. A knowledge of the primary structure of an enzyme is not, in itself, sufficient since the polypeptide chain will be folded and amino acid residues from widely separated parts of the chain will contribute to the active centre.

In 1965, Phillips and his associates described the tertiary structure of the enzyme lysozyme which they had obtained by X-ray structure analysis. The positions and orientations in space of all the 129 amino acid residues were clearly resolved. This enzyme occurs in many tissues and secretions and has an antibacterial action because it can cause the breakage of the polysaccharide chains which form part of the bacterial cell wall. A model of the structure showed a cleft running across the molecule into which a polysaccharide chain would fit neatly. It was possible to identify the R groups of amino acid residues which hold the substrate in place and the way in which amino acid residues cause the catalytic hydrolysis of the substrate. The R groups of the glutamate and aspartate residues which occur in positions 35 and 57 from the N-terminus of the polypeptide chain were shown to be intimately involved in the catalytic activity of the active centre.

Thus, a century after Liebig had proposed his controversial chemical theory of fermentation, an enzyme was revealed as a complex organic molecule of precisely known composition and structure, and the reaction mechanism of its function as a catalyst was understood and could be explained in terms of hydrogen ion migration and carbonium ion formation at known locations in the molecule. The active centre was revealed as a region of the molecule whose existence depends on the integrity of the structure of the whole molecule and was not just some active group borne on a protein carrier as Willstätter had suspected. Conditions which cause denaturation lead to a loss of enzymic activity.

When a compound has been isolated from living systems, the organic chemist is presented with two challenges—to find out its exact structure and to confirm this by a total synthesis in the laboratory. The biochemist also wants to know what the purpose of the compound is in the cell and how the cell made it. Success has been recorded in all these aspects with at least some enzymes but the first chemical synthesis of an active enzyme has only recently been achieved. The stepwise assembly of even a small protein molecule involves many hundreds of separate chemical operations. This brings problems of the purification of the desired product. For success, very high yields are essential at each stage or else the percentage yield of the final product becomes vanishingly small. These technical difficulties were overcome by Merrifield and his associates at the Rockefeller Institute in New York who developed a fine method which involved an automated and stepwise lengthening of the polypeptide chain while it was firmly anchored to a solid support. In 1971, they published full details of the synthesis of bovine ribonuclease A containing 124 amino acid residues. The synthetic product was indistinguishable from the natural enzyme by chemical, enzymic and immunological tests.

5.4. Prosthetic Groups and Coenzymes

(a) Aspects of structure

Some enzymes are simple proteins, (Chapter 4.2) consisting entirely of polypeptide chains. Examples of this type are the hydrolytic enzymes pepsin, trypsin, lysozyme and ribonuclease. More commonly, however, enzymes are conjugated proteins having some non-protein component associated, whose presence is essential for enzymic activity. This cofactor can take various forms and can differ in the firmness of its binding to the polypeptide chain. If the dissociation constant is so small that effectively all the polypeptide chains in solution retain their cofactors attached and the two remain together when the enzyme is isolated and purified, the enzyme is called a holo-enzyme and the cofactor is designated as a prosthetic group and can be thought of as an integral part of the enzyme molecule. The polypeptide portion is called the apo-enzyme, i.e.,

$$\text{holo-enzyme} = \text{prosthetic group} + \text{apo-enzyme}$$
$$\text{(active)} \qquad\qquad\qquad\qquad \text{(inactive)}$$

The binding between the prosthetic group and the apo-enzyme may be a covalent bond, so that there is no appreciable dissociation at all. Acetyl CoA carboxylase mentioned below is a good example as the biotin cofactor is united with the apo-enzyme by a covalent amide linkage. On the other hand, the bonds between the cofactor and polypeptide may be relatively weak (e.g. hydrogen bonds, ionic attractions, etc.) so that almost complete dissociation of the two occurs in solution during the isolation of the enzyme. The isolated enzyme will be found to be inactive until the cofactor is added back. If this separate cofactor is a small organic molecule, it is usually termed a coenzyme, important examples being NAD, NADP, co-carboxylase and pyridoxal phosphate. The distinction between a prosthetic group and a coenzyme is not absolute as sometimes the tightness of binding lies between the two extremes; for instance, the cofactor of D-amino acid oxidase (see later) can be removed by dialysis although it is regarded as a prosthetic group.

Enzyme activity is frequently dependent on the presence of metal ions; the commonest requirements being for Mg^{2+} although this can often be replaced in experimental systems by other divalent ions such as Mn^{2+} or Zn^{2+}. Less often, monovalent cations are required such as K^+ which may usually be replaced by NH_4^+ but not by Na^+. The size of the cation seems to be important in determining its ability to function as an activator. In some cases this may be due to the ion being used to bind the substrate to the active centre of the enzyme where a complex of the correct dimensions must be formed for reactivity to be possible. Mg^{2+} for instance, may serve to bind organic mono-esters of phosphoric acid through their negatively-charged phosphate groups to the active centres of phosphatases, the enzymes which can hydrolyze these esters. In other cases, the metal ion may complex with the substrate to form the true substrate for the enzyme. For instance, Mg^{2+} is a cofactor for creatine kinase since the true substrate is a magnesium salt of ATP rather than ATP itself. Common anions at physiological concentrations usually have little or no effect on enzyme activity, although one exception to this is shown by salivary amylase which hydrolyzes starch and is activated by Cl^- in the saliva.

While activating metals may function passively by maintaining the enzyme molecule in the best configuration for reaction or by facilitating a favourable substrate-enzyme complex, prosthetic groups and coenzymes enter actively into chemical reactions at the active centre of the enzymes. They usually act as intermediary carriers of electrons, hydrogen atoms or groups such as amino, carboxyl or acetyl.

A number of enzymes will now be described to illustrate points concerning prosthetic groups and coenzymes.

(b) Ascorbic acid oxidase

This is chosen as an example of an enzyme with metal atoms as prosthetic groups. The reaction catalyzed is the aerobic oxidation of ascorbic acid (vitamin C):

$$\text{Ascorbic acid} + \tfrac{1}{2} O_2 = \text{Dehydroascorbic acid} + H_2O$$

The pure enzyme is a blue protein as it contains 8 atoms of copper per molecule and these are firmly bound and cannot be removed by ion exchange resins or by extensive dialysis. Electron spin resonance studies have shown that the copper alternates between the cupric and cuprous states during reaction and is thus acting as an intermediate electron carrier. It is interesting to note that ionic copper itself has a catalytic effect on the oxidation of ascorbic acid but the activity is increased by more than a thousand times when the metal is combined with the apo-enzyme to give ascorbic acid oxidase.

(c) Acetyl coenzyme A carboxylase

This is an enzyme involved in the biosynthesis of fatty acids and catalyzes the formation of malonyl CoA from acetyl CoA and bicarbonate:

$$\begin{array}{ccc} \underset{\displaystyle \text{Acetyl CoA}}{\overset{\displaystyle CH_3}{\underset{\displaystyle CO \, S \cdot CoA}{|}}} + HCO_3^- + ATP & \rightleftharpoons & \underset{\displaystyle \text{Malonyl CoA}}{\overset{\displaystyle CH_2COO^-}{\underset{\displaystyle CO \, S \cdot CoA}{|}}} + ADP + P_i \end{array}$$

A molecule of ATP participates in the reaction and the energy associated with its breakdown to ADP and orthophosphate is trapped and is used to drive the overall carboxylation reaction. The involvement of the ATP in the reaction mechanism at the active centre of the enzyme is essential in order to avoid the loss of this energy as heat which is useless for driving biosynthetic reactions. The carboxylation reaction and the hydrolysis of the ATP must not be regarded as two separate processes going on simultaneously side by side but as a single complex reaction. The idea may be expressed less accurately by saying that the two reactions are tightly coupled, implying that energy transfer can occur without its liberation as heat. For convenience and clarity, the breakdown of ATP is often shown separately in reaction sequences (as in Fig. 5.1) but remember that in reality the ATP is intimately involved in the main reaction being displayed.

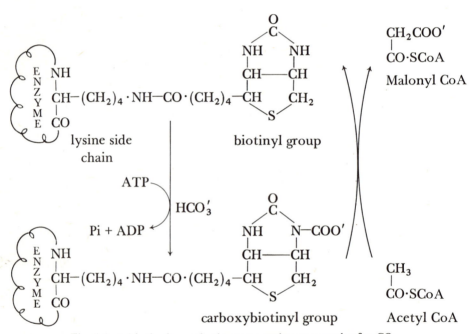

Fig. 5.1 A biotinyl prosthetic group acting as a carrier for CO_2.

The enzyme contains biotin as the prosthetic group and this is joined in an amide linkage to the side chain of one of the lysine residues. In the reaction involving ATP, the bound biotin combines with bicarbonate to give a carboxybiotinyl group which acts as a reactive carrier for the CO_2 which is then passed on to acetyl CoA, (see Fig. 5.1).

(d) D-amino acid oxidase

This catalyzes the oxidative deamination of various D-amino acids to the corresponding keto acids, oxygen being reduced to hydrogen peroxide in the process. The oxidase is a yellow enzyme as it contains flavin adenine dinucleotide as its prosthetic group which can exist in an oxidized and a reduced form (FAD and $FADH_2$). The prosthetic group first accepts two hydrogen atoms from the amino acid to give an imino acid (which spontaneously forms an ammonium salt of the keto acid by reaction with water) and subsequently passes the two hydrogen atoms on to oxygen and so regenerates the oxidized form of the prosthetic group, FAD.

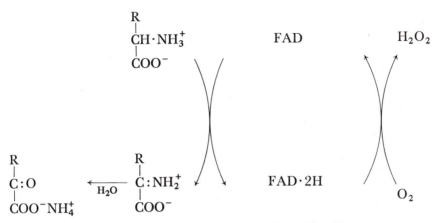

Fig. 5.2 Reaction catalyzed by D-amino acid oxidase.

The structure of the oxidized form of FAD is shown in Fig. 5.3, but it is sufficient to remember that the molecule is built up from the units shown in Fig. 5.3a and that the flavin contains a $-N:C-C:N-$ chain which can accept two hydrogen atoms as in Fig. 5.3b.

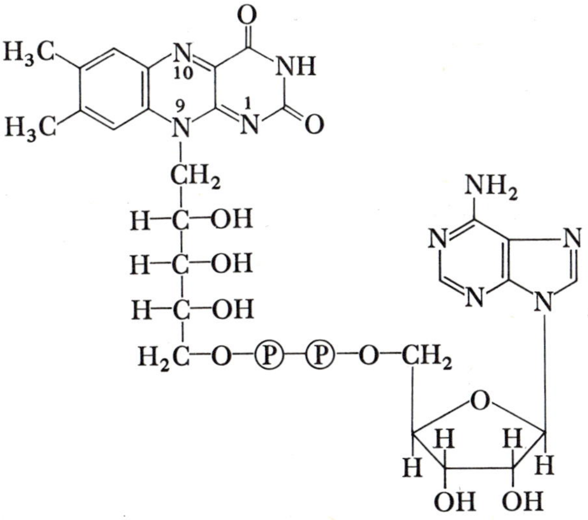

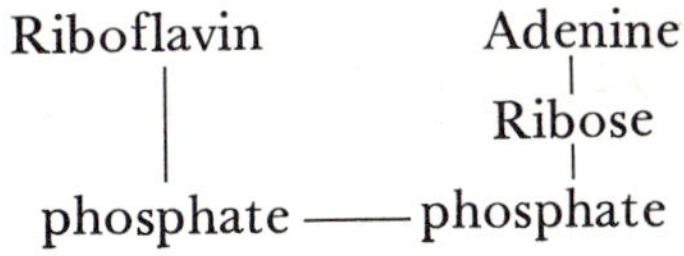

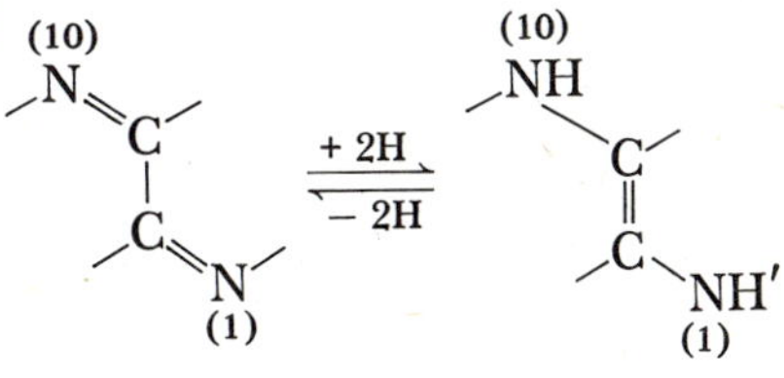

Fig. 5.3 Flavin adenine dinucleotide.

After the FAD has accepted the two hydrogen atoms it can only pass them on to some other molecule which will form a complex with this particular enzyme and in the living cell, oxygen is probably the only suitable acceptor. The fate of the hydrogen is, therefore, extremely restricted because the FAD·2H is bound to the enzyme as a prosthetic group. This should be contrasted with the situation (see below) in which a coenzyme such as NAD accepts the hydrogen, as there is the possibility here that it can move away from the enzyme where it was reduced and can participate in a great variety of reactions at numerous other enzymes which use the same coenzyme.

Many flavoprotein enzymes also contain metal atoms. The xanthine oxidase in milk is a good example as each molecule of enzyme contains 8 atoms of iron, 2 atoms of molybdenum and 2 molecules of FAD. In some flavoproteins, a smaller prosthetic group is found, riboflavin phosphate (often known as flavin mononucleotide, FMN) which functions in a similar manner to FAD. The flavin prosthetic groups are not usually bound covalently to the polypeptide and some reversible dissociation can occur. The FAD can be removed from D-amino acid oxidase, for instance, by dialysis against phosphate buffer at pH 7·6.

(e) NAD-dependent dehydrogenases

There are a large number of enzymes involved in oxidation/reduction reactions of specific substrates which use the same coenzyme, nicotinamide adenine dinucleotide, NAD, as a hydrogen carrier. The structure of this important coenzyme is shown in Fig. 5.4, reduction taking place at the nicotinamide end of the molecule as indicated in Fig. 5.4. The enzymic reaction with a substrate XH_2 can be written:

$$XH_2 + NAD^+ \rightleftharpoons X + NADH + H^+$$

Note that the formation of one molecule of reduced coenzyme is accompanied by the production of a hydrogen ion, i.e., one equivalent of acid is produced per mole of substrate oxidized. The reaction may also be written as

$$XH_2 + NAD \rightleftharpoons X + NADH_2$$

and this is probably the most convenient form for general use as the shorthand form. $NADH_2$ serves to remind us that the reaction involves the transfer of two hydrogen atoms, although chemically it is not quite correct. But NAD^+ is not correct either, as the net charge on the molecule as a whole at physiological pH is a single negative one and not a single positive one, but as the phosphate group will always be fully ionized and unchanged during reaction, it is convenient to ignore these negative charges. Both NADH and $NADH_2$ are acceptable abbreviations for the reduced form. The coenzyme was originally called Co-enzyme 1 (Co 1) and later diphosphopyridine nucleotide, DPN and these names may be encountered in older publications but NAD is preferred by international agreement.

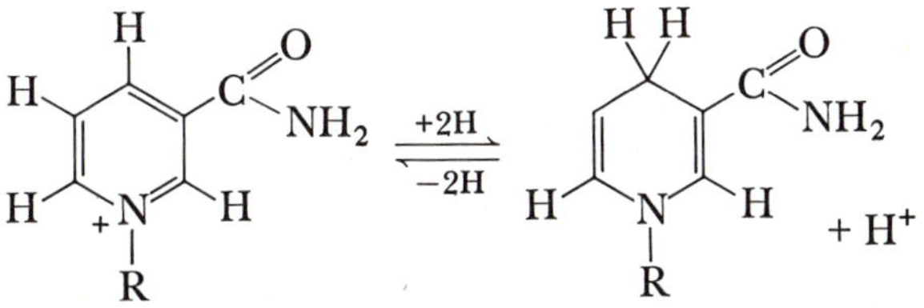

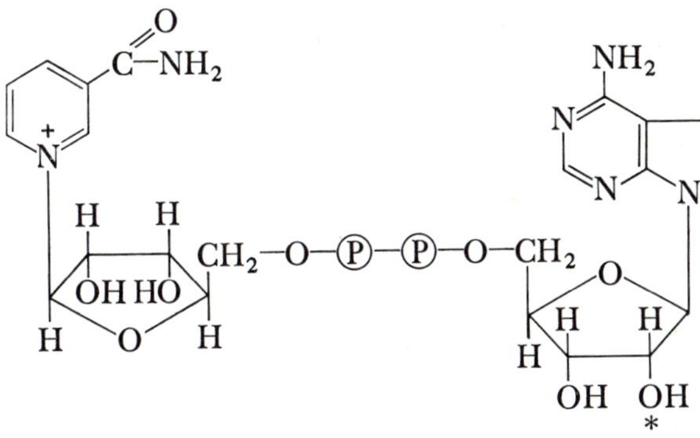

Nicotinamide   Adenine
    | |
ribose   ribose
    | |
phosphate —————————— phosphate

Fig. 5.4 Structure of oxidized form of NAD. There is an extra phosphate group in NADP at the position marked *.

Some enzymes use a slightly different coenzyme in which there is an additional phosphate group on the ribose attached to the adenine, (see Fig. 5.4). This is known as nicotinamide adenine dinucleotide phosphate, NADP and functions in a similar manner. It will be seen later that $NADPH_2$ is largely involved in biosynthetic reactions. Since these co-enzymes are mobile (i.e., not permanently attached to an apo-enzyme), they can serve to link two reactions if the two enzymes involved use the same coenzyme, e.g.,

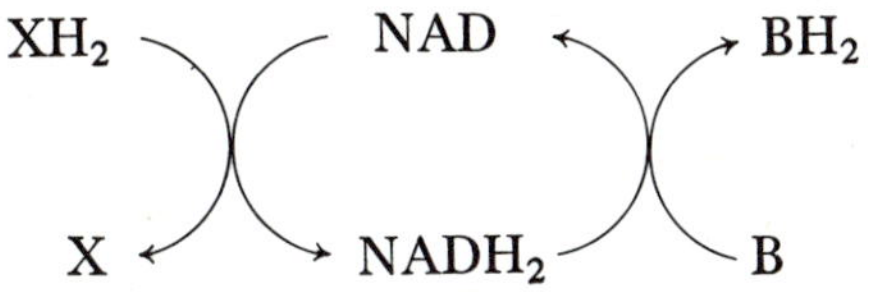

i.e., overall reaction is

$$XH_2 + B \rightleftharpoons X + BH_2$$

A good example of this linking of reactions will be met later in the discussion of glycolysis (Chapter 6).

These examples will suffice to illustrate typical properties of prosthetic groups and coenzymes. One general point to note is that several compounds which are recognized as vitamins by the nutritionist, are found to be components of prosthetic groups or coenzymes. Biotin, riboflavin and nicotinamide mentioned above are all well-known vitamins. Other examples will be found elsewhere in this book. A vitamin is an essential dietary constituent which has to be supplied in trace amounts because the body is unable to synthesize it. The total amount of enzymes needed by the cell is small because they are catalysts and this accounts for the quite small quantities of vitamins required. In the absence of such a vitamin, some key enzyme systems cannot be adequately developed, metabolic pathways are disorganized and illness or death of the individual from a deficiency disease may result.

5.5. Stability

Denaturation of enzyme protein will lead to loss of activity since this is dependent on the correct conformation of the protein. When handling enzyme preparations in the laboratory, loss of activity can occur unless care is taken to avoid conditions favourable to protein denaturation. Particularly important factors are listed.

Temperature   As a general rule, enzymes should not be exposed to temperatures above body heat and it is usually best to carry out preparative work around 0°C. There are a few enzymes, however, which are cold-sensitive, such as the one in mitochondria which hydrolyzes ATP and which undergoes inactivation at 0°C but is stable at room temperature.

pH   Enzymes are usually most stable around pH 6-8 and are inactivated under conditions much more acid than pH 5 or more alkaline than pH 9. There are exceptions, such as pepsin which is stable at pH 1 but is inactivated at neutral pH.

Interfaces   Proteins can be denatured at surfaces such as air-water interfaces, so enzyme solutions should not be so vigorously shaken as to cause the formation of froth and bubbles. They should be mixed by gentle stirring and poured carefully from one vessel to another.

Drying   If dehydrated enzymes are required, the water should be removed *in vacuo* from frozen solutions, a method known as freeze drying or lyophilization. Evaporation of water from an enzyme solution at room temperature usually leads to complete loss of enzyme activity. Precipitation of enzymes by alcohol or acetone must only be done at low temperatures.

Chemical contaminants   Enzyme activity may be lost by the introduction of an impurity which acts as a poison. Heavy metal ions such as copper, lead and mercury should be avoided by the use of good grade reagents and if present, their adverse effect may be overcome sometimes by the addition of chelating agents such as ethylenediamine tetraacetate (EDTA).

Storage It should not be forgotten that enzymes are a favourable medium for the growth of bacteria, moulds and fungi, so frozen solutions or freeze-dried preparations are best kept under deep-freeze conditions. Some enzymes are stable as suspensions in a saturated solution of ammonium sulphate and are provided commercially as such.

5.6. Catalytic Properties

Three criteria can be applied to inorganic catalysts—they are unchanged at the end of the reaction, only a small amount is required and they do not affect the position of equilibrium of the reaction. These statements are true for enzymes which in addition show a most important and unique property in their high specificity. These criteria will now be discussed briefly.

(a) Enzymes are unchanged after reaction

This is true for the reaction being catalyzed, in that the substrate unites with the enzyme, the reaction takes place and the products leave the active centre so that the enzyme is again in its original form. In practice, there may be a slight loss of enzyme due to side reactions e.g. if the enzyme is a little unstable at the pH of the reaction.

(b) Small amounts only are required

Two examples will illustrate this; one part of crystalline rennin, an enzyme from the stomach of the calf, will clot about ten million parts of milk in ten minutes at $37°C$ and of course, still be active. One molecule of catalase will split about five million molecules of hydrogen peroxide into water and oxygen per minute, but, admittedly, this is an extreme case, and rates of up to ten thousand molecules of substrate split per minute per molecule of enzyme are more usual.

(c) The position of equilibrium is unchanged

A self-evident truth derived from the second law of thermodynamics, otherwise a perpetual motion machine could be constructed by making a system do work if the position of equilibrium moved forwards and backwards as enzyme was added or removed. If we have a reaction

$$A + B \rightleftharpoons C + D$$

then the forward and back reactions must both be catalyzed by the enzyme and the same amounts of the four reactants will be present at the end whether one starts with (A + B) or (C + D). Consider the reaction catalyzed by fumarase:

$$HOOC-CH:CH-COOH + H_2O \rightleftharpoons HOOC-CH_2\,CH(OH)-COOH$$

$$\text{fumaric acid} \qquad\qquad\qquad \text{malic acid}$$

$$\Delta G^\circ = -3.62 \text{ kJ } (-0.865 \text{ kcal})$$

Using the relationship between ΔG^0 and K_{eq} it will be found that the equilibrium constant K_{eq} is about 4, so that starting with either fumarate or with malate, the final mixture will contain both in the ratio 1 : 4 and it would be easy to demonstrate experimentally the reversibility of this enzyme reaction. But for a reaction as:

$$\text{Phospho-}enol\text{pyruvate} + \text{water} = \text{pyruvate} + \text{phosphate}$$

$$\Delta G^\circ = -62.0 \text{ kJ } (-14.8 \text{ kcal})$$

the value of K_{eq} is about 2.9×10^{10} so that the position of equilibrium would be so far over to the right that the reaction could be considered as quite irreversible in practice if not in theory.

(d) Specificity

An inorganic catalyst like platinum black will catalyze the reduction by molecular hydrogen of tens of thousands of compounds with widely different structures. Enzymes, on the other hand are usually highly specific, that is, the catalytic activity is only shown in a reaction involving a single substance or a fairly small group of related substances. Dixon and Webb in their classic book on enzymes have observed that "enzyme specificity is one of the most important biological phenomena, without which the ordered metabolism of living matter would not exist and life itself would be impossible".

An enzyme is said to show *high* specificity if its action is limited to a very small group of compounds and *absolute* specificity if there appears to be only a single compound which will serve as substrate. Reduced cytochrome *c*, for instance, is the only known substrate for cytochrome oxidase which thus shows absolute specificity. Many hundreds of different reactions may be going on simultaneously in the living cell and it is a great advantage for it to have enzymes of high specificity so that the metabolism of some particular substance can be controlled independently by regulating the activity of a highly specific enzyme. On the other hand, if a fairly wide range of related compounds will act as substrates, the enzyme is said to show *low* specificity. This is the case with pancreatic lipase which will facilitate the hydrolysis of a very large number of different neutral fats, and this is advantageous in the process of digestion as many different fats occur in the diet and one enzyme can thus deal with all of them.

(e) Stereochemical specificity

Most enzymes show stereochemical specificity. If a substance exists in D- and L-forms, i.e., contains an asymmetric carbon atom and has structures which are mirror images, then it is almost always found that only one form will act as the substrate for a particular enzyme. Take as an example, the interconversion of lactate and pyruvate.

$$
\begin{array}{c}
CH_3 \\
| \\
H- C-OH + NAD \rightleftharpoons \\
| \\
COO^-
\end{array}
\qquad
\begin{array}{c}
CH_3 \\
| \\
CO \quad + \quad NADH_2 \\
| \\
COO^-
\end{array}
$$

$$
\text{Lactate} \qquad\qquad\qquad \text{Pyruvate}
$$

The lactate dehydrogenase in muscle will use only L-lactate as substrate and D-lactate will not react. If the conditions are such that the reverse reaction is favoured, then optically inactive pyruvate will be reduced enzymically to optically active L-lactate, i.e., a centre of asymmetry will be formed. In contrast, using an inorganic catalyst, there would be no difference in the rate of reaction if D- and L-lactate were oxidized while the product of the reverse reaction from pyruvate would be a mixture of equal parts of D and L-lactate. A lactic dehydrogenase from some bacterial species has the opposite requirement, using only D-lactate. This stereospecificity is a characteristic feature of enzymes and an explanation for it may be pictured on the following lines. The enzyme itself, being a protein, is built up from asymmetrical units, i.e., L-amino acid residues, so that it is unlikely that the configuration of the active centre will be symmetrical. A close fit between the asymmetric substrate and the active centre is essential and so it is reasonable to expect that this will occur with only one isomer of the substrate. It is as though the active centre is a glove and the substrate a hand, when only one hand of the pair will fit neatly into the glove.

If a compound exists in *cis* and *trans* forms due to different configurations of groups about a double bond, then normally only one of these geometrical isomers will act as a substrate; e.g., fumarase is specific for fumarate, the *trans* isomer and there is no reaction with maleate which has the *cis* configuration.

$$
\begin{array}{c}
{}^-OOC-C-H \\
\| \\
H-C-COO^-
\end{array}
\qquad\qquad
\begin{array}{c}
H-C-COO^- \\
\| \\
H-C-COO^-
\end{array}
$$

$$
\text{Fumarate} \qquad\qquad\qquad \text{Maleate}
$$

An interesting exception to stereospecificity is shown by a small group of enzymes which will cause a change in steric configuration in the substrate. An alanine racemase found in some bacteria is an example of an enzyme which will catalyze the conversion of both the D and L forms of the substrate into a DL mixture.

Enzyme specificity can be investigated by examining a large range of closely related natural and synthetic molecules as substrates. It is necessary to use an enzyme which has been highly purified to avoid the results being ambiguous due to contaminating enzymes. Usually no reaction can be detected with some of the compounds tested as possible substrates but others may show a range of speeds of reaction from fast—a good substrate—to very slow, a poor substrate. Thus most of the D-forms of the common amino acids (excepting glycine and D-glutamate) are substrates for the D-amino acid oxidase from kidney but rates of oxidation vary widely, for example being in the ratio of $297 : 100 : 22 : 0.9$ for D-tyrosine, D-alanine, D-leucine and D-lysine respectively. Experiments with synthetic substrates show that there must be one hydrogen atom on the alpha atom which must, of course, have the D-configuration but the nature of the R group in the amino acid $R-CH(NH_3^+)COO^-$ is less crucial although it does affect the rate of reaction. One would regard this oxidase as an enzyme which showed moderately low specificity. Data from specificity studies can provide the basis for speculation about the nature of the structural complementarity of the active centre and for possible mechanisms of enzyme reaction.

One important consequence of enzyme specificity can be illustrated by considering the nutritive values of amylose, (the straight chain component of starch) and of cellulose. These are both non-branched polysaccharides built up from glucose units joined through the $1:4$ positions and differ in the linkage being α in amylose and β in cellulose. The digestive poly-saccharases in the human gut are specific for α-glucosidic linkages so that starch is readily broken down and supplies a large part of the calorie requirement in man, but cellulose, on the other hand, is unattacked because the enzymes cannot hydrolyze β-glucosidic linkages and thus the material has no nutritive value and the large energy content cannot be tapped. It is strange that higher animals have not evolved a suitable cellulase since cellulose is one of the most abundant naturally occurring substances. Some molluscs, fungi and protozoa, however, do possess a cellulase and can digest wood. Fructose is a valuable foodstuff and yet inulin which is built up from β-fructose units and occurs as the reserve polysaccharide in Jerusalem artichokes, onions, etc. cannot be digested by man since his gut does not possess an enzyme with the correct specificity for hydrolyzing the β-fructosidic bonds.

5.7. Enzyme kinetics

(a) Units and assay methods

The study of the velocity of enzyme reactions under various conditions forms the subject matter of enzyme kinetics. There are various ways of measuring the velocity and the results are usually expressed in terms of the change in concentration of the substrate in unit time, preferably as micromole per litre per minute (μM/min.). The velocity observed will depend upon various factors including the temperature of the reaction mixture and its pH, the concentration of the enzyme and of the substrate and the presence or absence of activators or inhibitors. The amount of the enzyme preparation which will convert one micromole of substrate per minute at 30°C under optimum conditions is said, by international agreement to contain *one unit of enzyme activity*. If an enzyme is purchased, then the number of units stated to be present indicates how much activity one is getting for one's money but is, by itself, no guide to purity which requires information about the *specific activity*. This is defined as the number of units of enzyme activity per mg of protein, so that a preparation with a high specific activity will normally be purer than one with a lower activity since inert protein is the commonest impurity. The presence of salts such as ammonium sulphate will not affect the value of the specific activity. During purification of an enzyme, the specific activity rises and if a constant maximum value is attained, this may indicate homogeneity with respect to protein but during the purification the total number of units of activity may fall very considerably owing to losses of enzyme at various stages.

When a reaction rate is being determined under particular circumstances, it is essential to keep all the conditions as constant as possible. The reaction mixture is normally incubated in a thermostatically controlled water bath while the pH is maintained at the desired value by using a buffer which must be adequate to take care of any small amounts of acid or alkali produced during the reaction (see Chapter 2). The total time of reaction should be fairly short—from a few minutes to two hours—or else there could be loss of enzymic activity through denaturation or bacterial contamination. The concentration of substrate is bound to decrease as the reaction proceeds, and as the rate of reaction under certain conditions can be appreciably changed by a fall in substrate concentration, the conditions of assay should be chosen so that only a small proportion of the substrate initially present is used up. Correction must always be made for any spontaneous non-enzymic breakdown of the substrate and for any impurities present which may affect the readings, and this is done by running proper controls in parallel with the experimental mixtures. The controls usually contain the same components except that the enzyme is omitted, and readings made on these controls are subtracted from those made on the experimental mixtures. Enzyme reaction rates are usually determined by incubating enzyme-substrate mixtures and measuring the appearance of a product of the reaction since it is clearly more accurate to measure the concentration of a product which rises from, say, zero to 5 μM rather than that of the substrate which may only fall from, say, 200 to 195 μM.

When following the course of an enzymic reaction, it is advantageous to use a technique (when possible) that does not interfere with the course of the reaction so that a series of readings can be taken at successive time intervals and the results plotted on graph paper. If the substrate or one of the products has a strong optical absorption at some particular wavelength, then the photoelectric spectrophotometer is usually the instrument of choice since one can follow the changes in concentration with time of the substrate or product, by measuring the optical density at a selected wavelength, this variable being directly proportional to concentration within limits. Expensive instruments are available in which the optical cell containing the reaction mixture is thermostatically controlled and the optical density readings plotted continuously on an automatic recording chart, compensation being made for the absorption of a control blank reagent mixture. As an example, if one wished to follow the action of an alkaline phosphatase from kidney, p-nitrophenyl phosphate would be a useful substrate since it is colourless but one product of hydrolysis, p-nitrophenol, has an intense yellow colour at alkaline pH, so that measurement of the increasing optical density at 410 nm readily enables the reaction to be followed and measured.

The spectrophotometric method is particularly useful with reactions in which NAD or NADP participate since the reduced forms of these coenzymes absorb strongly in the ultraviolet at 340 nm whereas the oxidized forms absorb only weakly at this wavelength. Thus malic dehydrogenase can be assayed using the reverse reaction:

$$\text{oxaloacetate} + \text{NADH} + \text{H}^+ \rightleftharpoons \text{malate} + \text{NAD}^+$$

and the removal of the NADH followed by observing the fall in optical density at 340 nm.

Manometry may offer an alternative method of continuous assay if the reaction can be so arranged that a gas is evolved or absorbed in a Warburg respirometer. For instance, an esterase acting on a neutral ester will produce acid, and if the reaction is carried out in a bicarbonate buffer under an atmosphere of carbon dioxide in the manometer, then the extra carbon dioxide liberated by the acid can be measured readily. Manometry is probably less popular now than it was formerly since spectrophotometric methods (where available) are usually more sensitive, specific and convenient but manometry can still be useful, particularly when dealing with opaque systems such as suspensions of bacteria, yeast or tissue.

Sometimes continuous measurement is not possible and samples have to be removed from the reaction mixture at different times and some measurement made. For example, in the Anson method for the assay of the proteolytic activity of trypsin, denatured haemoglobin (deep red in colour) is used as the substrate and samples of the incubation mixture are withdrawn at intervals. The enzyme is inactivated immediately by the addition of trichloroacetic acid which also precipitates unchanged haemoglobin and large polypeptides to give a colourless supernatant solution containing amino acids and small peptides which can be estimated by a colorimetric method. The value obtained increases with time and gives a measure of the progressive breakdown of the protein as peptide bonds are hydrolyzed by the trypsin.

(b) Effect of the concentration of the substrate and enzyme on the rate of enzymic reaction

If the rate of the reaction is measured in the presence of different concentrations of substrate but with a constant amount of enzyme, the results usually have the form shown in Fig. 5.5.

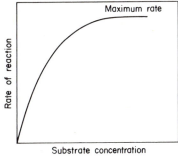

Fig. 5.5 Relationship between rate of an enzymic reaction and the substrate concentration.

At low concentrations of substrate, the rate is dependent on the concentration and increases as this is raised but eventually approaches a maximum value where it is almost independent of the concentration. The substrate is now said to be in excess and the enzyme to be saturated. When measuring the number of units of activity in an enzyme preparation, care must be taken that the substrate is in excess so that the enzyme is operating at its maximum rate.

If observations are made over an extended period on a mixture containing substrate and enzyme, a progress curve of the type shown in Fig. 5.6 may be obtained.

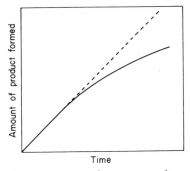

Fig. 5.6 Progress curve for an enzymic reaction.

At first, the product is formed at a fairly constant rate because the substrate is in excess and the enzyme is working at maximum speed, but later the rate of reaction falls off as the substrate concentration becomes rate-limiting, and finally no more product is produced because the substrate has been exhausted. Usually one is only interested in making observations in the linear region where the rate is constant and maximum, i.e., substrate in excess and observations are restricted to the period of time during which the rate of production of product is still linear.

If a series of observations are made using an excess of substrate and variable amounts of enzyme and working only in the linear region, results as indicated in Fig. 5.7 should be obtained.

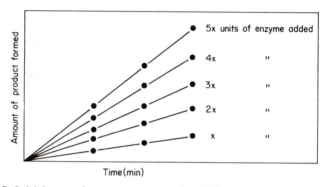

Fig. 5.7 Initial part of progress curves using different amounts of enzyme.

The rates of reaction are calculated from the slopes of the lines in Fig. 5.7 and a plot of these rates against amount of enzyme present should give a linear relationship (Fig. 5.8) i.e., the rate of reaction is directly proportional to the amount of enzyme present.

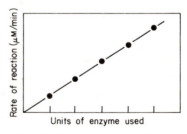

Fig. 5.8 Linear relationship between rate and amount of enzyme used.

In 1902, Henri put forward the hypothesis that a complex ES is formed between an enzyme E and its substrate S and that this complex can either dissociate to give its components again or undergo reaction and yield an end product P and regenerate the enzyme, i.e.,

$$E + S \underset{k_2}{\overset{k_1}{\rightleftharpoons}} ES \xrightarrow{k_3} E + P$$

This model has been used as the starting point for the derivation of an equation giving the relationship between two variables, the rate v at which product is formed and the concentration of the substrate [S] present. A number of simplifying assumptions have to be made in order to make the system amenable to mathematical analysis and different workers have made different assumptions. Michaelis and Menten assumed that the rate at which the complex ES reacted to give the product was very much slower than the rate at which it dissociated (i.e., k_2 is much greater than k_3) and that ES was in equilibrium with E and S. The relationship they deduced, known as the Michaelis-Menten equation is

$$v = \frac{V_{max} \, [S]}{K_m + [S]}$$

where v is the rate of formation of product when the substrate concentration is [S] and V_{max} is the maximum rate of reaction which can be attained when there is a large excess of substrate, while K_m is a constant, known as the Michaelis constant. Other workers, Haldane and Briggs, made different assumptions, namely that k_3 was not negligible compared with k_2 and that steady state kinetics were operative in which the rate of change of concentration of the complex ES was very small compared with the rate of change in substrate concentration and that a position of equilibrium was not attained in the first reaction. This alternative treatment yielded an equation identical in form with that of Michaelis and Menten, but the differing assumptions led to different meanings being attached to K_m.

Michaelis-Menten treatment, in which ES is in equilibrium with E and S	Haldane-Briggs treatment assuming steady state kinetics
$K_m = \dfrac{k_2}{k_1}$	$K_m = \dfrac{k_2 + k_3}{k_1}$

The second treatment is considered now to be the sounder and more general, and the Michaelis-Menten model is just a special case in which k_3 is very small compared with k_2. However, irrespective of the exact interpretation to be placed on K_m, this constant is a very valuable parameter to measure. For a given substrate, K_m has a fixed value for each enzyme (under constant conditions of temperature and pH) and can be used to characterize the enzyme. If two enzyme preparations catalyze the same reaction but show different Michaelis constants, then one can say that they contain different enzymes with similar activities. For example, malic dehydrogenase activity is shown by both mitochondrial and soluble fractions prepared from ox-heart, but the preparations show K_m values of $9 \cdot 9 \times 10^{-4}$ M and $5 \cdot 4 \times 10^{-4}$ M, so this tissue must contain two different malic dehydrogenases. It will be noted that K_m has the dimensions of a concentration. The enzyme concentration does not appear in the

Michaelis-Menten equation and this is very convenient as it is usually quite unknown and one can determine many of the properties of an enzyme in a crude preparation without isolation or without any knowledge of its purity or molecular weight.

Different values for K_m are found for different substrates with an enzyme and attempts have sometimes been made to correlate the values of K_m with the affinity of the enzyme for the various substrates. If Michaelis-Menten conditions hold, then K_m is the inverse of the equilibrium constant for the first reaction. If the enzyme has a strong affinity

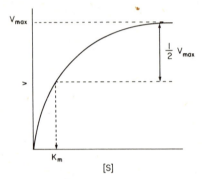

Fig. 5.9 Curve for Michaelis-Menten equation.

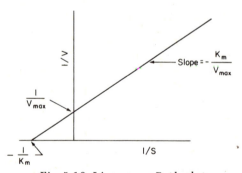

Fig. 5.10 Lineweaver-Burk plot.

for the substrate, formation of ES will be favoured and the equilibrium position will be well over to the right and the equilibrium constant will be high and hence K_m (its inverse) will be low. However, if k_3 is not negligible in comparison with k_2, then the Michaelis constant is not equal to the inverse of the equilibrium constant. The existence of a complex ES has been demonstrated experimentally using peroxidase and it has been found that k_3 is 26 times as large as k_2 and is thus quite the reverse of being negligible in comparison. As information about k_2 and k_3 is usually lacking, it is unwise to try and draw any conclusions about the relative affinities of the enzyme for different substrates from observed K_m values.

The Michaelis-Menten equation is the equation for a rectangular hyperbola (Fig. 5.9) in which the rate v rises as the substrate concentration is increased and approaches asymptotically a maximum value V_{max} but this is attained theoretically only at an infinite substrate concentration. The special case in which the observed rate is half the maximum is of interest. If one substitutes $v = \frac{1}{2} V_{max}$ into the Michaelis-Menten equation, it will be found to simplify to $K_m = [S]$ i.e., the Michaelis constant is equal to the concentration of substrate which causes the system to proceed at half the maximum rate.

Values of v for different substrate concentrations $[S]$ found experimentally usually give a satisfactory fit to a rectangular hyperbola. It may not be possible to work with high substrate concentrations owing to limits imposed by solubility, or if the substrate is very soluble (as with urea) then the enzyme may show inhibition by excess substrate and the rate falls off instead of increasing asymptotically. If experimental points are plotted and V_{max} determined, then K_m can be read off by observing the substrate concentration at $v = \frac{1}{2} V_{max}$ (Fig. 5.9). In practice, however, this is not a very satisfactory procedure as there may be considerable uncertainty in deciding where the curve would level off when extrapolated to find V_{max} and also, since the points will show some scatter due to experimental error, it may be difficult to decide where to draw the curve for a best fit. These difficulties can be avoided by a simple mathematical device due to Lineweaver and Burk. The Michaelis-Menten equation can be rearranged into the form

$$\frac{1}{v} = \frac{1}{V_{max}} + \frac{K_m}{V_{max}} \cdot \frac{1}{[S]}$$

If $1/v$ and $1/[S]$ are considered as the variables, the equation is now of the mathematical form $y = a + bx$ which is the equation of a straight line. It is now fairly easy to draw (or calculate) the straight line which gives the best fit, and if this is extrapolated back to where it cuts the x-axis the intercept gives $-1/K_m$. Alternatively, K_m can be found using the slope which is K_m/V_{max} and the intercept on the y-axis which is $1/V_{max}$ (Fig. 5.10). When determining K_m, it should be remembered that most of the rates are being measured in the region where the rate is highly sensitive to substrate concentration and will alter as substrate is used up. To minimize this effect, the observations made on appearance of product with time should be made over a period during which not more than about 5% of the substrate is used up. If the progress curve is found nevertheless to show an observable curvature, then the true rate is obtained by drawing a tangent to the curve at zero time. The initial almost linear region in Fig. 5.6 corresponds to a rate near V_{max} where the rate is almost independent of concentration.

Those who wish to study enzyme kinetics at greater length will find that the following is a useful book for the purpose: *Enzyme kinetics. A learning program for students of biological and medical sciences* by H. N. Christensen and G. A. Palmer (1967) W. B. Saunders Company. It includes two different derivations of the Michaelis-Menten equation.

(c) The effect of pH on the rate of reaction

When the rate of an enzymic reaction is determined in the presence of different concentrations of hydrogen ions, it is usually found that activity is shown only over a limited range of pH, but within this range the curve is roughly bell-shaped and shows a single maximum at a point known as the optimum pH for the enzyme (Fig. 5.11). This value is a characteristic

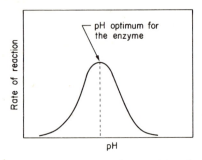

Fig. 5.11 Variation of rate with pH.

property of the enzyme although it may vary slightly with the buffer system used and frequently falls in the region of pH 6-8, a reflection perhaps of the selection by evolution of enzymes which work best in the physiological pH range of living cells. A noteworthy exception is shown by pepsin which has a pH optimum around 2·0 (and indeed is unstable and inactive above pH 6) but it functions in the stomach where HCl in the gastric juice provides a pH close to this value. The enzyme arginase, on the other hand, has a pH optimum around 10 but this extreme alkalinity is impossible in mammalian liver cells so the enzyme cannot function at its optimum pH *in vivo*.

The effect of a change in pH on an enzyme molecule is to alter the state or degree or ionization of acidic and basic groups (e.g., the —COOH in the side chain of glutamate or aspartate, the imidazolium nitrogen in histidine, the —SH of cysteine, the terminal NH_2 of lysine, etc.). One would expect that at least some groups—particularly those around the active centre— would have to be in a particular ionized or non-ionized state for effective combination to occur between enzyme and substrate and for the complex so formed to undergo reaction. Consideration of the following very simple and hypothetical model indicates that a maximum might be expected in the pH-activity curve. Suppose there are only two key ionizable groups at the active centre and that these are those associated with a histidine and a cysteine residue. There are three possible states:

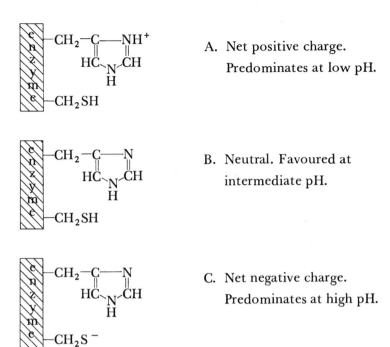

A. Net positive charge.

Predominates at low pH.

B. Neutral. Favoured at

intermediate pH.

C. Net negative charge.

Predominates at high pH.

Fig. 5.12 Three ionized forms of a simple model containing histidine and cysteine residues.

The pKs of these two groups (if unaffected by neighbouring charges) are about 6·9 and 9·0, so the distribution of the three forms at different pH will be as in Fig. 5.13. If it is assumed that enzymic activity is shown only by form B, then one might expect the enzyme to show maximum

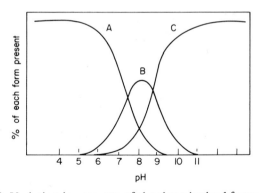

Fig. 5.13 Variation in amounts of the three ionized forms with pH.

activity under conditions where this form was present in the greatest amount—around pH 8. The simple model suggests a reason for the bell-shaped curve observed with enzymes but the real situation with an enzyme may be much more complicated owing to multiple ionizable groups which may interact with each other and there may be the added complication that the charge on the substrate may also change with pH, affecting its binding power.

If the two pKs in the model are close together, a sharp maximum is observed but if they are well separated, a broad-topped peak is produced. A wide variation in peak shapes is found with different enzymes.

(d) Effect of temperature on the rate of reaction

The rate of most chemical reactions is approximately doubled or trebled for each $10°C$ rise in temperature and this is true for enzymic reactions up to about 40-50°C but above this, thermal denaturation of the enzyme occurs leading to reduced activity and the rate falls and most enzymes show complete loss of activity if kept at 60°C. The temperature corresponding to maximum rate is not fixed but depends on the experimental conditions employed in assay, particularly the length of time of incubation. The effect of temperature on the rate of enzymic reactions is not very relevant to mammalian metabolic reactions *in vivo* since they normally always occur at temperatures quite close to blood heat. In major open heart surgery, the temperature of the patient may be reduced deliberately by cooling in order to slow down the enzymic reactions and reduce the demand for oxygen by metabolizing brain tissue.

5.8. How Enzymes Work

It is a familiar observation that many chemical reactions do not take place spontaneously at room temperature even though they are energetically favoured and there would be a liberation of energy (ΔG negative). The reason for non-reactivity can be explained on the transition state theory which postulates that before X and Y can react to give product Z, it is necessary for them to pass through a transition state in which they form an activated complex (X . . Y) which can then undergo further reaction to give the products. This complex however, possesses a higher free energy than that which X and Y together usually have, so that the extra amount—the activation energy—has first to be acquired although the reaction will eventually release a larger amount (Fig. 5.14). It is as though the molecules have to get over an energy hump. An analogy would be a ball lying in a depression on the top of a hill and it cannot roll down unless it is first given enough energy to take it over the lip of the depression.

The molecules in a population do not all possess the same energy but due to chance collisions which may be favourable or unfavourable, acquire different energies distributed over a range which is almost symmetrical about the mean (Fig. 5.15). The number with energies appreciably higher than the mean, falls off very rapidly as this difference increases. Only those molecules which have at least enough energy to allow the transition

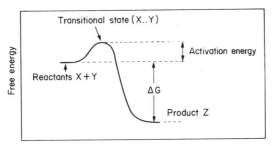

Fig. 5.14 The energy hump between reactants and product.

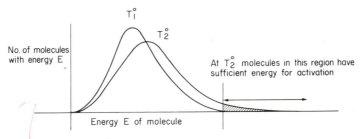

Fig. 5.15 Energy distribution among a population of molecules at a temperature T_1° and at a higher temperature T_2°.

state complex to be formed have the possibility of reacting should they collide in a favourable orientation. If one considers the simpler case of a monomolecular reaction in which a molecule decomposes and where there is no added complication about favourable orientation, then the molecules here must also first get into a transitional state (the relative positions and movements of some constituent atoms may differ from normal) and to do so they must first acquire the necessary higher free energy. If this activation energy is fairly high—e.g., 120 kJ/mole (29 kcal/mole)—then it will be so rare for a molecule to acquire this amount of extra energy by collision, that the reaction would be extremely slow, taking about a year to be half completed. If the activation energy was lower—say, 60 kJ/mole (14 kcal/mole)—then molecules with sufficient energy would be much more common and the reaction would be half over in one thousandth of a second. The rate of reaction is thus extremely sensitive to the magnitude of the activation energy (Table 5.1). A reduction of only 20 kJ from 120-100 kJ/mol is seen to make all the difference between a reaction which is proceeding almost too slowly to be detected and one which would be half over in an afternoon. The problem that has now to be answered is: how can a reaction be caused to proceed at a convenient and useful rate although normally at room temperature the rate may be indistinguishable from zero because of the high activation energy required? One solution would be to increase the energies of the molecules by raising the temperature of the reactants, the Bunsen burner often being

156

TABLE 5.1. Dependence of the time for half reaction on the energy
of activation for a monomolecular reaction

Energy of activation kJ per mole	Time for half reaction at 25°C for a monomolecular reaction
40	4×10^{-7} s
60	1×10^{-3} s
80	3·6 s
100	3 h
120	1·1 years

the chemist's best friend. Increased temperature causes the energy distribution curve to shift to the right (Fig. 5.15) so that there are now many more molecules with the necessary activation energy and the reaction proceeds much faster. This device of speeding up reactions by raising the temperature is denied to living cells and they make use of an alternative method, catalysis by enzymes. Catalysis solves the problem of getting over a high energy hump by reducing its height.

Instead of the products being formed from the reactants in a single reaction with a high activation energy, a catalyst enables the change to be carried out in a series of subsequent reactions, all of which have lower activation energies. Enzymes acting as catalysts achieve this by combining with the reagents and so enable different mechanisms of reaction to function. In the simplest case, the overall reaction is broken down into two, in the first the enzyme-substrate complex is formed and in the second, this reacts to give the products and to regenerate the enzyme. There may be successions of different complexes, as has been shown, for example, with peroxidase where four transient complexes involving the enzyme and one of the substrates, hydrogen peroxide, have been shown spectroscopically. Elucidation of the exact mechanism of action of a particular enzyme is a fascinating but difficult problem as the detailed structure of the active centre of the enzyme and how the substrate binds to it must be known. Only then can the complicated sequence of events be unravelled, probably involving the movement of electrical charge, formation of hydrogen bonds, involvement of small ions at strategic points, and so on leading to different complexes and reactions until the products are finally formed.

It was mentioned above that molecules have to have the correct orientation to each other at collision for the formation of the activated transitional complex. In free solution, a large fraction of collisions may be unproductive because of unfavourable orientation, although the necessary energy is available. An enzyme may be able to increase greatly the number of favourable collisions by bringing the reacting molecules together at the active centre in the correct orientation for reaction. This positioning effect may be able to increase the rate of reaction by a thousandfold. The consequence of a lowering of the height of the activation energy hump may be many orders of magnitude greater than this and a combination of the two effects acting together may account for the ability of enzymes to produce enormous increases in the rate of reactions—examples are known of enzyme-catalyzed reactions proceeding 10^{10}-10^{14} times as fast as uncatalyzed ones (Fig. 5.16). One cannot fail to be impressed by the ingenuity of living systems in evolving such beautiful and efficient devices. Few man-made tools are able to speed up a job to this extent.

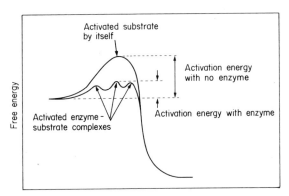

Fig. 5.16 Reduction in height of the activation energy hump by an enzyme.

5.9. Enzyme Inhibitors

The rate of an enzyme reaction may be greatly reduced, or the reaction stopped, by the addition of some reagent which may be without action on some other enzymes. There are thus specific inhibitors for enzymes and their study is important for a number of reasons:

(a) Importance of inhibitors

(i) Elucidation of metabolic pathways The product of one enzymic reaction is usually the substrate for another enzyme and a series of such reactions form a metabolic pathway in which the original substrate passes from one enzyme to another undergoing successive modification, rather like an engine block passing through a series of machine tools in a car factory. The conversion of the original substrate A into the observed product Z may involve many intermediate compounds, B, C, D . . . etc. and each stage requires its own specific enzyme:

$$A \xrightarrow[\text{Enzyme 1}]{} B \xrightarrow[\text{Enzyme 2}]{} C \xrightarrow[\text{Enzyme 3}]{} D \xrightarrow[\text{Enzyme 4}]{} E \dashrightarrow Z$$

The sequence of reactions may operate so smoothly that the various intermediate compounds may be present in such small amounts that only the final product Z can normally be detected. The elucidation of the intermediate stages can be difficult but inhibitors may prove to be a useful tool since if a specific inhibitor for, say enzyme 3, can be found, then C will accumulate since its further metabolism is blocked and it may be possible to isolate and identify it. Other inhibitors may block other enzymes and allow other intermediates to be detected, and eventually chemical considerations may allow the complete sequence of events in the pathway to be deduced. Glycolysis is a good example of a reaction pathway as the conversion of glucose to lactic acid by an extract of animal muscle actually occurs in eleven stages involving eleven different enzymes and ten intermediates.

Iodoacetate and fluoride are two inhibitors which were used to block this pathway at different points (see below) and helped in its elucidation.

(ii) Information about active centres of enzymes   Iodoacetate will react readily with thiol groups to form a stable sulphide:

$$X-SH + I\,CH_2COO^- = X-S-CH_2COO^- + HI$$

Certain enzymes are inhibited by iodoacetate (e.g., triose phosphate dehydrogenase in the glycolytic pathway) and this may indicate the presence of an essential $-SH$ group at the active centre of the enzyme and if blocked by iodoacetate, activity is lost.

If fluoride ion is found to inhibit an enzyme (e.g., enolase in glycolysis), the presumption is that it is removing Mg ions which are essential for the reaction mechanism to function at the active centre.

Some other enzymes, including trypsin, chymotrypsin and cholinesterase are strongly inhibited by certain organophosphorus compounds, such as diisopropyl phosphofluoridate (DFP) since they can esterify and block a key serine hydroxyl group at the active centre (Fig. 5.17).

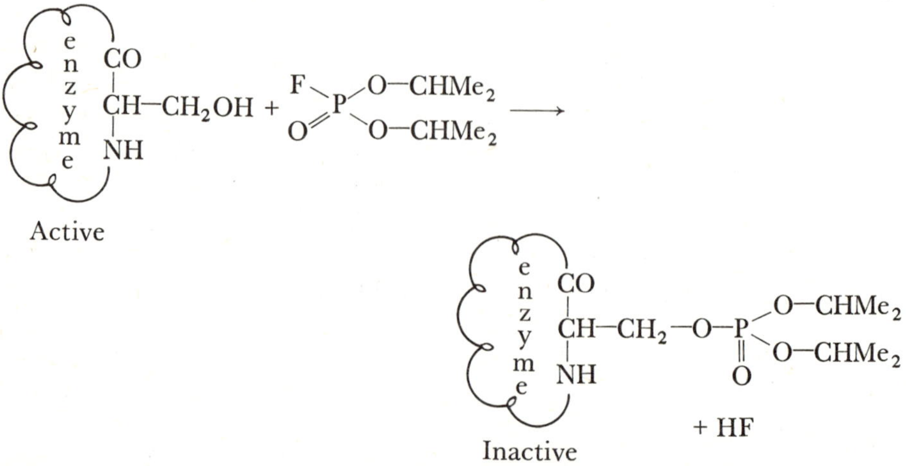

Fig. 5.17 Reaction of DFP with a serine residue at the active centre of an enzyme.

(iii) Explanation of toxic action   Two examples will illustrate this. Cytochrome oxidase is an essential enzyme in the respiration of all aerobic cells and if it is inhibited, the cell is prevented from using its oxygen supply. Cyanide is a very powerful poison since it easily inhibits this enzyme, blocks the respiration of brain cells in a few seconds and can lead to unconsciousness and death.

The organophosphorus compounds mentioned above are also highly toxic because they inhibit cholinesterase which plays a key role in the functioning of the central nervous system. Some of these compounds were developed as nerve gases during the Second World War and are among the most poisonous compounds known. Less toxic organophosphorus compounds of the same type have come into use as valuable insecticides and the occasional cases of poisoning of man and animals by these insecticides is due to their inhibitory action on cholinesterase.

(iv) Chemotherapy   One method of achieving a successful chemotherapeutic action against bacteria is to inhibit an enzyme which is essential for their growth but which plays little or no part in the metabolism of the infected mammalian host. The use of sulphanilamide is a good example of the exploitation of a vital difference in mammalian and bacterial metabolic pathways. Some bacteria require folic acid as an essential cofactor but are unable to absorb it ready-made and have to synthesize it, but mammals who also require folic acid, absorb this from their diet and lack the biosynthetic pathway. Sulphanilamide inhibits a stage in the bacterial synthetic pathway and prevents the growth of the organisms. The mechanism of this inhibition is indicated in the following section.

(b) Types of inhibition

Two main types may be distinguished, non-competitive and competitive inhibition. In the first type, the inhibitor does not resemble the substrate in chemical structure but is a reactive molecule which combines with some important group in the enzyme, iodoacetate and DFP mentioned above being good examples. Inhibition of the enzyme will not be an instantaneous process since it results from a chemical reaction between enzyme and inhibitor. The percentage inhibition of the enzyme observed will depend on the concentration of inhibitor used and on the length of time it is incubated with the enzyme before activity is assayed, but powerful inhibitors can produce almost one hundred per cent inhibition after quite a short time. Since a stable covalent bond is often formed, the enzyme will be permanently inactivated and the inhibition is said to be irreversible. This may not always be strictly true, as with some of the organophosphorus inhibitors, where a slow loss by hydrolysis of the alkyl phosphate group on the serine may lead to regeneration of active enzyme.

Just as a key may sometimes fit into a lock but be unable to work the mechanism because some detail of its shape is wrong, so a molecule which has a close structural resemblance to the true substrate may fit at the active centre of the enzyme and form a complex which, however, cannot react further, and so merely dissociates again. Suppose we have an enzyme working at its optimum rate converting substrate S into products, and we then add a compound I which forms a non-reacting complex EI. Two equilibria may be set up.

$$E + S \rightleftharpoons ES \; (\rightarrow products + E)$$
$$E + I \rightleftharpoons EI \; (no \; further \; reaction)$$

The number of molecules of enzyme available to carry out the first reaction will now be reduced because some of them will be temporarily unavailable as they are in the blocked form EI. The rate of the first reaction will thus be reduced and we should observe an inhibition, the extent of which will depend on (a) the affinity of the inhibitor I for the enzyme, since the greater this is, the more the second equilibrium will favour EI formation and leave less enzyme for the first reaction and (b) the relative amounts of S and I present since they are competing with each other for the same active centre. If the ratio of the concentration of substrate to that of inhibitor is increased, mass action effects will favour formation of ES at the expense of EI and so the inhibition will be less. Looked at in another way, the more substrate molecules there are for each inhibitor molecule, the less often will the latter manage to form a EI complex before a substrate molecule gets to a free active centre first and forms ES.

This type of competitive inhibition is reversible, since the complex EI readily dissociates and all the inhibitor may be removed by dialysis. Alternatively, the observed inhibition may be partially reversed by increasing the substrate concentration.

A classical example of this type of inhibition is shown by malonate in competing with succinate for succinic dehydrogenase. The distance between the two carboxyl groups is about the same in the two compounds and each will fit the active centre of succinic dehydrogenase, but only succinate has a grouping from which two hydrogen atoms may be removed (to give fumarate) while no dehydrogenation is possible with malonate (Fig. 5.18). The affinity of the enzyme for malonate, the inhibitor, is three times greater than that for its true substrate, succinate.

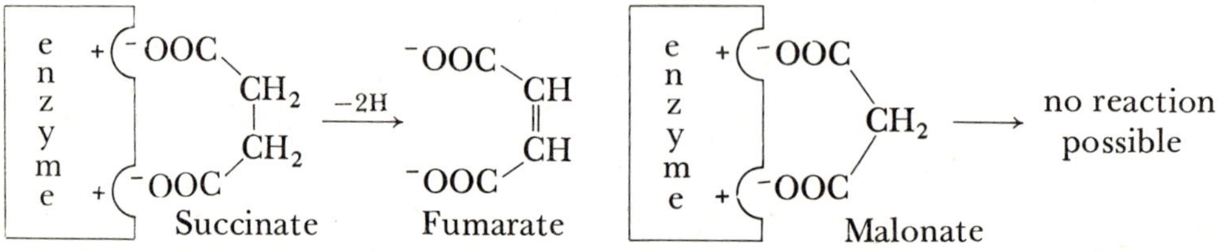

Fig. 5.18 Succinate and malonate at the active centre of succinic dehydrogenase.

Bacteria synthesize folic acid using p-amino-benzoic acid as a substrate at one stage. Sulphanilamide is a structural analogue and blocks an enzyme by taking the place of p-amino-benzoic acid but, since the inhibition is competitive, the effects of sulphanilamide on bacterial growth will be reduced by increasing the supply of the true substrate, p-amino-benzoic acid.

NH_2 NH_2

COOH SO_2NH_2

p-Amino-benzoic   Sulphanilamide
acid

Aminopterin is a structural analogue of folic acid (differing only in the replacement of an hydroxyl group by an amino group) and will compete with folic acid if administered and can so disorganize metabolism dependent on the participation of folic acid that aminopterin behaves as a highly toxic substance, the lethal dose for a rat being about 0·1 mg. Folic acid is a vitamin in man, so aminopterin is an example of an anti-vitamin.

An interesting case of competitive inhibition is shown by fluoro-acetate, $FCH_2 \cdot COO^-$ which is also highly toxic. The fluorine atom is not reactive and is larger than the hydrogen atom in acetate that it replaces but this difference does not interfere with the action of the first two enzymes involved in acetate metabolism. The organism treats fluoro-acetate as though it were acetate and first makes fluoroacetyl CoA and then fluorocitrate from it, but when the cell tries to use this compound as a substrate for aconitase instead of citrate the fluorine atom blocks the enzyme mechanism, the compound acts as a competitive inhibitor and citrate accumulates because its further metabolism is prevented.

An example of competitive inhibition involving an oxygen carrier rather than an enzyme is that shown by carbon monoxide (see Chapter 4.7(c)). The iron atoms in haemoglobin molecules can react reversibly with either a molecule of oxygen or with one of carbon monoxide and if exposed to a mixture of the two gases, there will be competition for the iron atoms. The high toxicity of carbon monoxide is due to its affinity for haemo-globin being over 200 times greater than that of oxygen, so that quite a low concentration of carbon monoxide in air can produce a serious fall in the degree of oxygenation of the blood, through conversion of useful haemoglobin into useless and inactive carboxyhaemoglobin.

5.10. The Kinetics of Inhibited Enzyme Systems

(a) Competitive inhibitors

Figure 5.19a shows the manner in which the rate of an enzyme reaction may be expected to vary with substrate concentration in the presence of a

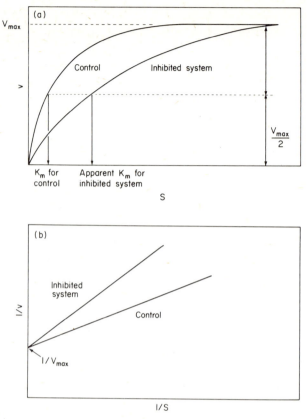

Fig. 5.19 (a) Relationship between the rate of an enzymic reaction and the substrate concentration in the presence of a competitive inhibitor. (b) The same, plotted in the Lineweaver-Burk form.

constant concentration of a competitive inhibitor. At high substrate concentrations, the effect of the inhibitor will be swamped so the maximum rate V_{max} of the control and inhibited systems will be the same. At lower substrate concentrations, the slower rate in the inhibited system will lead to a larger value for the "apparent K_m". The Lineweaver-Burk plots will have the form shown in Fig. 5.19b. The position of the straight line for the inhibited system relative to that for the control can be easily deduced by remembering (a) the two lines will cut the y-axis at the same point, the intercept being $1/V_{max}$ (b) the slope of the line for the control system is K_m/V_{max} so that for the inhibited system would be expected to be greater since the apparent K_m is larger and V_{max} unchanged.

The theoretical relationship for the inhibited system in the Lineweaver-Burk form is:

$$\frac{1}{v} = \frac{1}{V_{max}} + \frac{K_m}{V_{max}}\left(1 + \frac{[I]}{K_i}\right) \cdot \frac{1}{[S]}$$

where [I] = concentration of the inhibitor, K_i = dissociation constant for the complex EI (assuming that the inhibitor is not chemically modified by the enzyme).

The effectiveness of a competitive inhibitor is seen to depend both on its concentration [I] and on the firmness of its binding to the enzyme.

(b) Non-competitive inhibitors

Since a non-competitive inhibitor may react chemically with the enzyme, the amount of inhibition can often be observed to increase with time. In the simplest case, the rate of inhibition will be proportional to the product of the concentrations of enzyme and inhibitor in the incubation mixture. Figure 5.20 shows experimental data for the inhibition of

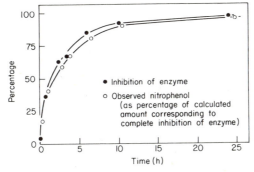

Fig. 5.20 Inhibition of chymotrypsin by diethyl p-nitrophenyl phosphate (10^{-3} M). One molecule of nitrophenol is liberated for each molecule of enzyme inhibited. (*Nature*, 1950, **166**, 784.)

chymotrypsin by diethyl p-nitrophenyl phosphate, a compound which, like DFP, reacts with a serine residue (see Fig. 5.17). Aliquots of the incubation mixture were withdrawn at different times and assayed to determine the percentage inhibition using excess substrate. If an aliquot of a partially inhibited enzyme is assayed using different substrate concentrations, curves as in Fig. 5.21 may be obtained. The maximum rate V_{max} will be lower in the inhibited system since some of the enzyme molecules will be completely inactive and the maximum rate depends on the amount of active enzyme. The uninhibited enzyme molecules present will, however, behave exactly as those in the control system and show the same K_m since the Michaelis-Menten equation is independent of enzyme concentration. One would anticipate that the slope of the Lineweaver-Burk line would be steeper than that of the control since K_m is the same but V_{max} less, and that it will cut the y-axis at a higher value than the line for the control since $1/V_{max}$ will be larger. It can be shown that the two lines intersect on the x-axis, and thus competitive and non-competitive inhibition can be distinguished by determination of the Lineweaver-Burk plots; the lines will meet on the y-axis for competitive and on the x-axis

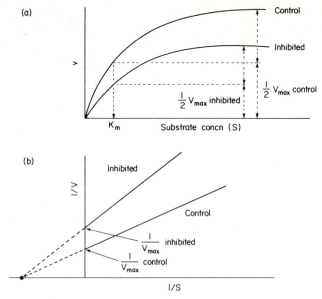

Fig. 5.21 (a) Relationship between the rate of an enzymic reaction partially inhibited by a non-competitive inhibitor and the substrate concentration. (b) The same, plotted in the Lineweaver-Burk form.

for non-competitive inhibition. This is a consequence of inhibited systems differing from controls as follows:

Type of inhibition	K_m	V_{max}
competitive	greater	same
non-competitive	same	smaller

5.11. Regulation

Although there are many hundreds of enzymic reactions going on simultaneously in a cell involving the synthesis, degradation or inter-conversion of a very large number of different compounds, the cell is nevertheless able to maintain the concentrations of these compounds at optimal values for efficient functioning. Materials such as glycogen and fat may be stored and mobilized only as required for biosynthesis or energy production. Precursors of macromolecules such as amino acids are produced only in the amounts required for immediate use. Metabolic pathways operate so efficiently that normally only trace amounts of intermediates are present. If an active enzyme is in contact with its substrate and all necessary cofactors are present, the reaction must proceed if the thermodynamic conditions are right since an enzyme has no will of its own to decide whether its action is required or not. If the phosphorylase and phosphatase in liver acted without restraint on liver glycogen, the blood sugar level would rise to a lethal value, but there are times when these enzymes must be active in order to increase the concentration of glucose in the blood. In this case, we should not be surprised to discover that some ingenious feed-back mechanisms are operating in which the glucose concentration can influence the enzymes involved in its formation or removal. The maintenance of compounds at optimal levels in the cell is a very complex process involving various mechanisms and some of these will now be briefly indicated.

(a) Law of mass action effect

If an active enzyme was present for a freely reversible reaction: $A + B \rightleftharpoons C + D$, then the concentrations of the components would be regulated by a mass action effect. If C or D was used up, the reaction would automatically move to the right and reduce the deficit. An example would be the reaction in muscle:

$$ATP + creatine \rightleftharpoons ADP + creatine\ phosphate$$
$$(creatine\ kinase)$$

where creatine phosphate is acting as a store of high energy phosphate groups. As ATP is converted into ADP during muscular work, the reaction will proceed from right to left to regenerate ATP at the expense of creatine phosphate. Later, as ATP becomes more plentiful, the reverse reaction will occur to reform creatine phosphate. This type of regulation is probably not of great importance as many enzyme reactions proceed in one direction only in practice or the product of one enzyme may be the substrate of the next enzyme and rapidly be removed so that a steady state is set up rather than a true equilibrium.

(b) Alteration in the amount of enzyme present

Some bacteria are very active metabolically and show a mean generation time of, say, 20 minutes. During this period every molecule present can be duplicated, the daughter cells being identical with the parent. Such active biosynthesis aids the rapid adaptation of bacteria to different environments. Cells of *Escherichia coli* grown on a medium containing glucose as the sole source of carbon and energy are initially unable to grow if transferred to a medium containing lactose instead of glucose; but after a lag phase measured in minutes only, they are found to contain an enzyme, galactosidase, which enables them to split the disaccharide lactose and resume growth. The amount of the enzyme in the cells may increase a thousandfold in a few minutes due to synthesis of new enzyme following a "turning-on" by lactose of the gene controlling the synthesis of this enzyme. This process is called enzyme induction (see Chapter 12.37).

In higher animals, the tempo of metabolism is slower and such short term changes are not possible. In a few cases, increases in the amount of a particular enzyme of the order of a hundredfold have been observed over several days. An interesting example of such an increase in the amount of enzyme is shown by assay of hydroxylases in the livers of animals treated with certain drugs. The hydroxylases are enzymes associated with the smooth endoplasmic reticulum and are often involved in the metabolism of foreign molecules such as drugs, alkaloids, carcinogenic compounds and insecticides. Administration of certain compounds, such as some barbiturates used as tranquillizing agents, may lead to a massive increase in the amount of smooth endoplasmic reticulum and a big increase in the amount of the enzymes present. If these enzymes are involved in detoxification processes (i.e., metabolism to less harmful or more easily excreted compounds), increased hydroxylase activity may be beneficial to the animal as the harmful compound could be more rapidly converted to a safer metabolite and the toxic effects reduced. An example of such a decrease in toxicity was demonstrated in experiments involving strychnine. A dose which killed 70% of a batch of rats was found to kill only 15% of a similar batch which had been treated for the four previous days with a barbiturate, thiopental. On the other hand, the hydroxylases may act on a foreign compound to convert it into a more toxic metabolite, a process that has been termed "lethal synthesis" and is the reverse of detoxification. An example of this is shown by the metabolism of the organophosphorus insecticide, Schradan, which can be converted by liver hydroxylases into a derivative many thousands of times more potent as an inhibitor of the cholinesterase of the nervous system and hence much more toxic. A dose of Schradan which would kill only 6% of a batch of normal rats was found to kill 75% of a batch predosed with thiopental. You should bear in mind the possibility that the administration of one drug may increase or decrease by several fold the toxicity or pharmacological response of another drug administered subsequently.

(c) Compartmentation

Enzymes may be specifically located inside or outside of organelles in the cell and this compartmentation can have a number of advantages. It may be desirable for enzymes to be kept away from their substrates as is the case with the enzymes in lysosomes. These organelles contain degradative enzymes as proteinases, ribonucleases and phosphatases which are potentially dangerous to the integrity of the cell but are safely confined within the lysosome by its membrane and thus out of contact with substrates present in the cytoplasm. Under various circumstances the lysosome membrane breaks, the enzymes are liberated and are able to effect the degradation of macromolecules. Compartmentation also makes it possible for the cell to carry out mutually incompatible processes at the same time, such as the degradation of fatty acids to acetyl CoA in mitochondria while the biosynthesis of other fatty acids is taking place in the soluble cytoplasm.

(d) Zymogens

Cells in the lining of the gastro-intestinal tract and in the pancreas synthesize and secrete proteolytic enzymes to facilitate digestion of proteins. The problem here is to prevent these enzymes destroying the cells which produce them. Control is achieved by the enzymes being synthesized in an inactive form—zymogens—which only become enzymically active after they reach the lumen of the tract whose walls are protected by mucus. Pepsinogen, for example, is an inactive protein secreted by the gastric mucosa into the stomach where it is converted into active pepsin by gastric HCl. This pepsin is then able to convert more pepsinogen into the active form so the process is autocatalytic. It has been found that 42 amino acid residues are split from the NH_2-terminal end of the pepsinogen molecule (Mol. wt. 40,400) to give pepsin (Mol. wt. 32,700) together with six peptide fragments. The active centre of the enzyme is only unmasked where it is safe for this to be done.

(e) Inhibition by the product of reaction

In a few cases, the immediate product of an enzymic reaction may inhibit the enzyme and so prevent an accumulation of product. The isoenzymes of lactic dehydrogenase in heart muscle are inhibited by lactate and so the reduction of pyruvate to lactate is restricted by the enzyme being "turned-off" as lactate builds up.

(f) Control by availability of an essential co-factor

A good example of this is the regulation of the ATP-ase activity of muscle, an enzyme which requires Ca^{2+} for activity. In the resting state, Ca^{2+} is absent from the sarcoplasmic space and the enzyme is inactive. When the muscle is stimulated by an electrical impulse down the nerve, Ca^{2+} is released from the sarcoplasmic reticulum into the sarcoplasmic space, the enzyme becomes active, ATP is broken down and muscle contraction takes place. A calcium ion pump then removes the calcium ions and the enzyme again becomes inactive and the resting state is restored.

(g) Competition for a common substrate

Where a compound can act as the substrate for more than one enzymic pathway, the appearance of one product may be controlled by factors which influence another pathway. For example, during glycolysis in muscle $NADH_2$ is produced and has to be reoxidized to NAD, and this can occur in either of two principal ways: (a) by being used to reduce pyruvate to lactate, or (b) by undergoing aerobic oxidation in the terminal respiratory pathway in which molecular oxygen acts as the final hydrogen acceptor.

The enzymes involved in (b) have a greater affinity for $NADH_2$ than does the lactic dehydrogenase used in (a) so that aerobic oxidation occurs preferentially when oxygen is present and the amount of lactate formed is greatly reduced. This switching of metabolism from anaerobic glycolysis to oxidative respiration also occurs with yeast cells and is known as the "Pasteur" effect after its discoverer.

(h) Feedback control

This is one of the most important regulatory mechanisms. Suppose the cell synthesizes a product P by a sequence of reactions such as:

$$A \xrightarrow[\text{enzyme 1}]{} B \xrightarrow[\text{enzyme 2}]{} C \xrightarrow[\text{enzyme 3}]{} D \xrightarrow[\text{enzyme 4}]{} \cdots P$$

where each stage is catalyzed by a separate enzyme. It is usually found that the thermodynamic conditions are such that the overall reaction proceeds only from left to right as some stages are effectively irreversible and an overall equilibrium position is not attained. If all the necessary cofactors are present and all enzymes active, then P would be synthesized so long as there was some A present; but the cell may well not want large amounts of the product and in order to match production with requirement, a cunning feedback mechanism often operates in which one of the enzymes in the pathway is inhibited by the product, so that as soon as this begins to accumulate, it automatically slows down the production line. The inhibition is reversible, and as the concentration of the product falls, the inhibitor-enzyme complex dissociates and synthesis of the product speeds up again. The inhibition usually takes place very early on the pathway before some irreversible stage. If, for instance, the stage B to C is irreversible and enzyme 3 was inhibited by the product, then the cell would accumulate a temporarily useless intermediate C; but if enzyme 1 is inhibited instead, then the starting product A can be diverted to other purposes.

A good example of feedback control is shown in the pathway used for the biosynthesis of cytidine triphosphate (CTP) from aspartic acid and carbamyl phosphate: as shown in Fig. 5.22. The final product, CTP inhibits the first enzyme, aspartate transcarbamylase, used in the multi-enzyme pathway and so controls its own synthesis. Even without studying the exact formulae and reactions it is easy to see that the final product, CTP, has little structural resemblance to those of the substrates. Thus one might expect that the mechanism of inhibition may well be quite different from the competitive inhibition discussed earlier (cf., succinate and malonate with succinic dehydrogenase) where substrate and inhibitor could compete for the same active centre because of their close structural resemblance. The idea developed that enzymes subject to regulation of the type shown by aspartate transcarbamylase might have two separate binding sites, one for substrate and one for the inhibitor. This has now been confirmed as the molecule of aspartate transcarbamylase can be

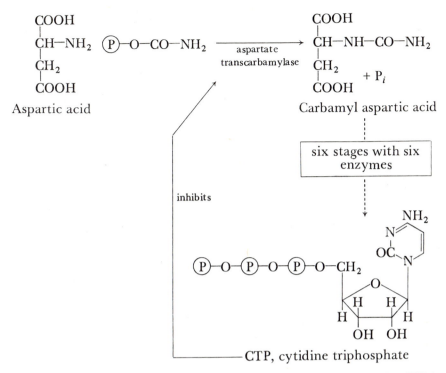

Fig. 5.22 The pathway for the biosynthesis of CTP can be inhibited by CTP itself acting on an allosteric enzyme at the beginning of the pathway.

dissociated into eight sub-units of two types. One shows enzymic activity but is unaffected by CTP while the other type of sub-unit will bind CTP but has no enzymic activity. A simplified picture of how this type of control works is indicated in Fig. 5.23. The enzyme molecule is to be considered as existing in two interchangeable configurations, one active

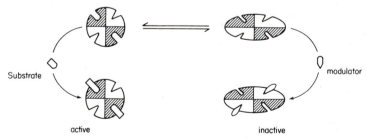

Fig. 5.23 Diagrammatic representation of the action of a modulator on an allosteric enzyme.

and one inactive. The first can bind substrate and is enzymically active but the other form binds the modulator (the molecule acting as an inhibitor) which stabilizes the molecule in the conformation in which the active centre is so distorted that the substrate will no longer fit and the enzyme is inactive. Enzymes of this type are known as allosteric or regulatory enzymes and can be of various types. Instead of the modulator inhibiting the enzyme as in the example above, it may activate it instead—i.e., the conformation binding the modulator is the one which will also bind the substrate. Some allosteric enzymes appear to have several sites at which different modulators can be bound: isocitric dehydrogenase, for example, is a key regulatory enzyme in the tricarboxylic acid cycle and is activated by ADP or NAD and inhibited by ATP and $NADH_2$, so that the amount of the enzyme in the active form will be controlled by the ratio of ADP to ATP and NAD to $NADH_2$. This enables the energy-producing cycle to be speeded up when energy rich compounds are running low and slowed down when they are plentiful.

Although definitive evidence for conformational changes in allosteric enzymes is still awaited, the plausibility of the concept has been strengthened by observations made on haemoglobin which may serve as a model for allosteric enzymes (see Chapter 4.7(c)). The molecule of haemoglobin is composed of four sub-units and when fully oxygenated will bind four molecules of oxygen to give oxyhaemoglobin. The entry of the first oxygen molecule enhances the binding of subsequent ones to the same molecule. Perutz has shown by X-ray analysis that the binding of oxygen (acting rather like an activating modulator) brings about a conformational change in the whole molecule in which there is relative rotation of the sub-units, an alteration in the distance between the haem groups and a change in the shape of the neighbourhood of the binding site for oxygen (analogous to an enzyme's active centre).

The increased affinity of haemoglobin for oxygen after the molecule has combined with the first O_2 causes the oxygen saturation curve to be sigmoid (see Section 2.7(b)).

(i) Chemical modification of the enzyme

Some key enzymes involved in metabolic control may exist in two forms, a phosphorylated and a non-phosphorylated one, which differ greatly in enzymic activity. The total activity of the enzyme in a tissue will depend on the ratio of these two forms and can be changed by altering this ratio, which is determined by the relative activities of two other enzymes—a specific protein kinase and a protein phosphatase—which carry out phosphorylation or dephosphorylation of the key enzyme. The activities of these two enzymes may in turn be under hormonal control. This topic is discussed more fully in Chapter 13.

(j) Isoenzymes

The development of sensitive methods for the electrophoretic separation of proteins during recent years has led to the discovery that many enzymes present in a single species and previously believed to be single entities, may be resolved into multiple forms, known as isoenzymes (or sometimes, as isozymes). The various forms may occur together or in different tissues (e.g., the acid phosphatases of human prostate gland and red blood cell differ) or even in different parts of the same cell (e.g., mitochondrial and cytoplasmic malate dehydrogenases). The five isoenzymes of lactic dehydrogenase (see Chapter 4.8) have been studied extensively as they can be separated readily by electrophoresis on agar, starch or polyacrylamide gels and made visible by suitable histochemical staining techniques. The molecule of lactic dehydrogenase is composed of four subunits of approximately the same size but of two types H and M, and all combinations of the two are possible i.e., H_4, H_3M, H_2M_2, HM_3 and M_4 leading to the five observed isoenzymes. The H subunit is more negatively charged at pH 7–9 than is the M subunit and when a mixture of the isoenzymes is separated by electrophoresis, the H_4 species will migrate fastest towards the positive electrode and M_4 be the slowest while the other three will be equally spaced between them since there is a constant change of net charge between adjacent structures.

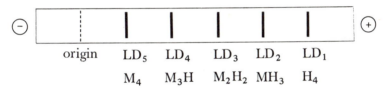

	origin	LD_5	LD_4	LD_3	LD_2	LD_1
		M_4	M_3H	M_2H_2	MH_3	H_4

Fig. 5.24 Diagram of electrophoretogram of lactic dehydrogenase isoenzymes.

Extracts from different tissues give characteristic patterns for the relative abundance of the five forms e.g., LD_1 predominates in the heart muscle while LD_5 predominates in skeletal muscle and liver. The pattern of the isoenzymes in plasma can sometimes give diagnostic information, e.g., a myocardial infarction (a "heart attack") leads to a massive increase in the amount of LD_1 which has been released from damaged heart cells, while infective hepatitis (jaundice) causes elevated levels of LD_4 and LD_5 following liver damage.

Isoenzymes may show different susceptibilities towards inhibitors and this may enable some metabolic control to be possible. For example, the isoenzymes of lactic dehydrogenase in heart muscle are inhibited by lactate and further production of this compound from pyruvate is prevented and the heart is compelled to generate energy aerobically. On the other hand, the isoenzymes in skeletal muscle are more resistant to lactate inhibition and enable the muscle to continue to reduce pyruvate to lactate and so generate energy by anaerobic glycolysis when there is insufficient oxygen available to meet an increased demand. A greater degree of exhaustion and lactate accumulation is permissible in skeletal muscle than would be desirable in heart muscle and the differing isoenzyme pattern may allow this to occur.

6

Carbohydrate Metabolism

6.1. The Relationship between Biosynthetic and Degradative Pathways

In this Chapter we shall be dealing in some detail with carbohydrate metabolism, and, since this is the first time we have discussed metabolism as such, it is important that we consider some of the principles that are involved.

Generally speaking, mature living organisms are characterized by a constancy of size, shape, weight and composition. This, however, is not a static condition but results from a balancing of numerous chemical reactions that are taking place. With a few notable exceptions (for example collagen) there is a constant breakdown, turnover and resynthesis of the macromolecular components of the organism, such as proteins, fats and carbohydrates. This situation has frequently been termed a state of Dynamic Equilibrium.

Superimposed upon and indeed part of this dynamic equilibrium, we have the metabolism of foodstuffs to provide the energy and the raw materials which allow living organisms to carry out their characteristic functions.

The sum total of these metabolic processes is referred to as Intermediary Metabolism although this is not a particularly useful term because of its rather wide definition. From what we have already said, we may divide the various reaction sequences into two types. Firstly, those that are degradative or Catabolic processes and secondly, those that are biosynthetic or Anabolic. The most important characteristic of catabolic

172

reactions is that they are exergonic, that is, they release energy to their surroundings. The free energy changes (see Chapter 9.8) that are involved are given a negative sign because there is a decrease in the energy content of the system.

Conversely, the characteristic property of anabolic or biosynthetic sequences is that by their very nature they are endergonic, that is, they require an energy input and the free energy change is said to be positive. However, they can only be deemed endergonic when considered in isolation since, from the principles of thermodynamics, for the reactions to proceed at all, they must be coupled to other reactions which themselves supply energy so that the overall, coupled processes, are no longer endergonic.

(a) The supply of energy to biosynthetic reactions

In living organisms, therefore, there is the constant need for the abstraction of energy from catabolic processes and its efficient transfer for use in biosynthetic reactions. In addition, energy is also required for the, perhaps more obvious, functions of muscular contraction, movement, secretion, absorption and nervous conduction. As we shall see, the compounds that function as energy carriers linking degradative and biosynthetic phases of metabolism are ATP and reduced NAD and NADP. The energy source for the other energy-requiring systems is exclusively ATP.

It is convenient to consider catabolism as taking place in at least two stages; other authors have defined more. Firstly, there is the breakdown of a large number of different macromolecules to their component units. These are relatively few in number and act both as the starting points for

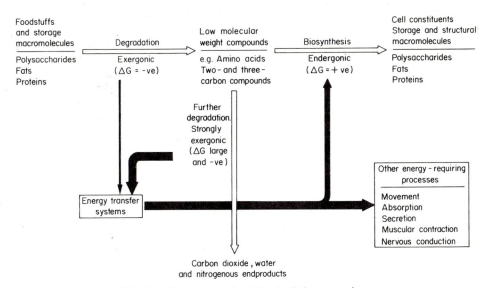

Fig. 6.1 Energy relationships in living organisms.

the synthesis of other macromolecules and also as the substrates for the reaction sequences leading to complete oxidation. Comparatively speaking, "stage one" reactions provide little usable energy, most of it being derived from the "stage two" processes. These are remarkably similar in all organisms. The overall relationship between energy-producing and energy-requiring systems is shown in Fig. 6.1.

Carbohydrate is particularly important with respect to the scheme shown, in that it provides the major (although not the most efficient) source of energy for living organisms. In man's food, this carbohydrate is present mainly as plant starch, the digestion of which is discussed in the next section. In addition to its significance in the diet, carbohydrate is stored in large amounts as glycogen and the most important energy substrate circulating in the blood-stream is glucose.

(b) The irreversibility of metabolic pathways

Until relatively recently it was tacitly assumed that the pathways by which macromolecules were synthesized were simply the *reverse* of the degradative ones. This is now known not to be the case in that different reactions, and consequently different enzymes, are involved in the two processes. This is made necessary thermodynamically since some of the free energy changes associated with certain individual reactions are such that these reactions are not readily reversed under physiological conditions. Moreover, as mentioned in Chapter 5.11 the enzymes required for biosynthesis and degradation, although present in the same cell, are often located in different cell organelles.

(c) Energy transfer systems

It is essential that we should understand now something about the nature of the energy transfer systems and the mechanisms that are involved in handling the energy derived from catabolic reactions. One of the important characteristics of biochemical degradation, when compared with chemical combustion, is that the former takes place in a series of discrete single-step reactions. These are enzyme-catalyzed and bring about small changes consisting, for example, of hydrolysis, isomerism, dehydration and decarboxylation and involving relatively few oxidation steps as such. The main advantages of this arrangement are two-fold: Firstly, the cell can exercise subtle and delicate control over the reactions that are taking place and secondly, energy can be obtained efficiently at a few points in the sequence in a useful form and with minimum dissipation in the form of heat.

(d) Energy-rich components

The compounds that are of fundamental importance in energy transfer are the organic esters and anhydrides of phosphoric acid. The hydrolysis of these compounds can be represented as follows:

$$R-O-PO_3H_2 + H_2O \rightarrow R \cdot OH + H_3PO_4$$

These derivatives of phosphoric acid can somewhat arbitrarily be divided into two groups, depending upon the extent of the free energy change that accompanies hydrolysis. These two groups are known as the low-energy and high-energy phosphate compounds, respectively (see Chapter 3.10). Examples from the first group are α-glycerophosphate and glucose 6-phosphate:

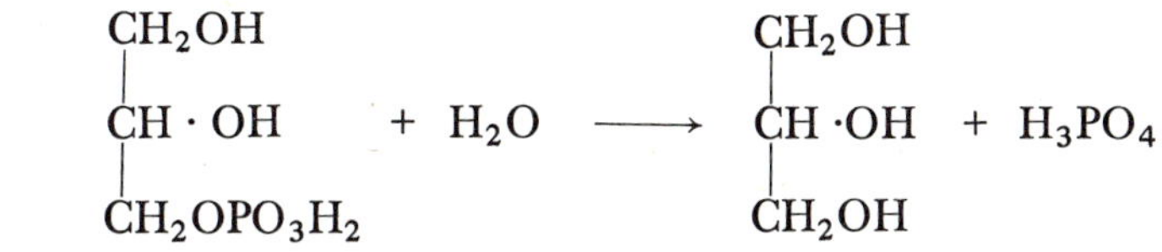

CH₂OH
|
CH · OH   + H₂O ⟶ CH ·OH + H₃PO₄
|
CH₂OPO₃H₂   CH₂OH

α-Glycerophosphate   Glycerol

$$\Delta G^{0\prime} = -9\cdot5 \text{ kJ } (-2\cdot3 \text{ kcal})$$

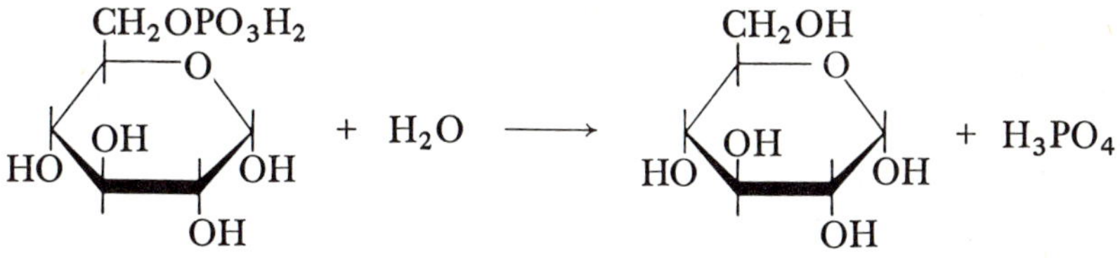

Glucose 6-phosphate   Glucose

$$\Delta G^{0\prime} = 13\cdot8 \text{ kJ } (-3\cdot3 \text{ kcal})$$

Phosphoenolpyruvate and creatine phosphate are important examples of high-energy phosphate compounds:

COOH   COOH
| |
C · OPO₃H₂ + H₂O ⟶ C=O + H₃PO₄
‖ |
CH₂   CH₃

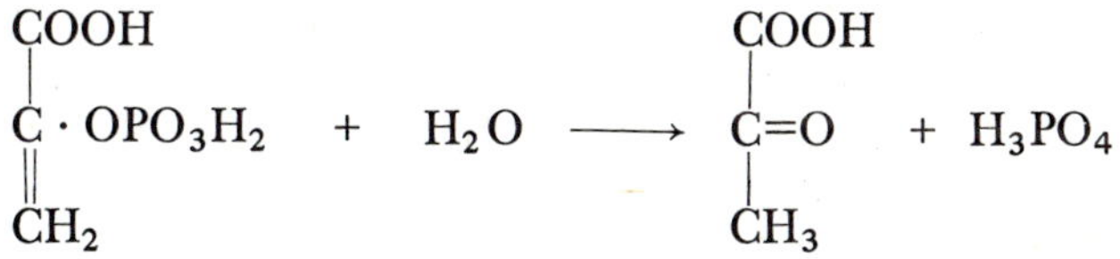

Phosphoenolpyruvic acid   Pyruvic acid

$$\Delta G^{0\prime} = -62\cdot0 \text{ kJ } (-14\cdot8 \text{ kcal})$$

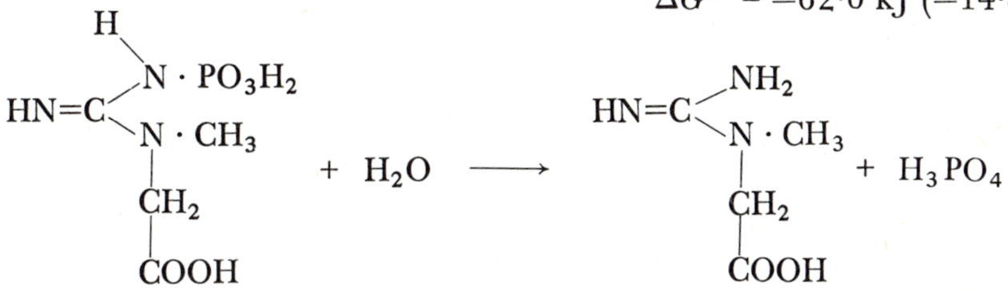

Creatine phosphate   Creatine

$$\Delta G^{0\prime} = -43\cdot0 \text{ kJ } (-10\cdot3 \text{ kcal})$$

$\Delta G^{0\prime}$ (sometimes referred to as $\Delta F^{0\prime}$) is the standard free energy change taking place under defined conditions of reactant and product concentrations and at a stated pH. Since the equilibrium constant of the reaction (K'_{eq}) is directly related to $\Delta G^{0\prime}$ by the equation:

$$\Delta G^{0\prime} = -2{\cdot}303\,RT\log K'_{eq}$$

it follows that reactions with the largest *negative* value of $\Delta G^{0\prime}$ will tend to proceed to completion to the greatest extent. Conversely, where $\Delta G^{0\prime}$ is large and positive most of the material will be present as unchanged reactants. The relationship between K'_{eq} and $\Delta G^{0\prime}$ is given in Table 6.1.

TABLE 6.1. The Relationship between Standard Free-Energy Changes and Equilibrium Constants at 37°C for the Hydrolysis of some Low-Energy and High-Energy Phosphates

	$\Delta G^{0\prime}$	K'_{eq}
Low-Energy Phosphate	kJ (kcal)	
α-Glycerophosphate	−9·5 (−2·3)	40·8
Glucose 6-phosphate	−13·8 (−3·3)	$1{\cdot}7 \times 10^2$
High-Energy Phosphate		
Phosphocreatine	−43·0 (−10·3)	$1{\cdot}9 \times 10^7$
Phosphoenolpyruvate	−62·0 (−14·8)	$2{\cdot}9 \times 10^{10}$

Undoubtedly, the most important high-energy phosphate compound of all is adenosine triphosphate (ATP). The full formula is given in Chapter 3.15; it can be represented here simply as

$$\text{ADENINE-RIBOSE}-O-\overset{\displaystyle O^-}{\underset{\displaystyle O}{P}}-O-\overset{\displaystyle O^-}{\underset{\displaystyle O}{P}}-O-\overset{\displaystyle O^-}{\underset{\displaystyle O}{P}}-O^-$$

from which it can be seen that three successive phosphate hydrolyses are possible. These are:

(i) $ATP + H_2O \rightarrow ADP$ (adenosine diphosphate) $+ H_3PO_4$

$$\Delta G^{0\prime} = -31{\cdot}0 \text{ kJ } (-7{\cdot}4 \text{ kcal})^*$$

(ii) $ADP + H_2O \rightarrow AMP$ (adenosine monophosphate, adenylic acid) $+ H_3PO_4$

$$\Delta G^{0\prime} = -31{\cdot}0 \text{ kJ } (-7{\cdot}4 \text{ kcal})$$

---

*It should be remembered that $\Delta G^{0\prime}$ is defined as the free energy change occurring under standard conditions, i.e., with molar concentrations of reactants. Since actual concentrations in the cell are difficult to determine, it is impossible to place an exact value on the amount of energy available from a given high-energy compound under physiological conditions. This, to a large extent, accounts for the discrepancies that are seen between the different free energy values given by various authors for the hydrolysis of compounds such as ATP.

(iii) $AMP + H_2O \rightarrow$ Adenosine $+ H_3PO_4$

$$\Delta G^{0\prime} = -14 \cdot 0 \text{ kJ} (-3 \cdot 4 \text{ kcal})$$

The first two of the hydrolyses are seen to involve high energy phosphate compounds. This store of chemical energy is of great importance in the promotion of energy-requiring systems, such as biosynthesis and muscular contraction. Close coupling of ATP with such systems allows much of the energy to be transferred and fulfil a useful function, whereas a simple hydrolysis of ATP to phosphate and ADP would result in the degradation of this energy into heat. Less commonly, AMP is produced from ATP but the phosphate of AMP is never removed for a similar purpose since the energy change is not large enough to perform a useful function.

In Chapter 5.4 the formation of malonyl coenzyme A was mentioned. This is a typical example of a biosynthetic reaction driven by ATP. Many such reactions can only use ATP and produce ADP as one of the end products, but there is an enzyme myokinase (adenylate kinase) which enables some ATP to be generated from this ADP and so, in effect, allows the energy associated with the two terminal phosphate groups of the original ATP to be used by the energy-requiring system. Myokinase catalyzes the reaction

$$2 \text{ ADP} \rightleftharpoons \text{ATP} + \text{AMP}$$

and occurs in various tissues, particularly in muscle where ATP is used to provide the energy for contraction.

(e) The significance of ATP

One of the main reasons why ATP is of such importance is that, although it is undoubtedly a high-energy phosphate compound, the free energy change associated with the hydrolysis of the terminal phosphate group lies somewhat below the free energy change accompanying the hydrolysis of most other high energy phosphate compounds. As a result, energy can be transferred to ADP from a number of donor compounds in the form of "high energy phosphate". The net free energy change associated with these reactions is the difference between the free energy changes which would normally be involved in the hydrolysis of the two high-energy compounds involved. Two examples are shown below; pyruvate kinase catalyzes the first reaction and phosphoglycerate kinase the second.

(i) 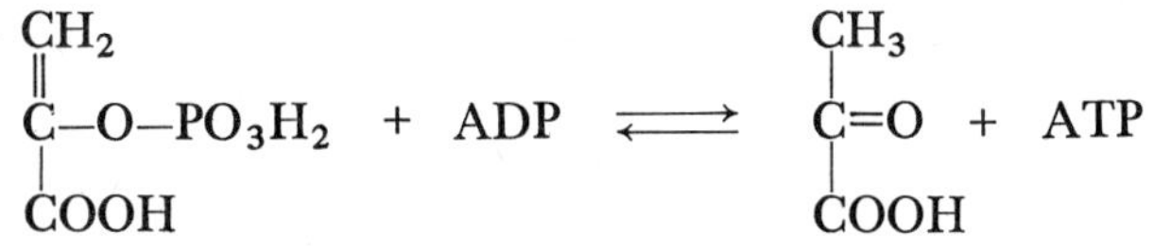
$$\begin{array}{ccc} CH_2 & & CH_3 \\ \| & & | \\ C-O-PO_3H_2 & + \text{ ADP} \rightleftharpoons & C=O & + \text{ ATP} \\ | & & | \\ COOH & & COOH \end{array}$$

Phosphoenolpyruvate   Pyruvate

$$\Delta G^{0\prime} = (-62) - (-31) = -31 \text{ kJ} (-7 \cdot 4 \text{ kcal}).$$

$$CH_2OPO_3H_2$$
$$CH \cdot OH \quad + \quad ADP \rightleftharpoons$$
$$CO \cdot OPO_3H_2$$

$$CH_2OPO_3H_2$$
$$CH \cdot OH \quad + \quad ATP$$
$$COOH$$

1,3-Diphosphoglycerate   3-Phosphoglycerate

$$\Delta G^{0\prime} = (-49) - (-31) = -18 \text{ kJ } (-4\cdot3 \text{ kcal})$$

Both these reactions take place during the metabolism of carbohydrate and bring about what is called substrate-level phosphorylation of ADP. The meaning of this will become clearer when other methods of ATP formation are discussed (see Chapters 6.12 and 9.10).

It will be seen that the free energy change accompanying the reaction catalyzed by pyruvate kinase is still large, resulting in a K'_{eq} value of 2×10^5 and making the reaction virtually irreversible under physiological conditions. The equilibrium constant of the phosphoglycerate kinase reaction (approx. 2×10^3) is low enough to allow the reaction to be reversed, especially where the products are rapidly removed by other enzymes.

Creatine phosphate is another high-energy phosphate compound that can react directly with ADP.

Creatine phosphate + ADP ⇌ Creatine + ATP

This reaction is catalyzed by creatine phosphokinase and it is particularly important in muscle where creatine phosphate forms a reserve store of readily available energy. Creatine is re-phosphorylated by ATP when the muscle is at rest.

A final group of high-energy compounds that will be encountered in the following sections are the esters of coenzyme A. Acetyl CoA and succinyl CoA are good examples. The hydrolysis of the latter, for instance, occurs as follows:

Succinyl CoA + H_2O → Succinate + CoA

$$\Delta G^{0\prime} = -31\cdot5 \text{ kJ } (-7\cdot5 \text{ kcal})$$

(f) The relationship between oxidation and ATP formation

Elsewhere in this book (Chapters 9 and 10), biological oxidation is dealt with in some detail. At this stage it is only necessary to understand in simple terms how ATP formation is linked to oxidation.

Perhaps surprisingly, only relatively rarely does biological oxidation occur by direct reaction of oxygen with the substrate. Most frequently, it is brought about by the removal of hydrogen which is transferred to one of a number of different coenzymes (see Chapter 5.4). The important coenzymes in this group are NAD, NADP and, to a lesser extent, FAD. The enzymes catalyzing oxidations of this type (dehydrogenases or oxidoreductases) are, of course, specific for one or other of these coenzymes; we may quote as an example, the oxidation of malate, catalyzed by malate dehydrogenase:

$$
\begin{array}{ccc}
\begin{array}{l}
\text{COOH} \\
| \\
\text{CH}_2 \\
| \\
\text{CH} \cdot \text{OH} \\
| \\
\text{COOH}
\end{array}
& + \text{NAD}^+ \rightleftharpoons &
\begin{array}{l}
\text{COOH} \\
| \\
\text{CH}_2 \\
| \\
\text{C=O} \\
| \\
\text{COOH}
\end{array}
& + \text{NADH} + \text{H}^+
\end{array}
$$

Malate   Oxaloacetate

With respect to energy metabolism, the reactions leading to the formation of NADH are by far the most important. It is during the reoxidation of NADH to NAD^+, linked to oxygen via the mitochondrial cytochrome system (Chapter 9), that ATP is synthesized from ADP. For every molecule of NADH oxidized in this way, and therefore for every pair of hydrogen atoms originally removed from the substrate, three molecules of ATP are formed. This process is known as oxidative phosphorylation.

NADPH differs from NADH in that it appears to be used primarily as the reductant in biosynthetic reactions rather than as an intermediate in ATP formation.

6.2. Dietary Aspects of Carbohydrate Metabolism

(a) The nature of dietary carbohydrate

Under normal circumstances in man, carbohydrate is the major component of the diet both in terms of bulk and with respect to foodstuffs meeting the daily calorie requirement. It is undoubtedly the most available and hence the cheapest form of food since it is so readily obtained in the form of starch from vegetable sources such as potato, yam, cassava and cereals. Sucrose, also of plant origin, forms a steadily increasing proportion of carbohydrate in the diet.

The average calorific value of carbohydrate has been shown to be about 17 kJ (4·1 kcal) per gram both by bomb calorimetry and by experiments involving direct measurement of heat production in animals fed a diet of known composition. The corresponding value for fat is 39 kJ/g (9·3 kcal/g) whereas for protein 22 kJ/g (5·3 kcal/g) is obtained by chemical combustion but only 17 kJ/g (4·1 kcal/g) by whole animal experiments. The lower physiological value is a consequence of the incomplete oxidation of nitrogenous compounds. Since carbohydrate might normally be expected to provide some 50-60% of the total calorie requirement, a balanced diet which provided 11,500 kJ (2750 kcal) per day would contain approximately 350-400 g carbohydrate.

In addition to starch and sucrose, carbohydrate is also present in the diet as cellulose, glycogen, glycoproteins and, particularly in the infant, as lactose. Cellulose cannot be digested by man as the enzyme necessary for its hydrolysis is not present. Ruminants and animals with a well-developed caecum (such as the rabbit) can utilize cellulose however, as symbiotic micro-organisms in the digestive tract can initiate its breakdown. Relatively small amounts of glucose and fructose and derivatives of other monosaccharides such as mannose and ribose may be ingested. Pentoses in general are quantitatively insignificant in food although they are present as components of nucleic acids in all plant and animal cells and as constituents of fruit and fruit juices. Pentosuria (the excretion of pentoses in the urine) may occur after the ingestion of large amounts of the latter. There is no dietary requirements for any specific carbohydrate since all the forms which are needed in the body are readily synthesized from glucose. Glucose itself can be synthesized from non-carbohydrate material, e.g., from pyruvate which could arise from the metabolism of certain amino acids. This process of formation of new sugar is termed gluconeogenesis (see Chapter 6.9).

(b) The digestion of carbohydrate

The digestion of carbohydrate compared with that of protein and lipid is an uncomplicated process. This is a reflexion of the relative simplicity and repetitive nature of the structures that are involved. The commonest linkage that is found is the α-1,4 glucosidic bond, the hydrolysis of which is catalyzed by the enzyme α-amylase.

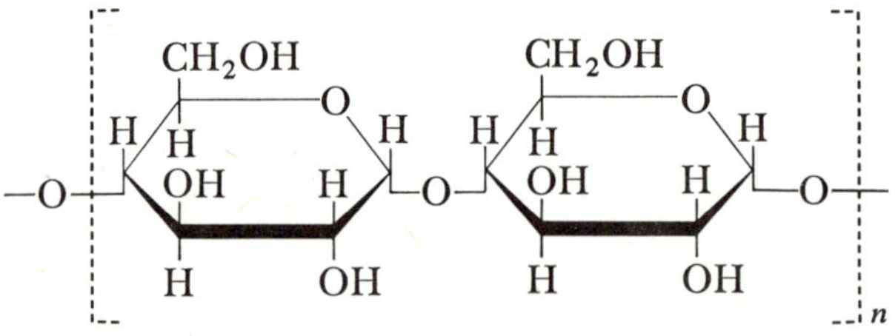

Fig. 6.2 Repeating α-1,4 linked glucose residues of amylose.

α-Amylase occurs both in saliva and in pancreatic secretion. It requires Cl^- and Ca^{2+} ions for full activity and brings about the hydrolysis of all α-1,4 glucosidic bonds, apart from that of the disaccharide maltose. Its action on a linear polysaccharide such as amylose produces mainly maltose but because of the random nature of its attack, some glucose is also formed. It has no effect on α-1,6 glucosidic bonds so that the products of the hydrolysis of the branched polysaccharides amylopectin and glycogen (see Chapters 3.12 and 6.3), in addition to maltose and glucose, are small oligosaccharides and the disaccharide isomaltose, which have the α-1,6 linkage still intact.

Hydrolytic enzymes acting on macromolecules may either remove residues starting at one end of the chain (*exo-*) or hydrolyze linkages within the chain (*endo-*). α-Amylase is an example of an endo-hydrolase. In the next chapter, exo- and endo-peptidases will be mentioned, enzymes which hydrolyze peptide bonds in proteins.

The extent of carbohydrate digestion brought about by salivary amylase is uncertain since the food is in the mouth for only a short period of time. The pH optimum for amylase activity is in the region of 7·1 so that digestion initiated by the enzyme will cease once the food is in contact with the acidic contents of the stomach. Individuals who do not have the ability to secrete salivary amylase do not appear to be at any disadvantage with respect to carbohydrate digestion.

There are no enzymes in gastric juice which digest carbohydrate so that the only hydrolysis taking place in the stomach will be the small amount that is catalyzed by gastric HCl. Most of the carbohydrate digestion taking place in the alimentary tract occurs in the small intestine and is brought about mainly by pancreatic α-amylase. A severe impairment in the secretion of pancreatic α-amylase results in the appearance of undigested starch granules in the faeces. Pancreatic α-amylase, both in terms of its mechanism of action and its physical properties, is indistinguishable from salivary amylase.

The other important function of the small intestine with respect to carbohydrate digestion is to bring about the hydrolysis of maltose and isomaltose produced by α-amylase action, and the hydrolysis of a number of disaccharides ingested in the food. Isomaltase catalyzes the formation of two molecules of glucose from one of isomaltose, and maltase has a similar action on maltose. Sucrase (or invertase) hydrolyzes sucrose to equal quantities of glucose and fructose. Lactase (β-galactosidase), the enzyme catalyzing the hydrolysis of lactose to glucose and galactose, is particularly active in the small intestine during infancy. None of these disaccharidases (disaccharases) is actually secreted into the lumen of the small intestine. They function while remaining attached to the brush border of the mucosal cells lining the intestinal tract.

(c) Abnormalities of disaccharide digestion

Occasionally there may be an impaired ability to digest one or more of the disaccharides that are formed by α-amylase action or which are present in the diet. This defect is characterized usually by diarrhoea which is caused by the bacterial fermentation of the undigested disaccharide in the large intestine. The inability is normally congenital (an inborn error of metabolism) and the individual is said to "show intolerance" for the particular disaccharide. Isomaltose intolerance has been described in several instances and, surprisingly, these patients also exhibit sucrose intolerance. Maltose intolerance has been demonstrated but much less frequently.

In contrast, inherited lactase deficiency has been encountered in a large number of infants. The patient suffers from severe diarrhoea and malnutrition from birth, and must be placed on a lactose-free diet. A second form of lactose deficiency is sometimes found in adults. This is an acquired enzyme deficiency and the mechanism by which it occurs is not understood.

(d) The absorption of carbohydrate

As a result of the digestive processes described above, carbohydrate is hydrolyzed to the level of monosaccharides, chiefly hexoses, before absorption occurs. This absorption takes place mainly in the mid part of the small intestine. The process, in common with many others that involve the transport of metabolites into or out of cells, is not one which relies solely upon simple diffusion. Transport frequently occurs against a concentration gradient showing that energy is required. The mechanism is said to be one of "Active Transport" and systems of this kind usually show varying degrees of specificity for the substances transported (see Chapter 11).

In the case of absorption of sugars from the small intestine, galactose appears to be the most readily absorbed hexose followed by glucose and fructose with the latter probably entering by simple diffusion; pentoses are taken up less quickly. The sugars, once absorbed by the mucosal cells, pass into the portal blood system and are carried to the liver.

6.3. Glycogen: its Structure, Properties and Function

Although the main characteristics of the structure of glycogen have already been described (Chapter 3.12), it will be necessary to emphasize these again, in greater detail, now that we are discussing the metabolism and physiological significance of this compound.

Glycogen is the storage form of carbohydrate in the body and it is extremely well-suited to perform this function. It is a polysaccharide, consisting entirely of glucose residues, which are joined mainly by α-1,4 glucosidic bonds but with residues being connected also by α-1,6 linkages. It has a particle weight which varies within the approximate range $0 \cdot 5$-5×10^6 so that such structures will consist of some 3000-30,000 glucose units; even larger sizes are not uncommon. The ratio of α-1,4 linkages to α-1,6 is from ten to twelve so that glycogen has a structure that branches, on average, every ten to twelve residues. There is only one "reducing" (see Chapter 3.6) or free C-1 hydroxyl group per molecule; the remaining chains have glucose units with free C-4 hydroxyl groups, i.e.,

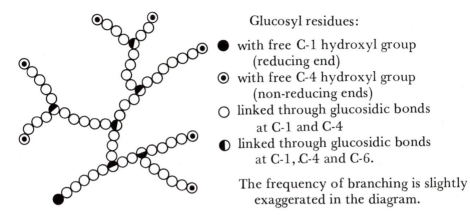

Glucosyl residues:

● with free C-1 hydroxyl group
    (reducing end)
◉ with free C-4 hydroxyl group
    (non-reducing ends)
○ linked through glucosidic bonds
    at C-1 and C-4
◑ linked through glucosidic bonds
    at C-1, C-4 and C-6.

The frequency of branching is slightly
exaggerated in the diagram.

Fig. 6.3 Glycogen.

they have "non-reducing ends" (Fig. 6.3). Since the chains of glucose residues are tightly folded, and not extended as shown in the diagram, the resulting structure tends to be spherical and extremely compact. So much so, that with the higher molecular-weight species the extent of branching, and thus the size of the molecule, becomes limited by steric hindrance.

Another property of the structure is important. Because of the regular degree of branching, in any given molecule up to rather more than 50% of the mass may be exterior to the outer branch points (the α-1,6 linkages). This means that from 5% to 10% of the total glucose units will be present as terminal residues having free C-4 hydroxyl groups.

The advantages of a storage molecule having the structure described above are manifold: firstly, because of its large size, minimum osmotic effects result from the storage of a large amount of material. Secondly, the high degree of branching confers solubility and at the same time allows the structure to take up the minimum amount of space. Thirdly, since glycogen is synthesized and degraded from the ends of the chains terminating in residues with free C-4 hydroxyl groups, there are many more growth and mobilization points within the molecule than would be present in a linear structure of the same size.

Although glycogen is present to some extent in most tissues two separate, distinct stores are recognizable. They are in liver and striated muscle, respectively, with each of these stores being mobilized for different purposes. The function of the glycogen stored in the liver is to act as a reservoir of carbohydrate so that, between meals, the concentration of glucose in the blood may be kept constant, normally within the range 4-5 mmol/l. In contrast to this, glycogen stored in muscle is not converted to blood glucose. It is used only for the provision of energy within the muscle itself. Specifically, it acts as the fuel to sustain muscular contraction when the demand for energy is greater than that which can be met from the glucose that circulates in the blood. The interrelationships of glycogen and glucose metabolism are shown in simplified form in Fig. 6.4.

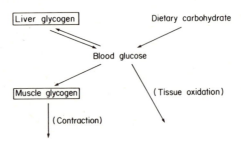

Fig. 6.4 Interrelationships of glycogen and glucose metabolism.

In liver, the glycogen concentration may vary from an average value of 6-8% of the wet weight shortly after a meal (in experimental animals on carbohydrate diets it may reach 15%), to well below 1% after fasting for 24 hours. In skeletal muscle the normal concentration range is around 0·5-0·8% and this is not affected by moderate periods of fasting. It is decreased by strenuous exercise of the particular muscle and restored, on resting, from blood glucose. In heart muscle, the concentration of glycogen only falls under abnormal conditions.

It should be noted that because of the relative bulk of the two storage sites, much more total glycogen is stored in muscle than in liver. The actual amounts vary from individual to individual but they are in the region of 150-250 g and 100-125 g, respectively, for the adult male. The glycogen stored in the liver is therefore sufficient for about a quarter of the day's total energy requirements. When this store is depleted, the body must draw upon its much more extensive reserves of lipid and begin to break down protein to meet the various energy demands made upon it.

Conveniently, when we consider the metabolism of glycogen, we find that the pathways of biosynthesis and degradation, although differing from each other, are identical for muscle and liver glycogen with the single exception that the latter gives rise to free glucose.

6.4. The Degradation of Glycogen and the Formation of Blood Glucose (Glucogenesis)

Glycogen, because of its size and structure, cannot be transported within the body but must be broken down in the cells in which it is stored; the mobile form of carbohydrate in the blood is glucose. Also because of its structure, a rather complex series of reactions is required in the initial stages of glycogen degradation.

The complete breakdown of glycogen to monosaccharide units takes place in three stages, catalyzed respectively by phosphorylase, oligo-1,4→1,4 glucantransferase and amylo-α-1,6 glucosidase. It is possible that the last two activities are properties of a single enzyme.

(a) Phosphorylase

The phosphorylases present in muscle and liver are of different genetic origin. They have very similar properties but they differ with respect to structural changes which are associated with their control. The reaction catalyzed by phosphorylase is the successive phosphorylytic cleavage of the terminal α-1,4 glucosidic bonds at the non-reducing ends of the glycogen chains. The enzyme from all sources requires Mg^{2+} for activity and also contains pyridoxal phosphate, the role of which is not known. Some forms of phosphorylase are allosterically regulated by AMP although this activation is superimposed upon other, more complex, control mechanisms.

Glycogen (n residues)

$+ H_3PO_4 \xrightarrow{Mg^{2+}}$

Glucose 1-phosphate   Glycogen (n-1 residues)

The reaction catalyzed by phosphorylase

Phosphorylase cannot break the α-1,6 linkages of glycogen and indeed its action along the outer chains of the molecule ceases when four glucose residues remain exterior to each branch point.

(b) Oligo-1,4→1,4-glucantransferase and amylo-α-1,6 glucosidase: the debranching process

During glycogen catabolism, the oligotransferase catalyzes the transfer of a terminal maltotriosyl or maltosyl unit from a shortened outer chain of the molecule to another similar chain. The transfer may be inter- or intra-molecular and exposes the single α-1,6 linked glucose residue (Fig. 6.5).

The α-1,6 glucose unit which remains is then removed hydrolytically

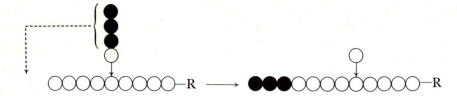

Fig. 6.5 The intra-molecular transfer of a maltotriosyl residue catalyzed by oligo 1,4 → 1,4 glucantransferase.

by the α-1,6 glucosidase and gives rise to free glucose. In this way, debranching is completed at the outer branch points and phosphorylase is able to act again, this time on the new α-1,4 linked chains which are produced. The overall mechanism is summarized in Fig. 6.6.

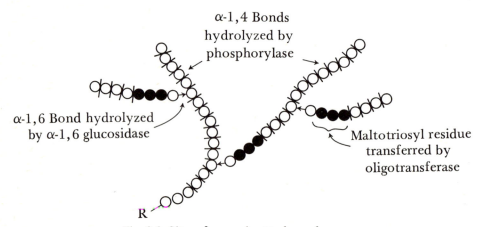

Fig. 6.6 Sites of enzymic attack on glycogen.

(c) The control of glycogen degradation

Since glycogen is a major storage compound, which is metabolized for energy purposes and the maintenance of the correct concentration of glucose in the blood under fasting conditions, it is to be expected that its degradation will be subject to sensitive control mechanisms that reflect the various demands made upon the body. One can see that in the case of muscle glycogen, extremely rapid degradation may be required when the muscle changes from a state of rest to one where vigorous contraction is taking place. The variation of phosphorylase activity is the mechanism by which the extent of degradation of glycogen is controlled. This is an efficient process since phosphorylase can be considered to be the "first enzyme" in a metabolic pathway (see Chapter 5.11).

Phosphorylase from skeletal muscle can exist in two forms which are conveniently referred to as the "active" *a* form and the "inactive" *b* form.

The active *a* form is so called because it is active without AMP whereas the inactive *b* form shows no activity in the absence of this nucleotide. The *a* form of the enzyme has been shown to be a tetramer with a molecular weight of 370,000 while the *b* form is a dimer which is half this size.

The interconversion of the two forms of muscle phosphorylase (Fig. 6.7) is itself under enzymic control. The formation of active *a* from

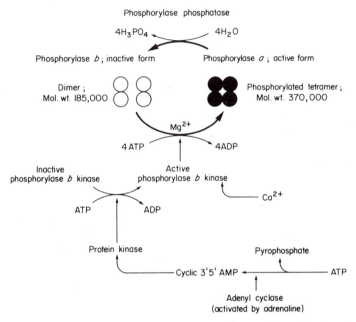

Fig. 6.7 The interconversion of inactive and active forms of muscle phosphorylase.

inactive *b* is catalyzed by phosphorylase *b* kinase and requires four molecules of ATP for each molecule of tetramer that is formed; the reaction brings about the phosphorylation of a serine residue in each of the four subunits. The reverse process, the conversion of active tetramer to inactive dimer is catalyzed by phosphorylase phosphatase which removes the four phosphate groups as inorganic phosphate.

Phosphorylase *b* kinase, which requires Ca^{2+} for full activity, is itself produced from an inactive form by protein kinase that is allosterically activated by cyclic $3',5'$ AMP (see Chapters 3.16 and 13.4). Cyclic AMP is formed from ATP by adenyl cyclase, a membrane-bound enzyme which, in skeletal muscle, is activated by the hormone adrenaline (epinephrine).

The overall activation of muscle phosphorylase thus consists of a cascade mechanism triggered off by adrenaline secreted by the adrenal medulla and dependent upon Ca^{2+} ions which are released from the sarcoplasmic reticulum; Ca^{2+} ions also play an important, direct role in muscle contraction.

Phosphorylase of heart muscle is subject to the same control system as the one described for skeletal muscle. Liver phosphorylase has slightly different control properties in that, although it undergoes an "inactive ↔ active" interconversion which is dependent upon phosphorylation ↔ dephosphorylation, a tetramer is not formed, neither does the inactive species show activity in the presence of AMP. The adenyl cyclase of liver is activated by glucagon, a hormone secreted by the pancreas in response to a low blood glucose level, as well as by adrenaline.

The advantages of a cascade control system of the kind described above are perhaps twofold. Firstly, because of the intermediate enzymic processes, there is obviously considerable amplification of the initial signal; and secondly, "branched" control mechanisms are possible whereby a number of different target sites may be affected by the one primary stimulus. This is particularly important with respect to cyclic AMP which is an effector (activator or inhibitor) for many reactions taking place within the cell (see Chapter 13.2). For example, as well as increasing the rate of degradation of glycogen by causing the activation of phosphorylase it also acts to decrease the rate of glycogen synthesis by bringing about the inactivation of glycogen synthetase.

(d) Phosphoglucomutase

As we have seen, glucose 1-phosphate is the major monosaccharide derivative formed from glycogen. In the liver, the conversion of glucose 1-phosphate to glucose 6-phosphate is the next reaction leading to the formation of glucose. The same reaction occurs in muscle when glycogen is being broken down for energy release.

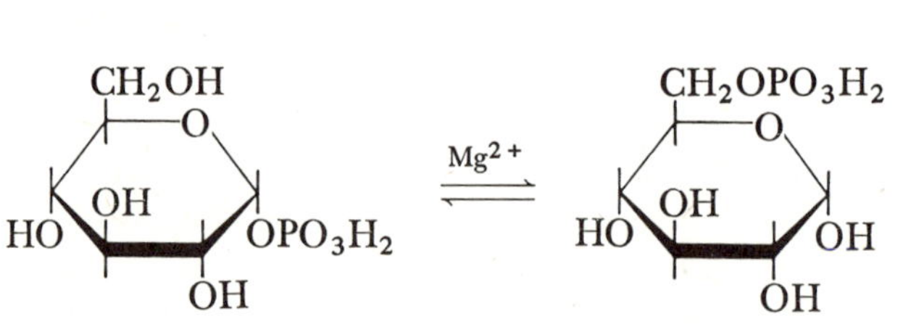

$$
\text{Glucose 1-phosphate} \quad \overset{\text{Mg}^{2+}}{\rightleftharpoons} \quad \text{Glucose 6-phosphate}
$$

Glucose 1-phosphate   Glucose 6-phosphate

This reaction is readily reversible and is catalyzed by phosphoglucomutase, an enzyme that requires Mg^{2+} as a cofactor.

(e) Glucose 6-phosphatase

The formation of free glucose in the liver is brought about by glucose 6-phosphatase. This enzyme, absent from muscle, catalyzes the hydrolytic removal of the phosphate group from glucose 6-phosphate.

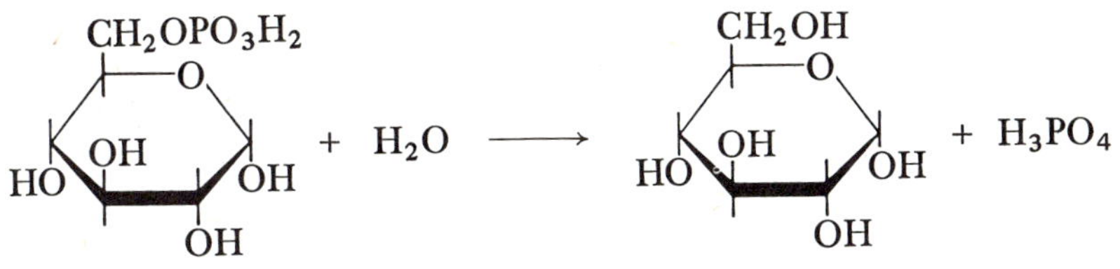

Glucose 6-phosphate   Glucose

6.5. The Synthesis of Glycogen from Glucose (Glycogenesis)

The pathways of glycogen synthesis from glucose are identical in muscle and liver.

(a) The phosphorylation of glucose: hexokinase and glucokinase

When glucose is being metabolized, whether for the synthesis of glycogen or for energy release, the initial reaction that takes place is one of phosphorylation. ATP is the phosphate donor.

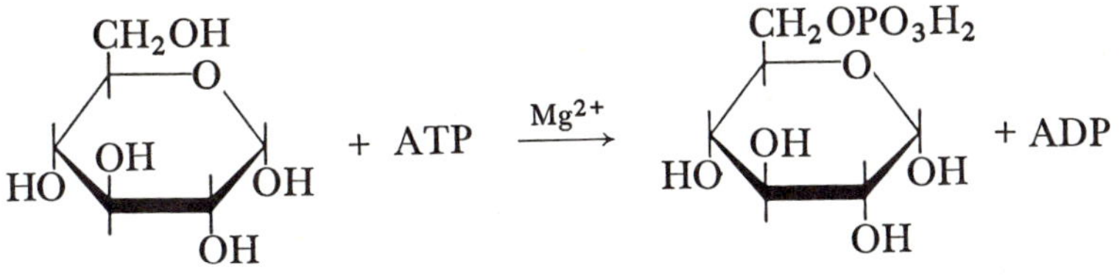

Glucose   Glucose 6-phosphate

There are, however, two enzymes which catalyze this reaction, both of them requiring Mg^{2+} ions for activity. Hexokinase, which is one of these, has a high affinity for glucose (K_m = 0·01 mM) and relatively low substrate specificity in that it will also phosphorylate mannose and fructose in the 6 position. The second enzyme, glucokinase, which is specific for glucose, has a much lower affinity for its substrate (K_m = 10 mM).

Only hexokinase is present in muscle. Both enzymes are found in liver but glucokinase is normally the more active of the two when the diet contains adequate amounts of carbohydrate. On fasting, or in diabetes, glucokinase activity virtually disappears but its activity in the diabetic rat can be increased by the administration of insulin.

Glucokinase therefore, under normal conditions, appears to be concerned exclusively with the synthesis of liver glycogen when the blood glucose concentration is high.

(b) The formation of glucose 1-phosphate from glucose 6-phosphate

The conversion of glucose 6-phosphate to glucose 1-phosphate is a simple reversal of the forward reaction and is catalyzed by phosphoglucomutase (see above).

(c) The synthesis of glycogen from glucose 1-phosphate

Although the synthesis of glycogen from glucose 1-phosphate can be demonstrated *in vitro* using phosphorylase, this reversal of the degradative reaction is unlikely to occur *in vivo* as the concentrations of glucose 1-phosphate and phosphate in the cell will be unfavourable for glycogen synthesis. The formation of glycogen *in vivo*, from glucose 1-phosphate, including the synthesis of α-1,6 glucosidic bonds, takes place in three stages. Energy is required in the form of nucleoside triphosphate for the formation of each glucosidic bond. The actual triphosphate that is used in the process is uridine triphosphate (UTP) although this is readily formed from UDP and ATP by nucleoside diphosphokinase, an enzyme that catalyzes the interconversion of nucleoside di- and triphosphate.

(i) URIDINE DIPHOSPHATE-GLUCOSE PYROPHOSPHORYLASE

The reaction of glucose 1-phosphate with UTP to give uridine diphosphate glucose (UDPG) and pyrophosphate is catalyzed by UDPG pyrophosphorylase.

$$\text{Glucose 1-phosphate} + \text{UTP} \rightarrow \text{UDPG} + \text{PP}_i \text{ (pyrophosphate)}$$

The pyrophosphate is rapidly hydrolyzed to two molecules of inorganic phosphate by the action of pyrophosphatase. This appears to be a reaction which is wasteful of energy since it is strongly exothermic. Nevertheless it constitutes an important mechanism for ensuring that reactions which produce pyrophosphate go to completion by the removal of an end product.

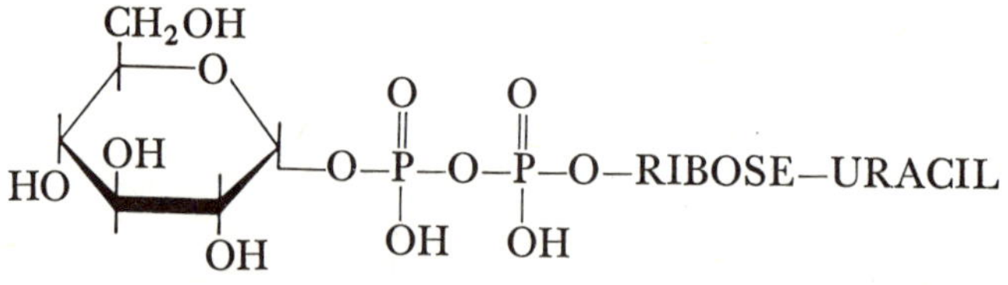

Uridine diphosphate-glucose (UDPG)

(ii) GLYCOGEN SYNTHETASE

Glycogen synthetase catalyzes the formation of the α-1,4 glucosidic bonds of glycogen by the stepwise extension of the non-reducing ends of the molecule.

$$\text{Glycogen} + \text{UDPG} \longrightarrow \text{Glycogen} + \text{UDP}$$
$$(n \text{ glucose residues}) \qquad (n + 1 \text{ glucose residues})$$

The rate of this reaction is very much dependent upon the size of the acceptor molecule in that the larger this is then the faster is the reaction.

Small molecules such as maltotriose and maltotetrose may act as primers for the system although the enzyme shows a much lower affinity for these substrates and the relative reaction rate is much reduced. It is even possible to demonstrate the complete synthesis of polysaccharide where there is no preformed acceptor molecule present.

TABLE 6.2. Relative rates of glycogen synthetase activity with acceptor molecules of different sizes

Acceptor molecule	Relative reaction rate
Glycogen	$1 \cdot 0$
Potato starch (solubilized)	$0 \cdot 6$
Maltotetrose to maltohexose	2×10^{-3}
Maltotriose	2×10^{-4}
Maltose	7×10^{-5}

This relationship between activity and primer size is probably physiologically significant in that it ensures that glycogen is preferentially synthesized at the sites where stores of the polysaccharide already exist. Old stores are added to, rather than new ones formed.

The overall pathway for the synthesis of the α-1,4 linked chains of glycogen from glucose 1-phosphate can, therefore, be represented as shown in Fig. 6.8.

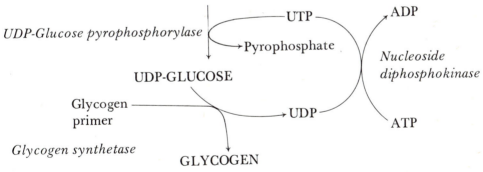

Fig. 6.8 The synthesis of α-1,4 glucosidic bonds of glycogen.

(iii) AMYLO-1,4→1,6-TRANSGLYCOSYLASE: THE BRANCHING PROCESS

The action of glycogen synthetase results in the formation of only the linear parts of the glycogen molecule. Branching of the structure is brought about by the branching enzyme, amylo-1,4→1,6-transglycosylase.

Glycogen synthetase first catalyzes the synthesis of an "extended chain" of α-1,4 linked glucosyl residues. When the chain is about ten or twelve residues long, the branching enzyme removes the terminal group of some six residues and transfers this unit to the 6 hydroxyl of a glucosyl residue in the same chain (Fig. 6.9) or in an adjacent chain. Chain elongation then occurs again by the action of glycogen synthetase.

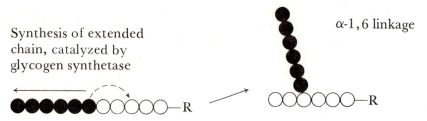

Fig. 6.9 The branching of glycogen.

(d) The control of glycogen synthesis

In many respects, the control of glycogen synthesis is similar to that of glycogen degradation and the two considered together should be seen as complementary processes (see Chapter 13.4).

Glycogen synthetase, like phosphorylase exists in two interconvertible forms. These are, respectively, the "active" I (independent) form and the "inactive" D (dependent) form. The terms independent and dependent refer to whether the synthetase is active or inactive in the presence of the allosteric effector, glucose 6-phosphate.

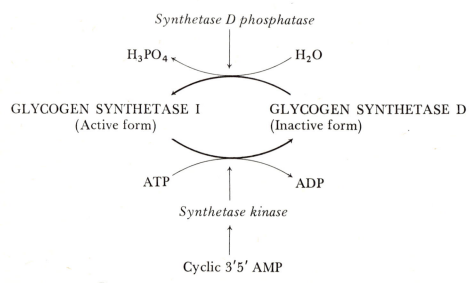

Fig. 6.10 The control of glycogen synthetase activity.

The interconversion of the two forms of glycogen synthetase is under enzymic control. However, in contrast to phosphorylase, glycogen synthetase is *inactivated* by the action of a kinase and *activated* by a phosphatase. The kinase, activated by cyclic $3',5'$ AMP, appears to be the same protein kinase that catalyzes the activation of phosphorylase b kinase. Again, therefore, there is an overall cascade control mechanism with the initial signal being the hormonal activation of adenyl cyclase. In this way, adrenaline in muscle, and glucagon and adrenaline in liver decrease the rate of glycogen synthesis at the same time that they increase that of glycogen degradation.

(e) The interconversion of glycogen and glucose: a summary

The reactions that we have discussed so far, leading to glycogen synthesis or bringing about glycogen degradation, can be summarized as shown in Fig. 6.11.

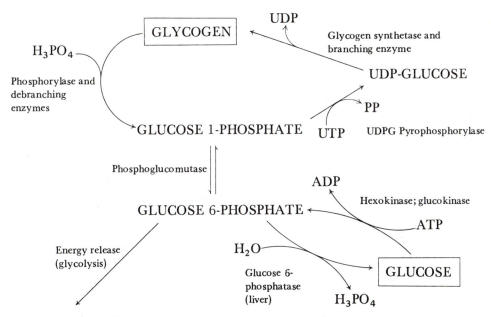

Fig. 6.11 Summary of glycogen and glucose interconversion.

(f) The direct formation of glucose from glycogen

The hydrolysis of glycogen to glucose occurs to some extent in all cells and is brought about by α-1,4 glucosidase. This enzyme is of lysosomal origin and like all the lysosomal enzymes it appears to have a "scavenging" role in the metabolism of the cell, in this case digesting unwanted glycogen. Also, in common with other enzymes of lysosomal origin, it has a low pH optimum (around pH 4) and because of this it is frequently referred to as acid maltase. It is not thought to be of significance in the mobilization of stored glycogen or in the metabolism of carbohydrate for energy release.

6.6. Glycogen-storage Diseases

A number of metabolic disturbances may occur which are characterized by either the over-accumulation of glycogen or the synthesis of "glycogen" with an abnormal structure. They are caused by genetic defects that are present in the enzymes that are concerned with the metabolism of glycogen. Collectively they are called the glycogen-storage diseases and the condition is known as glycogenosis. Brain damage during infancy is often caused by certain undetected glycogen storage diseases resulting in hypoglycaemia. These conditions are divided into distinct types (Table 6.3) and most of these can be related directly to a specific defect in one enzyme. The actual cause of Type VI glycogen-storage disease has not been identified. Frequently liver phosphorylase is low but this could be a secondary effect; it is probable that Type VI will be further subdivided.

TABLE 6.3. Glycogen-Storage Diseases

Classical Type	Common Name	Enzymic Lesion	Characteristics
I	Von Gierke's	Glucose 6-phosphatase	Poor mobilization of liver glycogen. Fasting hypoglycaemia.
II	Generalized or Pompe's disease	Lysosomal α-1,-4 glucosidase (Acid maltase)	Generalized, lysosomal accumulation of glycogen particularly in liver, heart, skeletal muscle and kidney. Enlargement of the heart and muscular weakness. Detected at approx. 3 months; usually fatal during 1st year.
III	Limit dextrinosis	Amylo-α-1,6-glucosidase or oligo-1,4$\rightarrow$1,4-glucantransferase (the debranching system)	Accumulation of glycogen with short outer chains. Hypoglycaemia common. Type A affects muscle and liver. Type B affects liver.
IV	Amylopectinosis	1,4$\rightarrow$1,6-glucosyltransferase (the branching enzyme)	Accumulation of liver glycogen with long chains and few branch points. Fasting hypoglycaemia. Usually fatal during first five years.
V	McArdles disease	Muscle phosphorylase	Early onset of muscular fatigue. Usually raised levels of muscle glycogen.
VI	Hers' disease	Not known. Probably more than one cause	Accumulation of glycogen in the liver. Sometimes accompanied by fasting hypoglycaemia.

In addition to the enzyme lesions listed in Table 6.3, the absence of liver glycogen synthetase has also been recognized, but strictly speaking, inactivity of this enzyme leads to a glycogen-deficiency disease rather than to a glycogen-storage disease. It is characterized by the presence of minimal amounts of glycogen in the liver and by profound hypoglycaemia between meals.

6.7. The Degradation of Glucose

The catabolism of glucose within cells for the specific purpose of ATP formation can be divided into two distinct stages. The first stage, which may proceed either aerobically or anaerobically, involves only the partial degradation of glucose and results in the formation of pyruvate or the accumulation of lactate respectively. The reactions that accomplish this, and which take place in the soluble cytoplasm of the cell, constitute the glycolytic pathway, or the Embden-Meyerhof-Parnas pathway, and the process itself is known as glycolysis.

The second stage, which must take place aerobically, occurs in the mitochondria of the cell and brings about the complete oxidation of the pyruvate that is formed by aerobic glycolysis. The pathway by which this occurs is the tricarboxylic acid (TCA) cycle, also known as the citric acid cycle or the Krebs' cycle. As will be seen, much more ATP is produced by this second stage of metabolism than by the first.

Although we have divided the metabolism of glucose into these two stages it must be emphasized that under aerobic conditions both are closely integrated to form a continuous catabolic pathway.

In addition to the metabolism of glucose by the combined action of glycolysis and the TCA cycle, other pathways of glucose degradation exist. These are collectively referred to as Alternative Pathways of glucose metabolism and one of them, the Pentose Phosphate Pathway, is particularly important in animal cells. The designation "Alternative" is somewhat misleading when applied to the pentose-phosphate pathway since it tends to suggest that the pathway may operate instead of glycolysis and the TCA cycle. This is not, of course, the case and both these pathways are essential for most cells and function side by side for different purposes.

6.8. The Glycolytic Pathway

The evidence for the individual reactions of glycolysis was obtained over a period of many years from studies of two completely different systems, namely, the fermentation of glucose by yeast and the production of lactic acid by various muscle preparations. The reactions occurring during both these processes, apart from the initial and final steps, are identical. They can be represented as follows:

(i) *Yeast fermentation*

$$\text{Glucose} \rightarrow 2C_2H_5OH + 2CO_2$$
$$\text{Ethanol}$$

(ii) *Anaerobic muscle contraction*

$$\text{Glycogen} \rightarrow 2CH_3CH(OH)COOH \text{ (per glucosyl residue)}$$
$$\text{Lactic acid}$$

Although yeast and muscle are able to effect extremely rapid glycolysis, it must be emphasized that most tissues, to varying extents, are able to derive energy by the anaerobic breakdown of glucose in this way.

TABLE 6.4. The rate of anaerobic glycolysis in various rat tissues.
(The units of rate are μmoles of lactic acid produced per hour per mg dry weight of tissue)

Tissue	Rate	Tissue	Rate
Retina	3·5	Placenta	0·7
Tumour cells			
(Jensen sarcoma)	1·6	Spleen	0·35
Kidney medulla	1.2	Kidney cortex	0·30
Embryo	1·0	Spermatozoa	0·25
Bone marrow	1·0	Liver	0·15
Brain	0·9	Erythrocytes	0·015

(a) The enzymes of the glycolytic pathway

The glycolytic pathway is shown in Fig. 6.12. Most of the intermediate steps are readily reversible with only three of them having free-energy changes so large that different reactions have to be used during the process of glucose synthesis (gluconeogenesis) from lactate or pyruvate.

The enzymes catalyzing the reactions of glycolysis and gluconeogenesis, with a few exceptions, fall conveniently into classes depending upon the type of reaction catalyzed. The kinases catalyze reactions involving the transfer of the terminal phosphate group of ATP; phosphatases remove a phosphate group from the substrate, hydrolytically, to give inorganic phosphate; mutases are involved in the transfer of phosphate within the molecule; isomerases catalyze the isomerization of the non-phosphate moiety of the substrate and dehydrogenases (oxidoreductases) bring about the transfer of two hydrogen atoms between substrate and hydrogen acceptor, in this case NAD. This simple classification is often useful to the student when relating the name of an enzyme to the reaction which it controls.

The individual reactions of the glycolytic pathway are catalyzed by the enzymes described below; the numbers correspond to those shown in Fig. 6.12. The conversion of muscle glycogen to glucose 6-phosphate has already been described and is omitted here.

(i) HEXOKINASE

Although the purpose of glycolysis is to produce ATP, an "energy debt" is initially incurred since ATP is itself required for two phosphorylation steps that take place before energy-producing reactions occur. Hexokinase, catalyzing the formation of glucose 6-phosphate from glucose and

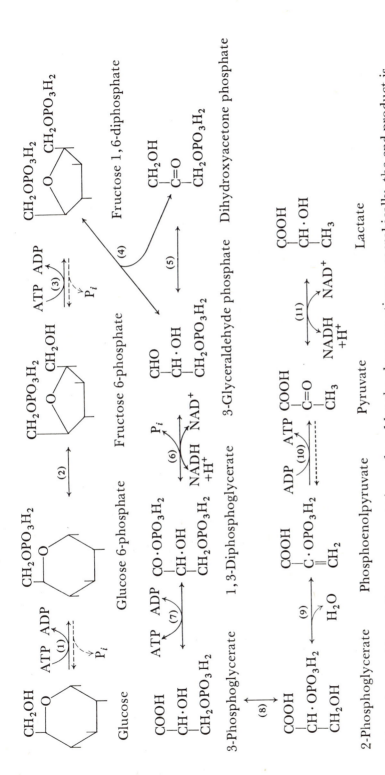

Fig. 6.12 The glycolytic pathway of glucose degradation.

(i) Under aerobic conditions, pyruvate is produced by the above reactions; anaerobically, the end-product is lactate.

(ii) The numbers alongside the reactions refer to numbered enzymes described in the text.

(iii) Unidirectional arrows indicate that different enzymes are involved when the pathway operates in reverse for the synthesis of glucose.

ATP, brings about the first of these phosphorylations, a reaction that is not directly reversible when gluconeogenesis is taking place.

Other properties of hexokinase, and the role of glucokinase in the phosphorylation of glucose in the liver, have already been discussed (see Chapter 6.5).

(ii) PHOSPHOGLUCOISOMERASE

This enzyme catalyzes the reversible formation of fructose 6-phosphate from glucose 6-phosphate; it is specific for these substrates. A similar enzyme, phosphomannoisomerase, catalyzes the interconversion of fructose 6-phosphate and mannose 6-phosphate. Since the latter is produced from mannose by the action of hexokinase, this sugar can be readily metabolized via the glycolytic pathway.

(iii) PHOSPHOFRUCTOKINASE

Phosphofructokinase requires Mg^{2+} and catalyzes the incorporation of a second phosphate group into the hexose molecule, using ATP as the phosphate donor. The products of the reaction are, therefore, fructose 1,6-diphosphate and ADP. The reaction is irreversible and a different enzyme, fructose 1,6-diphosphatase is required to bring about the formation of fructose 6-phosphate from fructose 1,6-diphosphate.

The control of phosphofructokinase is an important factor in the overall control of the glycolytic pathway. It is inhibited by citrate and by excess ATP, compounds which in many respects can be considered to be "end-products" of glycolysis, and activated by ADP and AMP (see Chapter 13.2).

(iv) ALDOLASE

Aldolase catalyzes the non-hydrolytic cleavage (an aldol cleavage) of fructose 1,6-diphosphate to give 3-phosphoglyceraldehyde and dihydroxyacetone phosphate. Although the equilibrium position of this reaction actually favours fructose 1,6-diphosphate synthesis, its progress in the forward direction is facilitated by the rapid removal of the products (the triose phosphates) by the enzymes catalyzing the subsequent reactions of glycolysis.

(v) TRIOSE PHOSPHATE ISOMERASE

Triose phosphate isomerase catalyzes the interconversion of 3-phosphoglyceraldehyde and dihydroxyacetone phosphate. Although the equilibrium position favours the formation of dihydroxyacetone phosphate, the reaction is readily reversible.

(vi) 3-PHOSPHOGLYCERALDEHYDE DEHYDROGENASE

The reaction catalyzed by 3-phosphoglyceraldehyde dehydrogenase is the only oxidation step that occurs during the whole of glycolysis. It is a rather complex reaction in that phosphorylation also takes place. 1,3-Diphosphoglycerate and NADH are formed from NAD^+, inorganic phosphate and 3-phosphoglyceraldehyde. It is therefore the first reaction in the pathway in which energy, in the form of NADH, is conserved in a utilizable form. The phosphate group which is incorporated into 1,3-diphosphoglycerate is also important since it becomes available for the formation of ATP in the following reaction.

(vii) PHOSPHOGLYCERATE KINASE

Phosphoglycerate kinase catalyzes the first of two reactions in glycolysis where ATP is formed. This occurs here by the transfer of the phosphate group from the 1 position of 1,3-diphosphoglycerate to ADP, leaving 3-phosphoglycerate as the other product of the reaction. The enzyme requires Mg^{2+} ions.

Since two molecules of 1,3-diphosphoglycerate take part in this reaction for every glucose molecule phosphorylated, the "ATP debt", incurred earlier by the action of hexokinase and phosphofructokinase, is cancelled out.

(viii) PHOSPHOGLYCEROMUTASE

3-Phosphoglycerate formed in the previous reaction is converted to an isomer, 2-phosphoglycerate, by the action of phosphoglyceromutase. This reaction is readily reversible and is similar to that catalyzed by phosphoglucomutase (see Chapter 6.5) in that Mg^{2+} ions are required and the transfer of a phosphate group occurs from one position of the molecule to another.

(ix) ENOLASE

Enolase catalyzes the dehydration of 2-phosphoglycerate to give phosphoenolpyruvate. There is a small, positive free energy change associated with this reaction so that the equilibrium position lies slightly in favour of 2-phosphoglycerate. Enolase requires either Mg^{2+} or Mn^{2+} for activity and it is very sensitive to inhibition by fluoride ions, particularly in the presence of added phosphate.

(x) PYRUVATE KINASE

Pyruvate kinase catalyzes the formation of pyruvate from phosphoenolpyruvate. The reaction is an important one in that the phosphate group is transferred to ADP to give ATP. Nevertheless, even though ATP is formed, the free energy change accompanying this reaction is strongly negative so that other enzymes are necessary to bring about the synthesis of phosphoenolpyruvate from pyruvate. Because of the unidirectional nature of the pyruvate kinase reaction, this step is also important with respect to the control of glycolysis. The enzyme is inhibited by NADH, citrate or ATP, the concentration of which will tend to increase as a result of aerobic metabolism in the cell when the need for the operation of the glycolytic pathway is much reduced.

(xi) LACTATE DEHYDROGENASE

The formation of lactate from pyruvate, brought about by lactate dehydrogenase, is only of significance under what is effectively an anaerobic condition. This results where the energy requirement of the cell is in excess of that which can be met by the oxidation of pyruvate in the mitochondria. This occurs because the rate of diffusion of oxygen into the tissues has become limiting. A good example is the condition that exists in rapidly contracting skeletal muscle. As glycogen is broken down to pyruvate, NADH is formed, and since all of this cannot be re-oxidized aerobically, the only way in which NAD can be recovered to allow further glycolysis to occur is for NADH to react with pyruvate to form lactate. The concentration of lactate in the blood at rest is normally about 0·5-1·5 mmol/l. During exercise it may rise to as high as 20 mmol/l. The accumulated lactate is subsequently used for the formation of glucose in the liver, by gluconeogenesis, and the depleted stores of muscle glycogen are replenished when muscle contraction ceases.

In most mammalian tissues up to five isoenzymes of lactate dehydrogenase (LDH) can be detected by electrophoresis or ion exchange chromatography. Their relative concentrations vary depending upon the

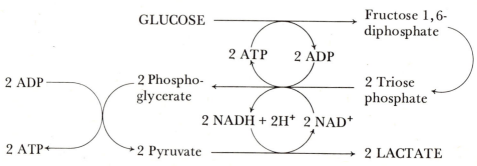

Fig. 6.13 Anaerobic glycolysis leading to the formation of lactate.

function of the particular tissue or organ where they are found. Consequently, each organ tends to have a characteristic pattern of LDH isoenzymes. The structural explanation of the existence of these isoenzymes, their physiological significance and clinical usefulness have been described in Chapter 5.11.

The overall balance of anaerobic glycolysis in mammalian cells, where lactate is the end-product, is shown in Fig. 6.13. During yeast fermentation, the formation of ethanol is directly analogous to the formation of lactate in mammalian cells in that the process serves to recycle NAD^+. The reactions are as follows:

(i) $CH_3 \cdot CO \cdot COOH \xrightarrow[\text{carboxylase}]{Mg^{2+}, \text{TPP}} CH_3CHO + CO_2$

   Pyruvate   Acetaldehyde

(ii) $CH_3CHO + NADH + H^+ \xrightarrow[\substack{\text{alcohol} \\ \text{dehydrogenase}}]{} C_2H_5OH + NAD^+$

                                                          Ethanol

Because of its role as the coenzyme in the carboxylase reaction, thiaminepyrophosphate (TPP) is sometimes referred to as co-carboxylase.

(b) The energy yield of glycolysis

An examination of Fig. 6.13 allows us to calculate the amount of ATP that is formed during glycolysis. Anaerobically, where lactate is the end-product, there is a net gain of two molecules of ATP for each molecule of glucose that is metabolized. Under aerobic conditions, in addition to these two molecules of ATP, the two molecules of NADH produced by the phosphoglyceraldehyde dehydrogenase reaction are equivalent to six molecules of ATP when their oxidation occurs via the cytochrome system of the mitochondria (see Chapter 9). The total yield of ATP aerobically from the reactions of glycolysis is therefore eight molecules per each glucose metabolized.

Where glycogen is the substrate for glycolysis then, as glucose 6-phosphate is formed without the utilization of ATP, the net yields of ATP per glucosyl residue metabolized are three and nine, respectively, under anaerobic and aerobic conditions.

6.9. Gluconeogenesis and Glyconeogenesis

Gluconeogenesis is defined as the synthesis of glucose from non-carbohydrate precursors and glyconeogenesis is the corresponding synthesis of glycogen from the same sources. Gluconeogenesis is essentially the reverse of glycolysis in that most of the intermediates and many of the enzymes are common to both pathways. The substrates for gluconeogenesis are lactate, pyruvate and the carbon skeletons of deaminated amino acids, many of which enter the pathway via oxaloacetate.

The main site of gluconeogenesis is the liver although some other organs, particularly the kidney, can also synthesize glucose this way. The enzymes that are specific for gluconeogenesis are virtually absent from skeletal muscle so that the lactate that is produced in the muscle during contraction must be first carried to the liver by the blood before it can be used for the synthesis of glucose. The important relationships between glycolysis, gluconeogenesis and the reactions that maintain a steady-state concentration of glucose in the blood are shown in Fig. 6.14.

(a) Reactions specific to gluconeogenesis

In the glycolytic pathway, leading from glucose to pyruvate, we saw that there were three reactions which had free energy changes such that they could not be readily reversed under physiological conditions. These reactions are, respectively, the formation of glucose 6-phosphate from glucose, fructose 1,6-diphosphate from fructose 6-phosphate and phosphoenol pyruvate from pyruvate. During gluconeogenesis, these steps are reversed as follows:

(i) PHOSPHOENOLPYRUVATE FROM PYRUVATE

Two pathways have been postulated to account for the formation of phosphoenolpyruvate from pyruvate. Both are very similar and involve the intermediate formation of one or more four-carbon compounds. It is now agreed that the most important route in the liver first involves the carboxylation of pyruvate, with CO_2, to form oxaloacetate.

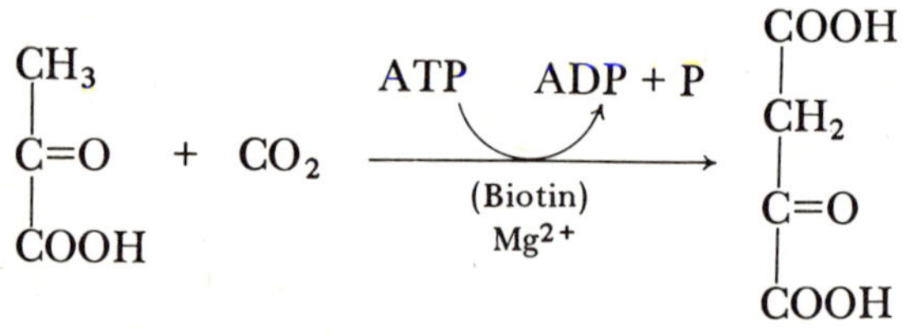

$$
\begin{array}{ccc}
CH_3 & & COOH \\
| & ATP \quad ADP + P & | \\
C{=}O \quad + \quad CO_2 \quad \xrightarrow[\substack{(Biotin) \\ Mg^{2+}}]{} & & CH_2 \\
| & & C{=}O \\
COOH & & | \\
& & COOH
\end{array}
$$

Pyruvate   Oxaloacetate

The enzyme catalyzing this reaction, pyruvate carboxylase, is a mitochondrial one and requires biotin as a coenzyme and ATP. It is also allosterically activated by acetyl coenzyme A.

The oxaloacetate so formed, ultimately reacts with GTP to give phosphoenolpyruvate.

$$
\begin{array}{ccc}
COOH & & \\
| & & CH_2 \\
CH_2 & CO_2 & \| \\
| \quad + \quad GTP \quad \xrightarrow{} & C{-}OPO_3H_2 \quad + \quad GDP \\
C{=}O & & | \\
| & & COOH \\
COOH & &
\end{array}
$$

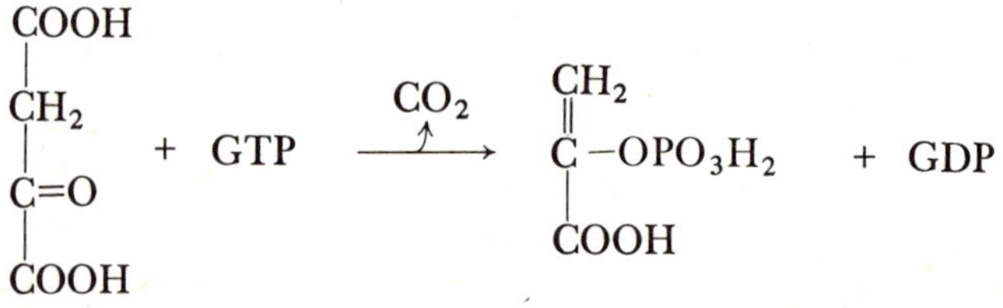

Oxaloacetate   Phosphoenolpyruvate

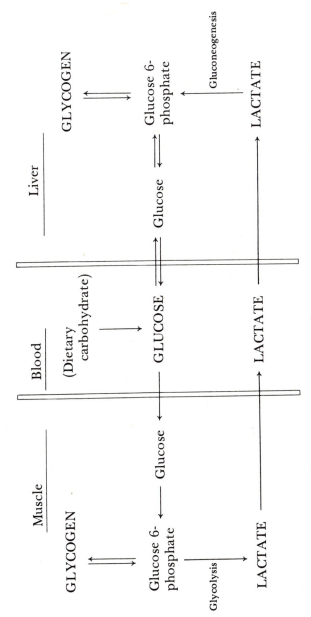

Fig. 6.14 The interrelationship of muscle glycolysis and liver gluconeogenesis.

However, the enzyme catalyzing this reaction, phosphoenolpyruvate carboxykinase, is found predominantly in the soluble cytoplasm and the mitochondrial membrane is impermeable to oxaloacetate. The mechanism by which net transfer of oxaloacetate across the membrane occurs, involves the participation of malate dehydrogenase, present as separate enzymes (quite different proteins) on either side of the mitochondrial membrane.

$$
\begin{array}{c}
\text{COOH} \\
| \\
\text{CH}_2 \\
| \\
\text{CH} \cdot \text{OH} \\
| \\
\text{COOH} \\
\text{Malate}
\end{array}
+ \text{NAD}^+ \rightleftharpoons
\begin{array}{c}
\text{COOH} \\
| \\
\text{CH}_2 \\
| \\
\text{C=O} \\
| \\
\text{COOH} \\
\text{Oxaloacetate}
\end{array}
+ \text{NADH} + \text{H}^+
$$

Malate, which is readily formed from or converted to oxaloacetate, passes freely out of the mitochondrion. The complete reaction sequence is shown in Fig. 6.15.

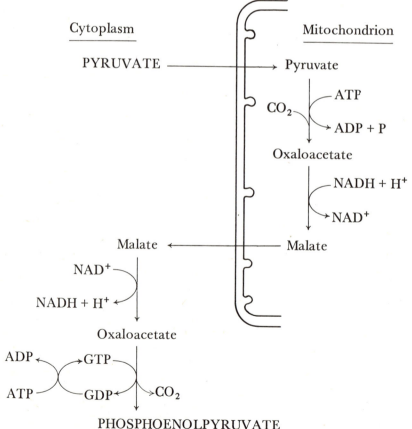

Fig. 6.15 The formation of phosphoenolpyruvate from pyruvate.

This results in the following overall reaction: Pyruvate + 2ATP → Phosphoenolpyruvate + H_3PO_4 + 2ADP. At the same time, two hydrogen atoms are transferred from mitochondrial NADH to cytoplasmic NAD.

It was originally thought that malate was formed directly from pyruvate by a reductive carboxylation reaction catalyzed by the malic enzyme

$$\text{Pyruvate} + \text{NADPH} + \text{H}^+ + \text{CO}_2 \rightleftharpoons \text{Malate} + \text{NADP}^+$$

However, although this enzyme is a soluble cytoplasmic one, all the evidence available suggests that the reaction normally functions in the "reverse" direction to bring about the formation of NADPH from NADP. It therefore appears to have no role in gluconeogenesis.

(ii) FRUCTOSE 6-PHOSPHATE FROM FRUCTOSE 1,6-DIPHOSPHATE

This reaction is catalyzed by fructose 1,6-diphosphatase and involves the hydrolytic removal of the phosphate group as inorganic phosphate.

$$\text{Fructose 1,6-diphosphate} + H_2O \rightarrow \text{Fructose 6-phosphate} + H_3PO_4$$

(iii) GLUCOSE FROM GLUCOSE 6-PHOSPHATE

This reaction is very similar to the previous one and is catalyzed by glucose 6-phosphatase.

$$\text{Glucose 6-phosphate} + H_2O \rightarrow \text{Glucose} + H_3PO_4$$

The role of this enzyme and its significance in bringing about the formation of blood glucose in the liver have already been described. It carries out this function whether the glucose 6-phosphate is an intermediate in glycogen breakdown or formed by gluconeogenesis.

6.10. The Metabolism of Hexoses other than Glucose

Apart from glucose, the two most important hexoses derived from dietary carbohydrate are fructose and galactose. The first of these arises mainly from sucrose while the other is a constituent of lactose, the carbohydrate of milk. The metabolism of both sugars takes place mainly in the liver with that of galactose being particularly important during infancy.

(a) The metabolism of fructose

Although fructose can be phosphorylated in the 6 position by hexokinase, and so enter the glycolytic pathway directly, the major pathway for its metabolism consists of the formation of the 1-phosphate by a specific fructokinase that requires ATP. The fructose 1-phosphate so formed is then cleaved by an isoenzyme of aldolase to give glyceraldehyde and dihydroxyacetone phosphate. Most of the glyceraldehyde formed in this way is phosphorylated to 3-phosphoglyceraldehyde although a small amount may be oxidized to glyceric acid before phosphorylation occurs. Whichever route is followed, the product is an intermediate of the glycolytic pathway.

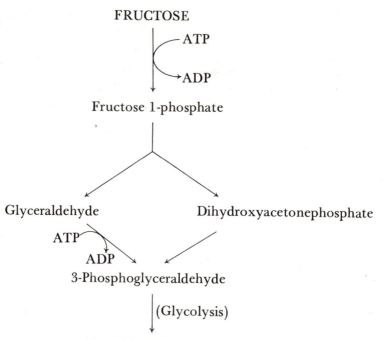

Fig. 6.16 The metabolism of fructose.

Two hereditary metabolic defects are known in the above pathway of fructose metabolism. The first, a deficiency of fructokinase, is harmless except that the resulting fructosuria may be mistaken for glucosuria and diabetes suspected. Despite the fact that some fructose metabolism occurs via the action of hexokinase, the concentration of this sugar in the urine after a meal may be as high as 100 mmol/l.

In the second case, hereditary fructose intolerance, the aldolase that cleaves fructose 1-phosphate is inactive. Fructose 1-phosphate accumulates in liver, intestine and kidney, all of which have relatively high activities of fructokinase. The main symptoms of this deficiency are nausea, vomiting and abdominal cramps. These are relieved by omitting fructose from the diet.

(b) The metabolism of galactose

Galactose is first phosphorylated in the 1-position by the action of a specific galactokinase using ATP as the phosphate donor. The enzyme is particularly active in the liver but is present also in the kidney and brain.

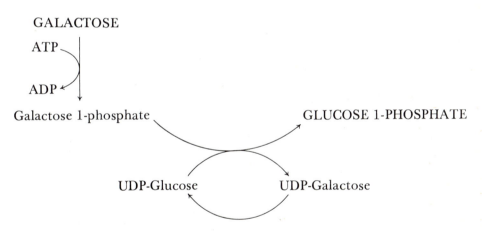

The galactose 1-phosphate then reacts with uridine diphosphate glucose (UDPG) to give UDP-galactose and glucose 1-phosphate. This reaction is catalyzed by uridyl transferase (phosphogalactose uridyl transferase).

UDP-glucose + galactose 1-phosphate $\rightleftharpoons$

UDP-galactose + glucose 1-phosphate

UDP-glucose is then formed from UDP-galactose by an epimerization reaction around carbon 4 of the galactose moiety. The complete reaction sequence is shown in Fig. 6.17.

Fig. 6.17 The formation of glucose 1-phosphate from galactose.

Galactosaemia is an hereditary disease which occurs because of a defective uridyltransferase. It is characterized by a high level of galactose and galactose 1-phosphate in the blood. It is detected in infancy because of the high lactose content of the diet. If not treated by the removal of galactose from the food, mental retardation occurs, there is enlargement of the liver and death frequently results. Only children are affected since in later life there is the development of an alternative pathway for galactose 1-phosphate metabolism.

Galactose 1-phosphate + UTP $\rightarrow$ UDP galactose + PP$_i$

This bypasses the uridyltransferase reaction.

6.11. The Metabolism of Glycerol

It is convenient to consider the metabolism of glycerol at this point since it is closely linked to the reactions of glycolysis and gluconeogenesis. Glycerol is a component of many lipids and quantitatively it is important as a derivative of neutral fat (triglyceride). It is readily converted to dihydroxyacetone phosphate by the action of two enzymes (i) glycerokinase and (ii) α-glycerophosphate dehydrogenase.

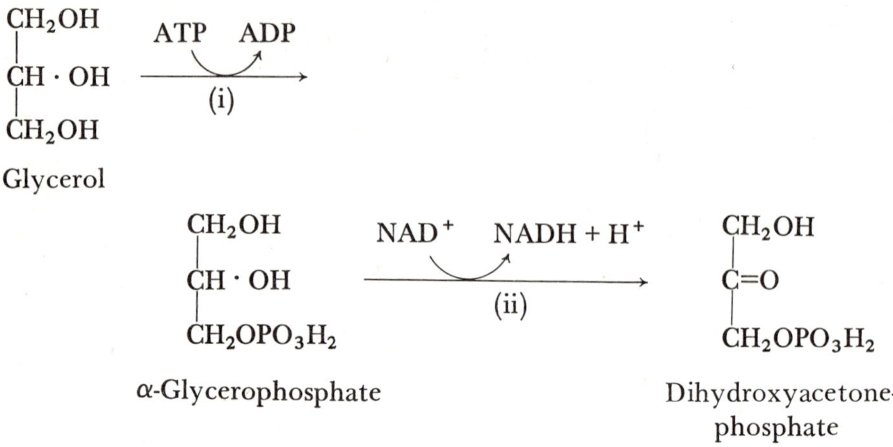

$$
\begin{array}{l}
CH_2OH \\
| \\
CH \cdot OH \\
| \\
CH_2OH
\end{array}
\quad
\xrightarrow[\text{(i)}]{ATP \quad ADP}
$$

Glycerol

$$
\begin{array}{l}
CH_2OH \\
| \\
CH \cdot OH \\
| \\
CH_2OPO_3H_2
\end{array}
\quad
\xrightarrow[\text{(ii)}]{NAD^+ \quad NADH + H^+}
\quad
\begin{array}{l}
CH_2OH \\
| \\
C{=}O \\
| \\
CH_2OPO_3H_2
\end{array}
$$

α-Glycerophosphate   Dihydroxyacetone-
                                                         phosphate

Glycerol may be synthesized by virtually a reversal of this pathway except that the phosphate group is removed from α-glycerophosphate by the action of α-glycerophosphatase, giving glycerol and inorganic phosphate.

6.12. The Aerobic Metabolism of Pyruvate: the Tricarboxylic Acid Cycle

We have seen that during the process of anaerobic glycolysis, glucose (or glycogen) is broken down and lactate accumulates. If oxygen is introduced into an isolated system metabolizing glucose in this way, two things happen. Firstly, the accumulated lactate disappears, and secondly, there is a marked decrease in the rate at which glucose is metabolized. This phenomenon was first described by Pasteur and it is consequently known as the Pasteur Effect. The reason why the Pasteur Effect occurs is that, in the presence of oxygen, the oxidation of the end-product of aerobic glycolysis (pyruvate) can go to completion. Since this process leads to the formation of a large amount of ATP then much less glucose needs to be metabolized to meet the energy requirements of the cell (see Chapter 13.2).

Whereas the reactions of glycolysis take place in the cytosol or "soluble" fraction of the cell, the oxidation of pyruvate occurs in the mitochondria. Mitochondria, when carefully isolated, can be shown to carry out three characteristic functions which are physiologically closely interrelated and which are relevant to the contents of this Chapter.

(i) They oxidize pyruvate to CO_2 with the formation, predominantly, of NADH.

(ii) They re-oxidize NADH to NAD^+ by transferring hydrogen to molecular oxygen by electron transport via the cytochrome system.

(iii) They couple the oxidation of each molecule of NADH to the formation of 3ATP; this is known as oxidative phosphorylation.

Much more will be said about electron transport and oxidative phosphorylation in Chapter 9.

The cyclic nature of the reaction sequence bringing about the oxidation of pyruvate to CO_2, the tricarboxylic acid cycle, was established by Krebs and Johnson in 1937. The cycle as postulated then has remained essentially unaltered to this day except that, as various enzymes have been purified and characterized, much has been learnt about the mechanisms of the separate steps. The cycle constitutes a particularly important series of reactions because, in many ways, it is the focal point of all the degradative pathways that are occurring in the cell. Not only is it able to oxidize pyruvate, obtained from carbohydrate, but it can also oxidize the end-products of fat and protein metabolism. Because of this, a large proportion of the ATP that is formed in the cell is synthesized by the mitochondria as a consequence of the operation of this cycle.

Most of the enzymes that catalyze the reactions of the tricarboxylic acid cycle are situated in the lumen of the mitochondria while others are firmly attached to the inner mitochondrial membrane. Some of the individual reactions of the cycle are also duplicated outside the mitochondria, but where this occurs, the reactions in the cytosol are catalyzed by completely different enzyme proteins. For example, we have already seen how two malate dehydrogenases effect the net transport of oxaloacetate out of the mitochondria during the early stages of gluconeogenesis. Enzyme duplication of this kind, on either side of the mitochondrial membrane, is not confined solely to enzymes of the tricarboxylic acid cycle. Others, particularly some of the transaminases (see Chapter 7) are found distributed in this way.

(a) The enzymes and coenzymes of the tricarboxylic acid cycle

The reactions which together comprise the tricarboxylic acid cycle are shown in Fig. 6.18. Although the cycle can be said to begin with the synthesis of citrate, it is convenient for us to consider here the formation of acetyl CoA from pyruvate. It must, however, be emphasized that acetyl CoA is not the only compound that enters the cycle. Any one of the

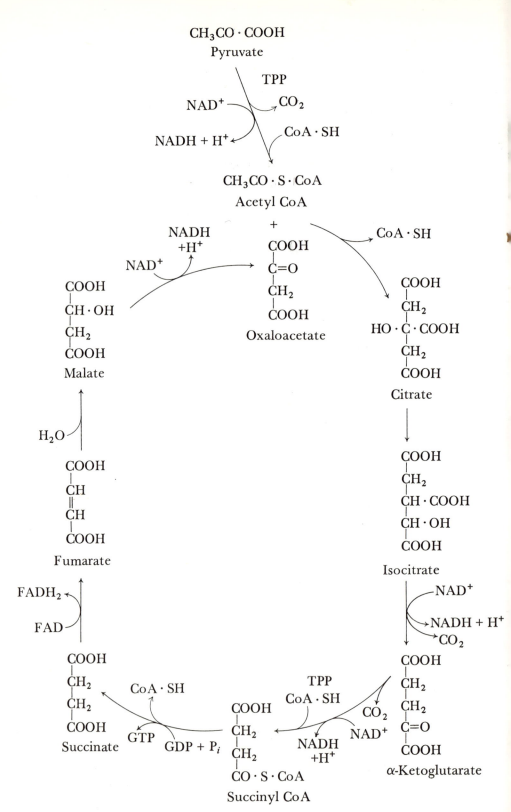

Fig. 6.18 The tricarboxylic acid cycle.

intermediates that is also an end-product of some other catabolic pathway can serve as an access point to the reactions of the cycle.

A number of different coenzymes are required by the enzymes catalyzing the reactions of the tricarboxylic acid cycle. Some of these have been described elsewhere, particularly NAD, NADP and FAD (see Chapter 5.4) but others are encountered for the first time. If we consider them as a group we find that all of them are derivatives of the B vitamins. This is shown in Table 6.5 where the particular B vitamin is listed, together with the coenzyme derived from it, and the type of reaction with which the particular coenzyme is concerned. These coenzymes, of course, are involved in many reactions in the cell and are not restricted to those of the tricarboxylic acid cycle. The list shown is only a partial one in that other B vitamins, namely folic acid and vitamin B_{12}, also function as coenzymes but not in the reactions discussed in this section.

TABLE 6.5. B Vitamins which are constituents of coenzymes for the enzymes of the tricarboxylic acid cycle and associated reactions

Vitamin	Coenzyme form	Function
Nicotinamide (niacin)	NAD; NADP	Hydrogen acceptors
Riboflavin (B_2)	FAD; FMN	
Thiamine (B_1)	Thiamine pyrophosphate (TPP)	Oxidative decarboxylation (Also transketolase reactions)
Lipoic acid* (Thioctic acid)	Lipoic acid	Oxidative decarboxylation
Pantothenic acid	Coenzyme A	"Activation"
Biotin	Biotin	Carboxylation (CO_2 fixation)
Pyridoxine (B_6)	Pyridoxal phosphate	Transamination, etc.

* Lipoic acid is included as a B vitamin although a specific requirement for it in the diet has not been shown. It is, however, required as a growth factor by many micro-organisms.

All these coenzymes play an "acceptor" or "transfer" role of one kind or another when they take part in a particular reaction. Their structures and the reactions that they undergo are shown in Fig. 6.19. Only TPP, lipoic acid and coenzyme A are given here since the remainder are described elsewhere.

(i) THE FORMATION OF ACETYL CoA FROM PYRUVATE

The oxidation of pyruvate to acetyl CoA, and the transfer of hydrogen to NAD^+, can be divided into a number of separate reactions. Although three different enzymes are involved in the overall reaction, these enzymes are closely associated with each other to form a structurally ordered complex which also contains the coenzymes that are required. This enzyme complex, consisting of many molecules of each protein can be

THIAMINE PYROPHOSPHATE (TPP)

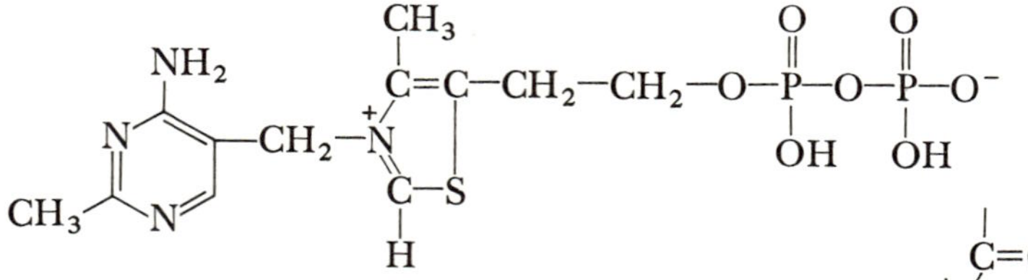

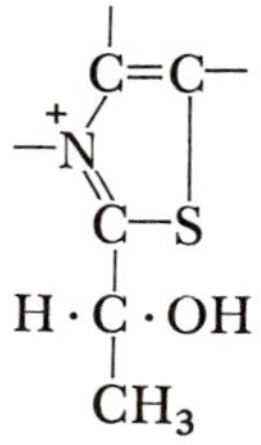

Hydroxyethyl thiaminepyrophosphate $(CH_3 CH(OH)TPP)$

ENZYME BOUND LIPOATE (Lipoic acid)

Oxidized lipoate

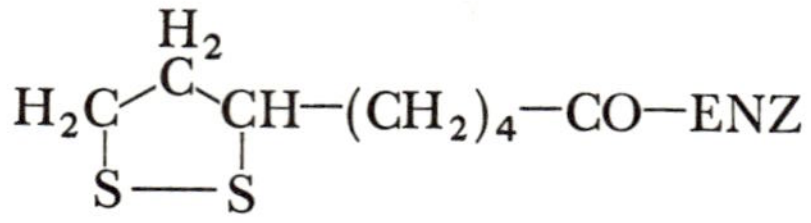

Reduced lipoate

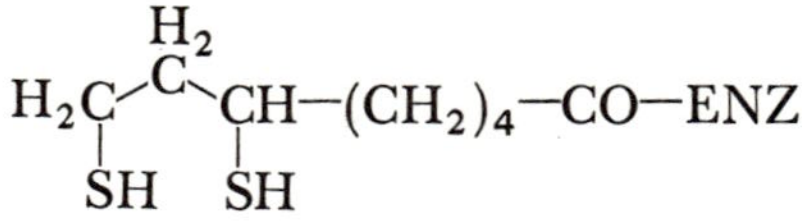

Acetyl lipoate

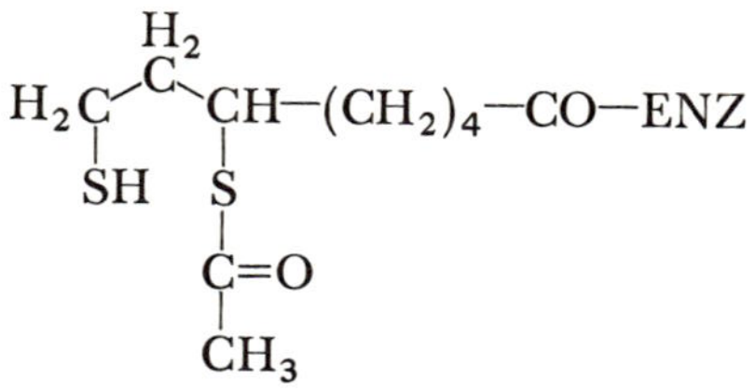

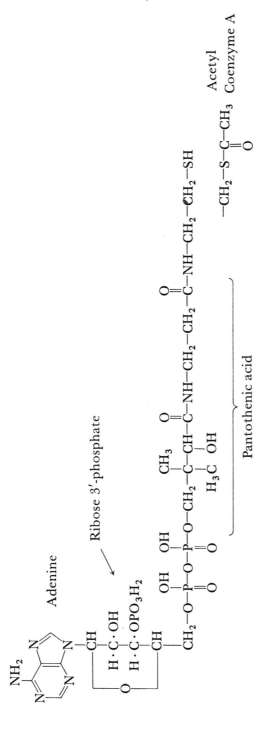

Fig. 6.19 Formulae of thiaminepyrophosphate, lipoic acid and coenzyme A.

resolved into its separate components and these components will re-assemble to form an active complex once more.

The formation of acetyl CoA from pyruvate is directly analogous to the formation of succinyl CoA from α-ketoglutarate. The process is one of oxidative decarboxylation and in each case loss of CO_2 from an α-keto acid is followed by the formation of a coenzyme A derivative. The enzyme systems catalyzing these two interconversions are, respectively, the pyruvate dehydrogenase complex and the α-ketoglutarate complex.

If we take the formation of acetyl CoA as being typical of both reactions, then the various stages can be represented as follows although the intermediate forms that are shown remain firmly bound to the proteins of the complex:

(i) $CH_3COCOOH + TPP \rightarrow CH_3CHOH-TPP + CO_2$

     Pyruvate   TPP complex

(ii) $CH_3CHOH-TPP$ + Oxidized lipoate $\rightarrow$ Acetyl lipoate + TPP

(iii)   Acetyl lipoate + $CoA \cdot SH \rightarrow CH_3CO \cdot S \cdot CoA$ + Reduced lipoate

         Coenzyme A   Acetyl CoA

The reduced lipoate is finally reoxidized by dihydrolipoyl dehydro-genase, an enzyme which has FAD as a firmly bound prosthetic group and which is itself a component of the dehydrogenase complex.

(iv)   Reduced lipoate + $NAD^+ \rightarrow$ Oxidized lipoate + $NADH + H^+$

The sum total of these reactions is therefore

Pyruvate + CoASH + $NAD^+ \rightarrow$ Acetyl CoA + CO_2 + $NADH + H^+$

The free-energy change associated with this conversion is large and negative ($\Delta G^{0\prime} = -34$ kJ; $-8 \cdot 0$ kcal) so that the reaction is effectively irreversible. This has important physiological consequences since it follows that acetyl CoA cannot be carboxylated to pyruvate and thus serve as a substrate for gluconeogenesis. Therefore carbohydrate cannot be synthe-sized from acetyl CoA or from fat, which is degraded mainly via acetyl CoA.

(ii) THE FORMATION OF CITRATE

To enter the tricarboxylic acid cycle, acetyl CoA must react with oxaloacetate. The enzyme catalyzing this step is citrate synthase, originally called the condensing enzyme; the products of the reaction are citrate and coenzyme A.

Citrate synthase is inhibited by NADH and ATP so that these meta-bolites, which result from the operation of the cycle, act as "negative feedback" inhibitors of their own synthesis.

Monofluoroacetate ($CH_2F \cdot COOH$) has long been recognized as a potent inhibitor of the tricarboxylic acid cycle and for a number of years it was used with some success as a rat-poison. Its lack of selectivity, however, and particularly its high toxicity towards farm and domestic animals, led to the discontinuation of its use for this purpose. Its mechanism of action, leading to the accumulation of citrate in all tissues, is described in Chapter 5.9.

(iii) THE ACONITASE REACTION

Aconitase catalyzes the formation of L-isocitrate from citrate by what is effectively the intramolecular rearrangement of a molecule of water, without the intermediate formation of free *cis*-aconitate. Preparations containing aconitase will, however, convert *cis*-aconitate to citrate and isocitrate and in fact the equilibrium mixture contains all three compounds. The enzyme is activated by Fe^{2+} ions.

$$
\begin{array}{ccc}
\begin{array}{l}
*COOH \\
| \\
*CH_2 \\
| \\
HO \cdot C \cdot COOH \\
| \\
CH_2 \\
| \\
COOH
\end{array}
&
\xrightleftharpoons{\;H_2O\;}
\left[
\begin{array}{l}
*COOH \\
| \\
*CH_2 \\
| \\
C \cdot COOH \\
\| \\
CH \\
| \\
COOH
\end{array}
\right]
\xrightleftharpoons{\;H_2O\;}
&
\begin{array}{l}
*COOH \\
| \\
*CH_2 \\
| \\
H \cdot C \cdot COOH \\
| \\
CH \cdot OH \\
| \\
COOH
\end{array}
\\
\text{Citrate} & \text{\textit{Cis}-aconitate} & \text{Isocitrate}
\end{array}
$$

The aconitase reaction: The asterisks show the carbon atoms originally derived from the acetyl moiety of acetyl CoA.

Although citrate is superficially a symmetrical molecule, it is attacked asymmetrically during the aconitase reaction in that "dehydration" always occurs across what were originally carbon atoms 2 and 3 of oxaloacetate. For this to happen, citrate must be able to react with the enzyme in one way only, and having done so, form a complex which makes asymmetric dehydration possible. The formation of a complex of this kind, which apparently confers asymmetry upon the citrate molecule, is possible if we postulate a minimum of three specific contact points between enzyme and substrate.

(iv) THE ISOCITRATE DEHYDROGENASE REACTION

Isocitrate dehydrogenase catalyzes the oxidation of isocitrate to α-ketoglutarate. The enzyme responsible for this reaction in the mitochondria, as a component reaction of the tricarboxylic acid cycle, is NAD-dependent and requires either Mg^{2+} or Mn^{2+} for activity. It is an important regulatory enzyme of the cycle since it is activated by ADP and inhibited by ATP and NADH. Its overall activity, and hence the rate of operation of the cycle are, therefore, very much influenced by the concentration ratios of ATP : ADP and NADH : NAD^+ (see Chapter 13.2).

Isocitrate dehydrogenase is also present in the cytosol but this enzyme is NADP-dependent and it is not subject to the same allosteric regulation as the mitochondrial enzyme. The isocitrate dehydrogenase in the cytosol is probably concerned with the provision of α-ketoglutarate for transamination and with the formation of NADPH needed for biosynthetic reactions.

The existence of the NAD-dependent enzyme was not appreciated until long after the NADP-dependent one had been discovered. This was mainly because of the much greater stability of the NADP-dependent enzyme and the fact that in consequence it was always the one found in enzyme preparations. This led to some temporary confusion since it was thought that the NADP-dependent enzyme catalyzed the oxidation of isocitrate during the operation of the tricarboxylic acid cycle in the mitochondria.

(v) THE α-KETOGLUTARATE DEHYDROGENASE REACTION

The similarities between the reactions forming succinyl CoA from α-ketoglutarate and those forming acetyl CoA from pyruvate have already been referred to. Thiamine pyrophosphate and lipoic acid function as coenzymes and dihydrolipoyl dehydrogenase catalyzes the transfer of hydrogen from reduced lipoate to NAD. The overall reaction, which is irreversible, can be written as follows:

$$\text{α-Ketoglutarate} + \text{CoASH} + NAD^+ \rightarrow \text{Succinyl CoA} + CO_2 + \text{NADH} + H^+$$

(vi) THE METABOLISM OF SUCCINYL CoA

Succinyl CoA, like acetyl CoA, is a high energy thio-ester (see Chapter 6.1). Whereas the energy associated with acetyl CoA is an important factor in driving the citrate synthase reaction in the direction of citrate, that of succinyl CoA can be conserved in the form of a nucleoside triphosphate. The enzyme catalyzing this reaction, succinyl thiokinase, requires inorganic phosphate and GDP.

$$\text{Succinyl CoA} + \text{GDP} + H_3PO_4 \rightleftharpoons \text{Succinate} + \text{GTP} + \text{CoASH}$$

The formation of a high energy nucleoside triphosphate in this way is another example of a substrate-level phosphorylation of which we have already mentioned two examples with respect to ATP synthesis. GTP readily gives ATP by transferring its terminal phosphate to ADP, a reaction catalyzed by nucleoside diphosphokinase. Succinyl CoA, as well as being converted to succinate by the reaction shown above, can also participate in CoA transfer reactions and in so doing be used to form the CoA derivatives of other carboxylic acids. The most important of these reactions is the following which is relevant to the metabolism of ketone bodies (see Chapter 8.6).

Succinyl CoA + Acetoacetate ⇌ Succinate + Acetoacetyl CoA

(vii) THE SUCCINATE DEHYDROGENASE REACTION

The oxidation of succinate to fumarate differs from the other oxidation reactions of the tricarboxylic acid cycle in that there is no requirement for NAD, the hydrogen acceptor being FAD. The enzyme catalyzing this reaction, succinate dehydrogenase, is firmly attached to the inner mitochondrial membrane and is, therefore, closely associated with the cytochrome system which also forms part of this membrane. The succinate dehydrogenase reaction is competitively inhibited by a number of dicarboxylic acids structurally related to the substrate. The most potent of these inhibitors is malonate which was used in some of the earliest experiments involving the reactions of the tricarboxylic acid cycle and has already been referred to as a competitive inhibitor in Chapter 5.9. If tissue homogenates, for example of muscle, are incubated with any cycle intermediate in the presence of malonate, then succinate accumulates. This was a key piece of evidence in establishing the cyclic nature of the pathway by which the various substrates are oxidized.

Although inhibition by malonate is of no physiological significance, that brought about by oxaloacetate probably is. Since oxaloacetate is produced in the last reaction of the tricarboxylic acid cycle it is possible that it can control its own rate of synthesis by inhibiting the succinate dehydrogenase reaction.

(viii) THE FUMARASE REACTION

Fumarase catalyzes the hydration of fumarate to give malate. The carboxyl groups of fumarate are in the *trans* position and water is added across the double bond; L-malate is formed.

(ix) THE OXIDATION OF MALATE TO OXALOACETATE

The final reaction of the tricarboxylic acid cycle is that catalyzed by malate dehydrogenase and results in the formation of oxaloacetate and NADH. The equilibrium position of this reaction strongly favours the synthesis of malate. However, the rapid removal of oxaloacetate, by its reaction with acetyl CoA to form more citrate, enables malate oxidation to occur.

Since oxaloacetate is recovered in this way, the cycle acts catalytically when it is oxidizing pyruvate or acetyl CoA. By this we mean that, in theory at least, the presence of a small amount of oxaloacetate can facilitate the oxidation of an infinite amount of pyruvate. The overall reaction when this occurs can be represented as follows:

$$CH_3CO \cdot COOH + 2H_2O + GDP + H_3PO_4 \rightarrow 3CO_2 + 10\ H + GTP$$

(b) The fate of the hydrogen atoms removed during the oxidation steps

When pyruvate is metabolized to CO_2, by the tricarboxylic acid cycle, there are five oxidation reactions which result in the removal of five pairs of hydrogen atoms that are ultimately oxidized to water via the electron transport system. NADH is formed in four of these reactions but the hydrogen from the remaining one, the oxidation of succinate, is transferred via the bound FAD of succinic dehydrogenase. The hydrogen from NADH similarly passes to the FAD of another flavoprotein, NADH dehydrogenase. The pathway of hydrogen from substrate to oxygen, during the combined operation of the tricarboxylic acid cycle and electron transport, is summarized in a simplified form, Fig. 6.20. A much more detailed description of electron transport is given in Chapter 9.

The reoxidation of one molecule of NADH, by the system shown above, gives rise to 3 ATP, but the transfer of the hydrogen from succinate, via FAD, results in the formation of only 2 ATP. We can thus calculate how much ATP is synthesized during the oxidation of pyruvate, as follows: 12 ATP are obtained from the oxidation of the 4 NADH molecules and 2 ATP are formed by the oxidation of the $FADH_2$ of succinate dehydrogenase. Since one high-energy phosphate is synthesized at the substrate level of the tricarboxylic acid cycle, by the action of succinyl thiokinase, a total of 15 ATP results from the oxidation of one molecule of pyruvate.

(c) The efficiency of energy conservation during the oxidation of glucose

We are now able to calculate the total amount of high-energy phosphate that becomes available from the complete, aerobic degradation of glucose. If we add the 8 ATP formed by anaerobic glycolysis (see Chapter 6.8) to the 30 ATP that are synthesized during the subsequent oxidation of two molecules of pyruvate, we find that 38 ATP are obtained per glucose molecule. This total of 38 obtained aerobically can be compared with the 2 that are formed by anaerobic glycolysis when glucose is converted to pyruvate and thence to lactate.

Very crudely, using the above figures, it is possible to attach numerical values to the efficiencies of the anaerobic and aerobic processes; surprisingly these efficiencies are not too dissimilar.

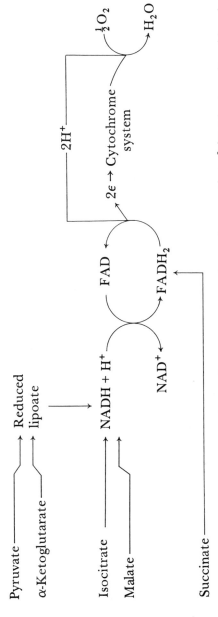

Fig. 6.20 The flow of hydrogen from substrate to oxygen during the operation of the tricarboxylic acid cycle.

Chemically, the conversion of glucose to either lactate or carbon dioxide and water, can be written as follows:

$$C_6H_{12}O_6 \rightarrow 2CH_3 \cdot CHOH \cdot COOH \quad \Delta G^{0\prime} = -197 \text{ kJ } (-47 \text{ kcal})/\text{mole}$$

$$C_6H_{12}O_6 + 6O_2 \rightarrow 6CO_2 + 6H_2O \quad \Delta G^{0\prime} = -2870 \text{ kJ } (-686 \text{ kcal})/\text{mole}$$

If we assume that, under physiological conditions, the energy available from the hydrolysis of ATP to ADP is about 35 kJ (8 kcal)/mole then anaerobically 2 x 35 kJ, and aerobically 38 x 35 kJ, become available during the metabolism of one mole of glucose. The efficiencies of the two processes are therefore 70 x 100/197 (35%) and 1330 x 100/2870 (46%), respectively.

It must be emphasized again that these are only approximate (probably minimum) values because of the uncertainty in calculating the free energy change associated with the hydrolysis of ATP when this occurs in the cell where the concentration of the various reactants is not known.

(d) The role of the tricarboxylic acid cycle in biosynthesis and its role in the metabolism of degradation products of fat and protein

We have seen that the tricarboxylic acid cycle can operate catalytically to bring about the oxidation of pyruvate to CO_2, thereby allowing relatively large amounts of ATP to be synthesized by oxidative phosphorylation. When functioning in this way the role of the tricarboxylic acid cycle is clearly to provide energy in a utilizable form. Many of the intermediates of the cycle, however, serve as starting points for the biosynthesis of a large number of metabolites, most of which are required for the further synthesis of macromolecules. It is not possible to consider these pathways in detail here, and anyway, the important ones will be described in other chapters; a few examples will suffice. Glutamate and aspartate are readily formed by transamination of α-ketoglutarate and oxaloacetate. Both of these amino acids are incorporated into protein but each of them can also be converted into other components; proline and hydroxyproline, for example, are synthesized from glutamate, and the pyrimidine bases from aspartate. Similarly, succinyl CoA is the starting point for the synthesis of porphyrins which are constituents of haemoglobin and the cytochromes.

One of the most important compounds with respect to biosynthetic pathways that originate in the tricarboxylic acid cycle is oxaloacetate. Apart from giving rise to aspartate, oxaloacetate is a key intermediate of gluconeogenesis. Compounds from which glucose can be synthesized are referred to as glucogenic substances, or as glycogenic substances since glycogen is readily synthesized from glucose. It follows, from what has been said above, that any precursor of oxaloacetate is glucogenic and that among these compounds are included all the tricarboxylic acid cycle intermediates and the substances from which they can be formed.

The two roles of the tricarboxylic acid cycle, its catalytic one for the oxidation of acetyl CoA and its role in biosynthesis, are to some extent in conflict since the latter results in the removal of oxaloacetate, the presence of which is essential for the oxidation of acetyl CoA to occur. The way in which this apparent paradox is resolved is that, quite simply, compounds other than acetyl CoA must enter the tricarboxylic acid cycle to compensate for the intermediates used for biosynthesis.

In this respect, the pyruvate carboxylase reaction (see Chapter 6.9) is particularly important. Moreover, since acetyl CoA is an allosteric activator for this enzyme there exists a physiological method for ensuring that acetyl CoA stimulates the synthesis of oxaloacetate which is required for the continued operation of the cycle. The relationship is as follows:

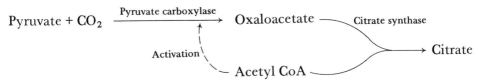

The pyruvate for this reaction originates from carbohydrate but also from a number of amino acids such as alanine and serine. Many other amino acids, for example aspartate, glutamate and histidine, can give tricarboxylic acid cycle intermediates directly; amino acids in both these groups are consequently glucogenic.

(e) Ketogenic substances

Under certain circumstances, the rate of formation of acetyl CoA in the liver may be such that the capacity of the tricarboxylic acid cycle to oxidize it is exceeded. When this occurs, there is an excess production of ketone bodies (acetoacetate, β-hydroxybutyrate and acetone). For this reason, acetyl CoA and compounds giving rise to acetyl CoA are termed ketogenic. The substance which is most strongly ketogenic is fat since this is degraded predominantly to acetyl CoA. Also ketogenic is the amino acid leucine while a number of other amino acids are partly ketogenic and partly glucogenic.

The relationship of the tricarboxylic acid cycle to glycolysis and gluconeogenesis and to the metabolism of the end products of protein and fat degradation is shown in Fig. 6.21. This elementary metabolic map emphasizes the importance of the cycle in the metabolism of a large number of compounds other than acetyl CoA.

(f) The oxidation of ethanol and acetic acid

Neither ethanol nor acetic acid are normally produced by the body but both compounds may occur to varying extents in the diet. The quantitative significance of the pathways for their metabolism will, therefore, tend to vary from person to person.

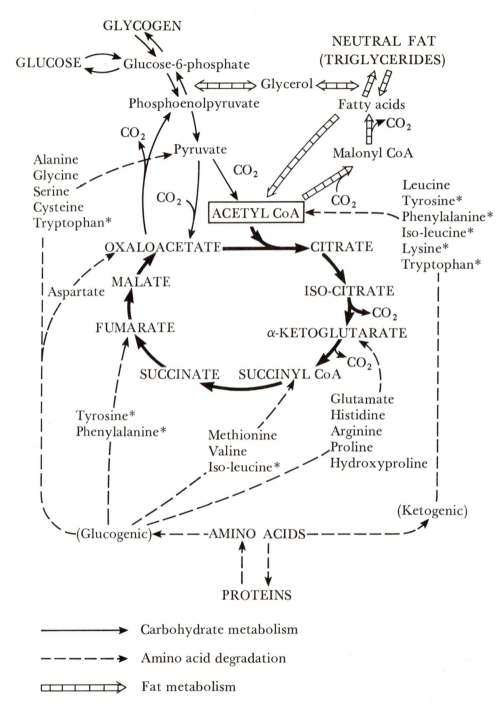

Fig. 6.21 The central role of the tricarboxylic acid cycle in intermediary metabolism.

Ethanol is first oxidized in the liver by alcohol dehydrogenase to acetaldehyde; it is thought that the formation of acetaldehyde accounts for many of the unpleasant side effects associated with excess consumption of alcohol. The acetaldehyde is oxidized to acetic acid.

$$CH_3CHO + H_2O + NAD^+ \rightleftharpoons CH_3COOH + NADH + H^+$$

Acetaldehyde   Acetic acid

The acetic acid formed in this way can then give acetyl CoA by the action of the enzyme acetate thiokinase. As shown below two stages in this reaction can be recognized although there are no free intermediates formed. The formation of acetyl-AMP is analogous to the formation of amino acid adenylates which occur as a preliminary to protein synthesis (see Chapter 12.16).

The overall reaction is identical to that bringing about the activation of long chain fatty acids prior to oxidation (see Chapter 8.5).

(i) $ATP + CH_3COOH \rightarrow CH_3CO\text{-}AMP + $ pyrophosphate

(ii) $CH_3CO\text{-}AMP + CoASH \rightarrow CH_3CO \cdot SCoA + AMP$

Sum $CH_3COOH + ATP + CoASH \rightarrow CH_3CO \cdot SCoA + AMP + $ pyrophosphate

Acetic acid   Acetyl CoA

(g) The metabolism of propionyl CoA

Propionyl CoA remains as the terminal fragment after fatty acids with an odd number of carbon atoms have been degraded by oxidation. However, fatty acids of this kind are relatively few and so little propionyl CoA is formed in this way. Rather more is obtained from the degradation of methionine and isoleucine; a closely related compound, methylmalonyl CoA is obtained from valine. Propionyl CoA is oxidized by the tricarboxylic acid cycle after it has first been converted to succinyl CoA.

The reactions are as follows:

(i)

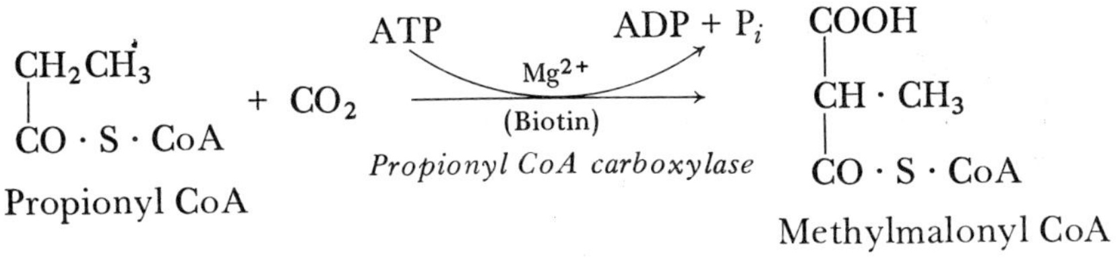

$$
\begin{array}{lcl}
\begin{array}{c} CH_2CH_3 \\ | \\ CO \cdot S \cdot CoA \end{array} + CO_2 &
\xrightarrow[\substack{(Biotin) \\ \textit{Propionyl CoA carboxylase}}]{\substack{ATP \quad ADP + P_i \\ Mg^{2+}}} &
\begin{array}{c} COOH \\ | \\ CH \cdot CH_3 \\ | \\ CO \cdot S \cdot CoA \end{array}
\end{array}
$$

Propionyl CoA   Methylmalonyl CoA

(ii)

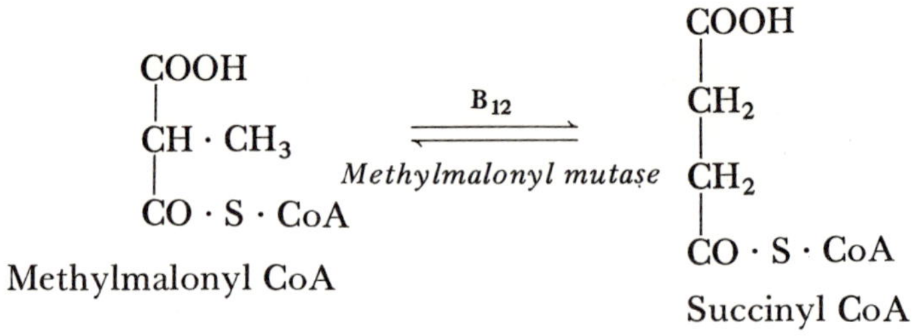

$$
\underset{\text{Methylmalonyl CoA}}{
\begin{array}{c}
\text{COOH} \\
| \\
\text{CH} \cdot \text{CH}_3 \\
| \\
\text{CO} \cdot \text{S} \cdot \text{CoA}
\end{array}}
\quad \underset{\textit{Methylmalonyl mutase}}{\overset{B_{12}}{\rightleftharpoons}} \quad
\underset{\text{Succinyl CoA}}{
\begin{array}{c}
\text{COOH} \\
| \\
\text{CH}_2 \\
| \\
\text{CH}_2 \\
| \\
\text{CO} \cdot \text{S} \cdot \text{CoA}
\end{array}}
$$

The second of the above reactions is of interest since it requires vitamin B_{12} as a co-factor. The non-absorption of this vitamin from the intestine, due to lack of intrinsic factor in gastric secretion, leads to pernicious anaemia. Although impairment of methylmalonyl mutase is not the cause of the clinical condition, patients suffering from pernicious anaemia excrete large amounts of propionic acid and methylmalonic acid in the urine.

6.13. The Pentose Phosphate Pathway of Glucose Oxidation

Earlier in this chapter (Section 6.7) it was emphasized that glucose can be metabolized to carbon dioxide other than by the combined action of glycolysis and the tricarboxylic acid cycle. In mammals, only one of these "alternative" pathways is of significance. This is the pentose phosphate pathway, also known as the hexose monophosphate shunt or 6-phosphogluconate pathway.

In most tissues glycolysis and the pentose phosphate pathway function side by side although the extent to which either of them operates depends firstly upon the particular tissue and secondly upon the nutritional state of the animal. For example, the glycolytic pathway is virtually the only means of glucose degradation in muscle but in liver, some 30% of the glucose may be oxidized by the pentose phosphate pathway; in mammary gland, adipose tissue, testis, adrenal cortex and leucocytes the contribution made by the pentose phosphate pathway may be even higher.

The two most important features of this pathway are, first, that all the enzymes concerned are in the cytosol of the cell and, therefore, oxidation is not dependent upon the mitochondria or the tricarboxylic acid cycle, and secondly, that the pathway produces NADPH rather than NADH or ATP.

To understand how the pentose phosphate pathway can bring about the oxidation of glucose to carbon dioxide, it is necessary to consider the metabolism not of one molecule, but of six. This is because the reactions of the pathway can be divided into two stages; the first, in which hexose is decarboxylated to pentose, and the second, during which six molecules of pentose undergo a rearrangement reaction to give five molecules of hexose. This can be summarized as follows, although as in the case of glycolysis it is the sugar phosphates that take part in the various reactions:

(i) 6 Hexose + $6H_2O \rightarrow$ 6 Pentose + $6CO_2$ + 24H

(ii) 6 Pentose $\longrightarrow$ 5 Hexose

Sum   Hexose + $6H_2O \rightarrow 6CO_2$ + 24H

The reactions of the first stage, those leading to decarboxylation, can perhaps be considered to be the most important since all the oxidation that occurs in the pathway takes place here. These are the reactions, therefore, that result in the formation of NADPH.

The remaining reactions, those which allow hexose phosphate to be resynthesized from pentose phosphate give the pathway its characteristic complexity. In describing the various transformations that take place during this stage, the formulae of the intermediates have been deliberately omitted since they are not considered essential to an understanding of the function of the pathway. It is important, however, that the student should know the number of carbon atoms in each compound so that the nature of the reactions can be appreciated.

The pentose phosphate pathway is shown in Fig. 6.22 starting with six molecules of glucose (or glucose 6-phosphate) and presented so that re-synthesis of five molecules of glucose 6-phosphate is shown. It should be realized that this is undoubtedly an oversimplification since in many tissues the enzymes of the pentose phosphate pathway and those of glycolysis will be present together; the kind of balance that is shown in the diagram is, therefore, probably never achieved. It is possible, for example, for much of the 3-phosphoglyceraldehyde formed from pentose to be oxidized to pyruvate by the glycolytic enzymes.

(a) The individual reactions of the pathway

(i) THE FORMATION OF PENTOSE-PHOSPHATE

The first reaction, specific to the pentose-phosphate pathway, is the NADP-dependent oxidation of glucose 6-phosphate which gives 6-phosphogluconolactone and NADPH. The enzyme catalyzing this reaction is glucose 6-phosphate dehydrogenase. The 6-phosphoglucolactone so formed can then hydrolyze spontaneously but slowly to the straight-chain sugar, 6-phosphogluconate; but the conversion is accelerated by a specific lactonase. 6-Phosphogluconate dehydrogenase catalyzes the second NADP-dependent oxidation reaction which also brings about de-carboxylation of the substrate to form a pentose phosphate, ribulose 5-phosphate.

Glucose 6-phosphate dehydrogenase of erythrocytes is of some interest

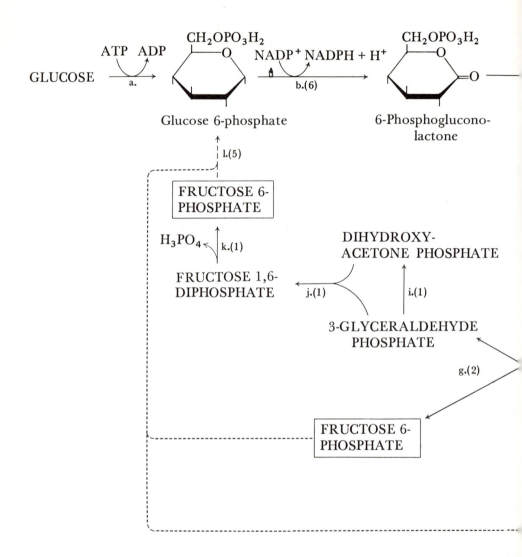

The letters e.g. b. refer to the enzymes listed below.

The figure after each letter e.g. (2) is the number of molecules proceeding via the reaction shown.

Enzymes specific to the Pentose Phosphate Pathway

b. Glucose 6-phosphate dehydrogenase
c. Gluconolactonase
d. 6-Phosphogluconate dehydrogenase
e. Phosphopentose epimerase
f. Phosphopentose isomerase
g. Transketolase
h. Transaldolase

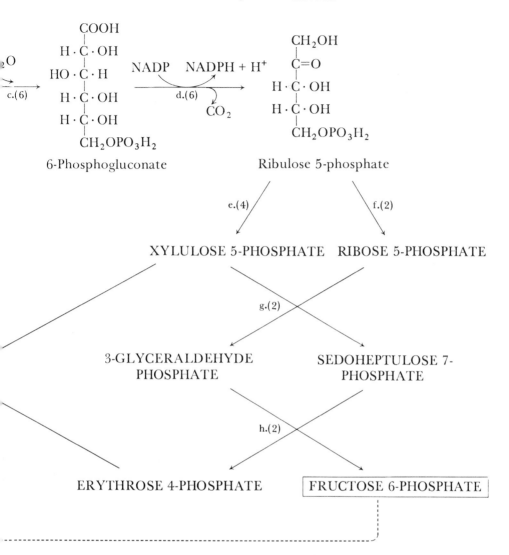

6-Phosphogluconate   Ribulose 5-phosphate

XYLULOSE 5-PHOSPHATE   RIBOSE 5-PHOSPHATE

3-GLYCERALDEHYDE   SEDOHEPTULOSE 7-
PHOSPHATE   PHOSPHATE

ERYTHROSE 4-PHOSPHATE   FRUCTOSE 6-PHOSPHATE

Enzymes of Glycolysis and Gluconeogenesis

a. Hexokinase
i. Triose phosphate isomerase
j. Aldolase
k. Fructose 1,6-diphosphatase
l. Phosphoglucoisomerase

Fig. 6.22 Reactions of the pentose phosphate pathway.

because a significant decrease in its activity leads to a disorder known as primaquine sensitivity. In this condition, which is genetically determined, the erythrocytes are sensitive to haemolysis by a wide range of compounds, particularly the antimalarial drug primaquine. This is due to a lowered concentration of reduced glutathione, the presence of which is necessary to maintain the integrity of the erythrocyte membrane. Reduced glutathione is normally synthesized by the action of glutathione reductase which uses NADPH produced by the glucose 6-phosphate dehydrogenase reaction.

(ii) THE RESYNTHESIS OF GLUCOSE 6-PHOSPHATE FROM PENTOSE-PHOSPHATE

Ribulose 5-phosphate can undergo two isomerization reactions, one catalyzed by phosphopentose epimerase and forming xylulose 5-phosphate, and the other catalyzed by phosphopentose isomerase, giving ribose 5-phosphate. Transketolase, an enzyme requiring thiamine pyrophosphate (TPP) and Mg^{2+} then transfers a terminal two-carbon fragment from one of these pentose derivatives (xylulose 5-phosphate) to the other. The products of the reaction are, therefore, a seven-carbon compound, sedoheptulose 7-phosphate, and a three-carbon compound, 3-phosphoglyceraldehyde.

A further transfer reaction then results in the recovery of the first molecule of hexose-phosphate. The enzyme catalyzing this reaction, transaldolase, transfers a three-carbon moiety from sedoheptulose 7-phosphate to 3-phosphoglyceraldehyde, thereby forming fructose 6-phosphate and the tetrose sugar, erythrose 4-phosphate.

In the final reaction, specific to the pathway, transketolase catalyzes a reaction between erythrose 4-phosphate and a second molecule of xylulose 5-phosphate. As in the earlier transketolase reaction, a two-carbon moiety is transferred from xylulose 5-phosphate but in this case the acceptor molecule is a four-carbon sugar; the products are therefore 3-phosphoglyceraldehyde and a hexose-phosphate, fructose 6-phosphate. The synthesis of fructose 1,6-diphosphate from 3-phosphoglyceraldehyde, and the conversion of the fructose derivatives to glucose 6-phosphate is accomplished by the enzymes of gluconeogenesis as shown in Fig. 6.12.

For every six molecules of glucose 6-phosphate metabolized by the pathway shown, the following overall reaction occurs:

6 Glucose 6-phosphate + $7H_2O$ + $12NADP^+$ → 5 Glucose 6-phosphate + $6CO_2$ + H_3PO_4 + $12NADPH$ + $12H^+$

The net reaction is therefore:

Glucose 6-phosphate + $7H_2O$ + $12NADP^+$ → $6CO_2$ + H_3PO_4 + $12NADPH$ + $12H^+$.

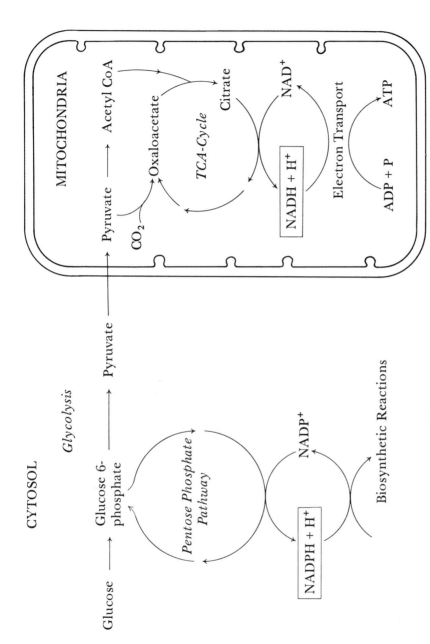

Fig. 6.23 The relationship between the pentose phosphate pathway, glycolysis and the tricarboxylic acid cycle; their respective roles in the formation of NADPH and NADH.

(b) The significance of the pentose-phosphate pathway

Although the pentose-phosphate pathway provides a mechanism for the synthesis and degradation of sugars other than hexoses, particularly pentoses, it is now generally accepted that its main function is to synthesize NADPH from $NADP^+$. In this respect, it stands in sharp contrast to the reactions of glycolysis and the tricarboxylic acid cycle which form NADH. Whereas the NADH of the cell is produced mainly in the mitochondria and used almost exclusively for the production of ATP, NADPH is synthesized primarily in the soluble cytosol where it appears to have a unique role in biosynthetic reactions.

Thus the pentose-phosphate pathway, in higher organisms at least, can be seen as a mechanism which primarily drives biosynthesis while the combined glycolysis-tricarboxylic acid cycle pathway is concerned with the capture of energy in the more widely usable form of ATP. This is summarized in Fig. 6.23.

Once again, the above presentation is probably an oversimplification since we know that reducing equivalents can cross the mitochondrial membrane and also that the NAD^+ and $NADP^+$ are not strictly compartmentalized as the scheme suggests. Nevertheless, there is much evidence in favour of the following hypothesis.

Firstly, where reducing power is needed in a biosynthetic reaction the appropriate enzyme usually has a specific requirement for NADPH. There are many examples but the reactions of fatty acid synthesis and steroidogenesis (including steroid hydroxylation) are perhaps the best.

Secondly, biosynthetic reactions, like the reactions of the pentose-phosphate pathway, take place outside the mitochondria.

Thirdly, the distribution of the pentose phosphate pathway within the tissues of the body is consistent with its having the role described above. For example, it is most active in tissues which are concerned with the synthesis of fatty acids and steroids, and it is also found that substances that stimulate lipogenesis (for example insulin) also stimulate glucose oxidation via the pentose phosphate pathway.

Because the reactions from ribulose phosphate to fructose phosphate are readily reversible, the pentose phosphate pathway also provides a mechanism for the non-oxidative synthesis of pentoses from hexoses.

7

Nitrogen Metabolism

7.1. Protein Digestion

The major part of the intake of nitrogenous compounds by man is in the form of proteins and so a brief account of their digestion is a convenient starting point for a discussion of nitrogen metabolism.

Except for a few rare and exceptional cases, intact protein molecules cannot be absorbed across the walls of the gastro-intestinal tract of adult animals and the protein has first to be broken down to its constituent amino acids or small peptides. Although proteins are complex structures of high molecular weight and are usually composed of some twenty different amino acids, the linkage between the amino acids is always the peptide bond —CO—NH— and breakage of these bonds by simple hydrolysis would lead to a mixture of amino acids. Thermodynamic conditions are favourable to this breakdown since there is a decrease of free energy of about 15 kJ per peptide bond hydrolyzed and theoretically a single enzyme specific only for the peptide bond should suffice for the digestion of proteins. In reality, the situation is a little more complicated as the digestive enzymes are somewhat more specific and may be susceptible to the size of the polypeptide, the location of the point of hydrolysis and to the nature of the amino acids making up the peptide bond (i.e., to the structure of the R side groups adjacent to the bond).

Although many millions of different proteins may be encountered in the diet, they can be digested by the concerted action of quite a small group of proteolytic enzymes—perhaps less than a dozen important ones.

231

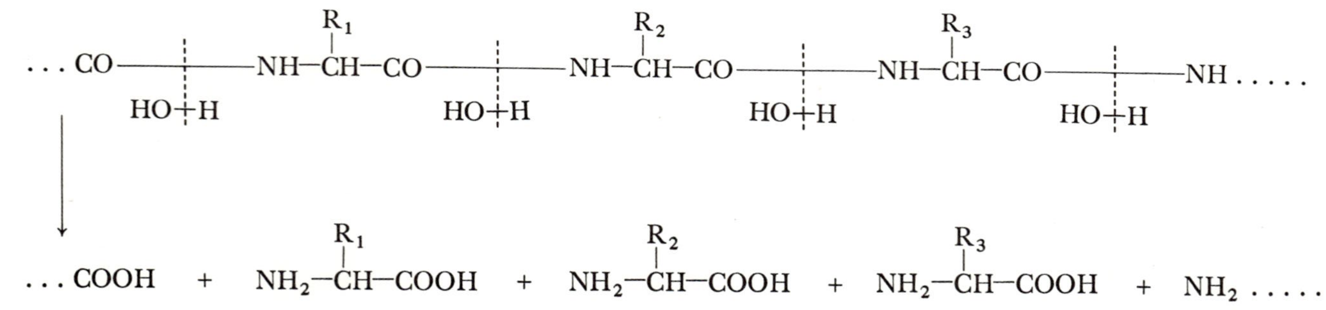

$$\cdots CO\underset{HO\,+\,H}{\underline{\quad\quad}} NH\text{-}CH\text{-}CO\underset{HO\,+\,H}{\underline{\quad}} NH\text{-}CH\text{-}CO\underset{HO\,+\,H}{\underline{\quad}} NH\text{-}CH\text{-}CO\underset{HO\,+\,H}{\underline{\quad}} NH\cdots$$

with side chains R_1, R_2, R_3 on the α-carbons.

$$\longrightarrow$$

$$\cdots COOH \;+\; NH_2\text{-}CH\text{-}COOH \;+\; NH_2\text{-}CH\text{-}COOH \;+\; NH_2\text{-}CH\text{-}COOH \;+\; NH_2\cdots$$

with side chains R_1, R_2, R_3.

or more correctly,

$$\cdots COO^- \;+\; {}^+NH_3\text{-}CH\text{-}COO^- \;+\; {}^+NH_3\text{-}CH\text{-}COO^- \;+\; {}^+NH_3\text{-}CH\text{-}COO^- \;+\; {}^+NH_3\cdots$$

with side chains R_1, R_2, R_3.

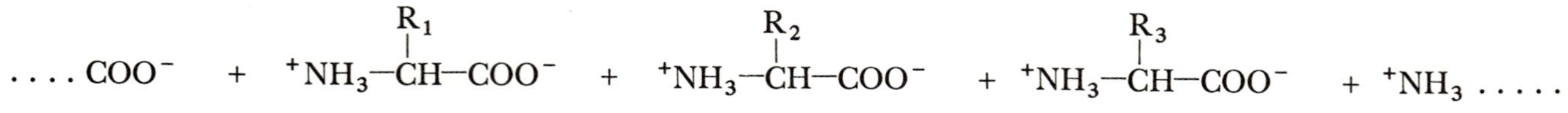

Fig. 7.1 Hydrolysis of a protein chain to individual amino acids.

They can be classified under two headings:

(a) Exo-peptidases which remove amino acids one at a time from the end of a peptide chain by hydrolyzing the terminal peptide bond to give a free amino acid and a polypeptide e.g., carboxy-, amino-, di- and tri-peptidases.

(b) Endo-peptidases which usually hydrolyze peptide bonds within the chain so that the products are smaller polypeptides e.g., pepsin, trypsin, chymotrypsin, elastin. (Care—*endo* means "within" and not "terminal"!)

Examples of the first type are carboxypeptidases A and B which are made in the pancreas and secreted into the gut in the pancreatic juice. A polypeptide has an amino group at one end (N-terminus) and a free carboxyl group at the other (C-terminus), and as the name suggests, a carboxypeptidase acts only at the C-end where it hydrolyzes off the terminal amino acid bearing the free $-COO^-$ group. The reaction leaves a new free $-COO^-$ group and the enzyme can then remove another amino acid and so on, nibbling its way, as it were, along the peptide chain, one amino acid at each bite.

$$\boxed{etc.}-NH-\underset{\underset{\displaystyle HO\vdots H}{|}}{\overset{\overset{\displaystyle R_2}{|}}{CH}}-CO\vdots NH-\overset{\overset{\displaystyle R_1}{|}}{CH}-COO^- \longrightarrow R_2 \qquad\qquad R_1$$

$$\boxed{etc.}-NH-\overset{\overset{\displaystyle R_2}{|}}{CH}-COO^- + NH_3^+-\overset{\overset{\displaystyle R_1}{|}}{CH}-COO^-$$

The three-dimensional structure of carboxypeptidase A is known and the active centre is a pocket in the enzyme into which the side chain of the C-terminal amino acid of the substrate fits, while the free carboxylate group is held in position by an arginine residue in the enzyme so that the last peptide bond of the substrate is correctly oriented for cleavage, a process involving an essential zinc atom in the enzyme (see Chapter 4.11). Carboxypeptidase A will remove the terminal amino acid most readily when this has an aromatic or branched-aliphatic side chain while carboxypeptidase B requires the terminal acid to be lysine or arginine. As we saw in Chapter 4 the pure enzymes are useful tools in studies on primary protein structure since if they are allowed to act for a short time only, the first free amino acid to appear will be the C-terminal one in the polypeptide. Both enzymes are synthesized as zymogens, inactive forms which are converted into the active ones in the gut by trypsin, another proteolytic enzyme.

The intestinal juice secreted by the lining of the small intestine contains aminopeptidases which act in a similar manner to the carboxypeptidases, except that they remove amino acids one at a time from the other end of the polypeptide chain, i.e., the N-terminus. The best known member of the group is called leucine aminopeptidase because peptides with N-terminal leucine residues are especially good substrates, but other terminal amino acids can also be liberated by hydrolysis. The intestinal juice also contains tripeptidases which remove one amino acid from tripeptides to give dipeptides which then act as substrates for dipeptidases present to give two free amino acids.

Exo-peptidases might be able to digest a protein molecule completely by hydrolytic attack from both ends but it could be a slow process and the importance of endo-peptidases is that they break down long peptide chains into small polypeptide fragments and thus greatly increase the number of free ends at which the exo-peptidases can react. The first proteolytic enzyme to act on dietary proteins is the pepsin in the stomach. This endo-peptidase has a fairly broad specificity for side chains but the peptide bonds most susceptible to hydrolysis are those formed between two amino acids which each bear hydrophobic side chains, such as phenylalanine, tyrosine and leucine. Pepsin thus begins the breakdown of protein chains to polypeptides of different sizes, depending on how long the enzyme has to act, and thereby causes some disintegration of the structure of the fabric of the food. It should be noted that pepsin is an exceptional enzyme in that the pH optimum is not near neutrality but at pH 2·0 and that the gastric mucosa secretes HCl so that the pH of the stomach contents is favourable for peptic activity. The enzyme is secreted as pepsinogen, an inactive zymogen, which is activated initially by HCl and then autocatalytically by pepsin itself (see Chapter 5.11(d)). It is possible that this zymogen may give rise in man to several closely related pepsins and to a pepsin-like enzyme called gastricsin which has a pH optimum of 3.

Rennin, an enzyme very active in clotting milk, may occur in the gastric juice of infants but is absent in the human adult. It occurs abundantly in the fourth stomach of the calf (which is used to make rennet, a preparation employed by cooks to convert milk into junket) where its function appears to be to coagulate milk in order to slow down its passage through the digestive tract. Rennin acts on milk casein, a phosphoprotein present as a soluble calcium salt, and converts it into paracasein whose calcium salt is insoluble and separates out as a curd. If the calcium ions are removed from milk, no curd is formed.

The flow of gastric juice, containing pepsinogen, HCl and mucus, is stimulated by a hormone gastrin which is released into the blood stream following the action of certain food constituents ("secretogogues") on the gastric mucosa. Gastrin is a polypeptide of known composition containing 17 amino acid residues and exists in a sulphated and an unsulphated form, the former being the more active. It is interesting to find that the gastrins of different species differ in composition except for the last five amino acids which are common to all and is where the activity resides. This pentapeptide is active and a chemical analogue—pentagastrin—is used experimentally to stimulate the secretion of gastric juice. Histamine is also a powerful stimulant if injected.

After leaving the stomach, the semi-digested food mass (the chyme) enters the duodenum and small intestine where it is mixed with three secretions—the pancreatic juice, the intestinal juice and the bile, all of which have a pH of 7 to 8 because they are rich in bicarbonate which neutralizes the gastric HCl in the chyme and brings its pH nearer to the pH optima of the digestive enzymes which are secreted in the pancreatic and intestinal juices. This titration of the acid appears to be controlled in the following manner:

(a) the acid in the chyme acts on the mucosa of the duodenum and causes a hormone, secretin, to be released from it into the bloodstream (possibly from a precursor molecule, prosecretin),

(b) the secretin is carried to the pancreas where it stimulates an increased flow of alkaline juice. The hormone disappears from the blood after a short time,

(c) as the acid is neutralized, secretin release is reduced, the stimulation of the pancreas decreases and the flow of alkaline juice slows down.

Secretin is a polypeptide of known structure with 27 amino acid residues and has been synthesized. The pancreatic juice liberated following stimulation by secretin is poor in enzymes although rich in bicarbonate but stimulation by a second hormone, pancreozymin, produces an enzyme-rich juice.

Three important endo-peptidases are now involved in the further digestion of protein—trypsin, chymotrypsin and elastase. All three are made in the pancreas as zymogens and are secreted in the pancreatic juice into the gut where activation occurs. Trypsinogen is activated initially by a specific enzyme, enterokinase, secreted by the intestinal mucosa but once free trypsin has been formed it can activate more trypsinogen and the process is autocatalytic as with pepsin and pepsinogen. The process of activation involves only the removal of a hexa-peptide from the N-terminal end of trypsinogen, a molecule containing 249 amino acid residues. Chymotrypsinogen and proelastase are converted to the active enzymes by fission of some peptide bonds by trypsin. The three endo-peptidases complement each other by having different specificities so that a wide range of peptide bonds are split, yielding a mixture of small peptides and some free amino acids. Elastase shows a fairly broad specificity but with a preference for peptide bonds formed from amino acids with small hydrophobic side chains as in alanine, glycine and serine. Trypsin, on the other hand, displays a rather narrow specificity for bonds involving the carboxyl group of arginine and lysine, (i.e., with basic side chains), while chymotrypsin mainly hydrolyzes peptide bonds which involve the carboxyl groups of the aromatic amino acids, tryptophan, tyrosine and phenylalanine.

Elastase was given its name because it will attack elastin, a protein occurring in yellow connective tissue and very rich in glycine and alanine and therefore a good substrate for elastase. Trypsin and chymotrypsin are without effect on this protein. The structures of these three enzymes show many points of resemblance in sequences of amino acids and positions of disulphide bridges and all have an essential serine residue at the active centre which is inhibited by DFP and other organophosphorus compounds. In addition to hydrolyzing proteins, they can also hydrolyze certain amide and ester linkages in synthetic substrates. It has been suggested that they have evolved from some common ancestral endopeptidase.

7.2. Absorption

The products of protein digestion are absorbed through the intestinal wall and released into the portal blood which carries them to the liver where they may be metabolized. Experiments have shown that amino acids of the naturally occurring L-series are absorbed much more rapidly than those of the unnatural D-series. If simple diffusion only was involved, then the L and D forms of an amino acid would be expected to be absorbed at the same rate—since they pass through a dialysis membrane equally fast. The difference in rate observed *in vivo* clearly shows that some selective transport system is working to pump the L-amino acids through the living intestinal mucosa. This matter is considered further in Chapter 11, p. 381.

It is generally agreed that intact protein molecules pass through the gut wall into the blood only extremely rarely and under special circumstances. Proteins acting as antibodies present in first milk in lactation of certain animals, as the horse, may be absorbed unchanged by the newly born offspring during the first day or so of life and thereby confer a passive immunity on the offspring. The allergic reactions to certain foodstuffs shown by a few people are caused by traces of particular proteins in the food passing into the circulation and stimulating the formation of specific antibodies which interact with the same dietary protein when it is subsequently absorbed. This gives rise to the symptoms of an allergic reaction.

It is probable that passage of macromolecules into the circulation takes place in the respiratory tract as well as in the gut. This is certainly true of certain viruses such as poliomyelitis. The exceptional nature of antibody formation to any of the innumerable proteins in the diet is a good indication that it is extremely rare for an intact protein to be absorbed.

Recent work has shown that many small peptides are taken up by the intestinal mucosa at a faster rate than that for their constituent amino acids and this suggests that there are active transport mechanisms for small peptides in addition to those for free amino acids. Good evidence for the existence of different transport mechanisms is supplied by observations on patients with certain inborn errors of amino acid transport. For instance, in Hartnup disease the rate of absorption of L-phenylalanine is almost zero but the dipeptide L-phenylalanyl-L-phenylalanine is well absorbed.

It seems probable that protein digestion in the lumen of the gut results in the production of free amino acids and small peptides which are then absorbed and the final stages of digestion—the hydrolysis of the small peptides into free amino acids—takes place within the columnar epithelial cells. Free amino acids then leave these cells and enter the bloodstream. There are still many unanswered questions, such as the relative amounts of amino acid and peptide absorbed, the site of hydrolysis of the small peptides (at the cell surface or intracellular?) and the mechanism of the transport systems, and much further research will be necessary to elucidate the true nature of the absorption process.

7.3. The Amino Acid Pool

It is convenient to regard the total amount of free amino acids in the blood and other body fluids as constituting a "pool" to which amino acids are constantly being added and removed so that the total amount remains fairly constant. The mean concentration of free amino acids in the blood is around 50 mg/100 ml, corresponding to a total amount in the blood of only about 2·5 g, yet digestion of proteins alone will add about 70 g of amino acids a day to the pool. The contents of the pool are changed, therefore, many times each day, but additions so closely balance removals that the mean level does not vary very much. The most important causes of addition and removal of amino acids to and from the pool are indicated in Fig. 7.2.

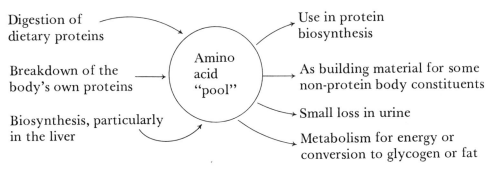

Fig. 7.2 The turnover of amino acids contributing to the amino acid pool.

The use of isotopically labelled amino acids has shown that the proteins in the body are in a state of flux being continually degraded and resynthesized again so that individual molecules have a finite life even though the total amount may be unchanged (see Chapter 12.2).

In an adult whose body weight is constant, the contribution to the pool of amino acids from tissue breakdown is exactly balanced by amino acids withdrawn for the synthesis of replacement proteins. The two amounts will not be in balance for a growing child or an expectant mother who are accumulating extra tissue protein, nor during lactation when considerable quantities of milk protein are produced. A female goat, for instance, may produce 300 g of milk protein a day derived from amino acids in a pool which never contains more than a few grams at any one time.

Amino acids are very frequently used in the biosynthesis of other molecules because they provide a wide range of useful carbon chains and rings and bear various reactive groups. Numerous examples will be encountered in this book, but a few are indicated in Fig. 7.3.

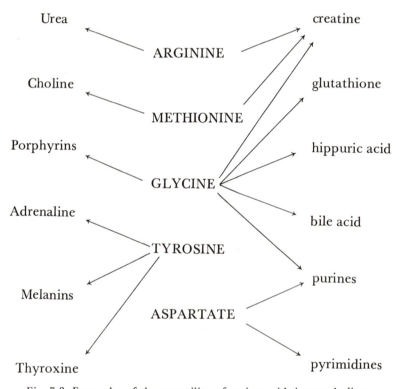

Fig. 7.3 Examples of the versatility of amino acids in metabolism.

There is a small loss of amino acids in the urine but this amounts to about 1·5 g a day which is small in comparison with the daily intake of amino acids.

The body has no way of storing amino acids which are surplus to immediate requirements (unlike fat which can be deposited in adipose tissue or sugar which can be converted into glycogen). All proteins appear to be synthesized in the body for specific purposes and none merely as a store for amino acid residues. The surplus is metabolized, therefore, completely to yield useful chemical energy and the concentration of amino acids in the blood is maintained fairly constant.

7.4. The Metabolism of Amino Acids

This is a fairly complicated story because there are many different amino acids and each can often be metabolized in more than one way. The mainstream of metabolism is that leading to complete degradation to carbon dioxide, water and ammonia. A person on a normal diet may produce about 1200 kJ (290 kcal) a day from the oxidation of 70 g of amino acid, that is, about 10% of his daily energy requirement. The minor pathways of metabolism are those which yield useful or indeed essential compounds but quantitatively only a very small amount of amino acid may be involved.

The first stage in the oxidation of amino acids is usually the removal of the amino group and its replacement by a keto group and since all the 20 amino acids (with the sole exception of proline) have the same general formula $R \cdot CH(NH_3^+)COO^-$, it is possible for the cell to use fairly general methods for the initial deamination, but the keto acids thereby produced show a wide range of structures (e.g., may be straight or branched chain aliphatic compounds, possess aromatic or heterocyclic ring systems, etc.) and each keto acid then has to be dealt with individually. The pathway may be quite complicated (e.g., 14 stages are required to degrade tryptophan to acetyl CoA) or much simpler as with glutamic and aspartic acids since these are fed into the tricarboxylic acid cycle in one step. The purpose of these pathways is to convert the amino acids into simpler compounds which can be metabolized by the cycle. This conversion involves degradation until pyruvate, acetyl CoA or some compound on the cycle (e.g., oxaloacetate, succinyl CoA) is produced.

The fate of the 20 common amino acids can be summarized:

Two
  on deamination form a component of the TCA cycle, i.e., aspartate and glutamate which yield oxaloacetate and α-ketoglutarate respectively.
Three
  are converted to succinyl CoA (also in the TCA cycle), i.e., methionine, isoleucine and valine
One
  asparagine, loses its amide group to give aspartate and then oxaloacetate
Four
  are first converted to glutamate, i.e., proline, glutamine, histidine and arginine

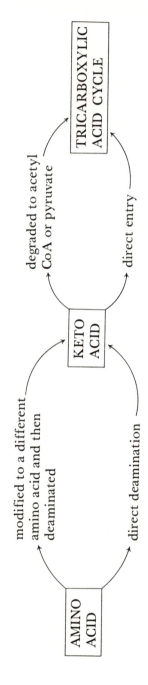

Fig. 7.4 Metabolic routes from amino acids to components of the tricarboxylic acid cycle.

Five

are degraded to pyruvate which can enter the TCA cycle after oxidation to acetyl CoA (i.e., cysteine, alanine, glycine, serine and threonine)

Five

give rise to acetoacetyl CoA which yields acetyl CoA and enters the TCA cycle i.e., tyrosine, leucine, phenylalanine, lysine and tryptophan.

The last five were sometimes described as "ketogenic" amino acids because, if fed to an untreated diabetic patient showing the symptoms of ketosis, they made his ketosis more severe. The explanation is that these amino acids produced more acetyl CoA, the precursor of the "ketone bodies", and this cannot be shunted into the pathway for glucose synthesis because the pyruvate to acetyl CoA reaction is not reversible. The other amino acids were called "glucogenic" since they gave rise to pyruvate (or some compound which could be converted into this) and might, therefore, be used for glucose synthesis as an alternative to oxidation to acetyl CoA and the further formation of ketone bodies could thereby be avoided. The distinction is now recognized as being ambiguous since some amino acids are both ketogenic and glucogenic; of the nine carbon atoms of tyrosine, for example, four could give rise to acetoacetate and three to pyruvate.

7.5. The Conversion of Amino Acids into Keto Acids

The two most important and general methods by which this can be achieved are oxidative deamination and transamination.

(a) Oxidative deamination

The three best known enzymes are D-amino acid oxidase, L-amino acid oxidase and glutamate dehydrogenase.

(i) D-AMINO ACID OXIDASE

This has already been described in Chapter 5.3(d) where it was given as an example of an enzyme containing FAD as a prosthetic group, and catalyzing the reaction:

$$R \cdot CH(NH_3^+)COO^- + O_2 + H_2O \rightleftharpoons R \cdot CO \cdot COO^- NH_4^+ + H_2O_2$$

Amino acid   Keto acid

The purpose of this enzyme is rather a mystery since its substrates are amino acids belonging to the unnatural D-series and are rather rare in nature, although some are synthesized by bacteria to form a part of their cell wall. It is possible that some D-amino acids derived from digestion of dead bacteria might be absorbed by the gut wall and their removal from the circulation might be advantageous. It has been claimed that the

D-amino acid oxidase level in germ-free rabbits was low but increased when the gut was allowed to acquire a bacterial flora.

The enzyme is located within special organelles in the cell—microbodies, also called peroxisomes—which also contain catalase, the enzyme which will decompose the hydrogen peroxide produced by the oxidase.

(b) L-amino acid oxidase

The reaction catalyzed is similar to that with the previous enzyme except that the substrates are the common L-amino acids. The prosthetic group is FMN instead of FAD. The enzyme is only weakly active in mammalian cells and is not considered to play any major part in amino acid metabolism, except possibly with lysine which does not appear to be a substrate for the alternative enzymes, but which can be oxidatively deaminated after prior conversion to the ε-acetyl derivative.

(c) Glutamate dehydrogenase

This is by far the most important of the three enzymes, not only because of the important role it plays in the metabolism of glutamate but also because it can result in the deamination of most other amino acids by coupling with transaminases (see later).

The reaction catalyzed is:

$$
\begin{array}{l}
\text{COOH} \\
|\\
\text{CH}_2 \\
|\\
\text{CH}_2 \quad + \text{NAD}^+ + \text{H}_2\text{O} \\
|\\
\text{CH}-\text{NH}_3^+ \\
|\\
\text{COO}^- \\
\text{Glutamate}
\end{array}
\rightleftharpoons
\begin{array}{l}
\text{COOH} \\
|\\
\text{CH}_2 \\
|\\
\text{CH}_2 \quad + \text{NADH} + \text{H}^+ + \text{NH}_4^+ \\
|\\
\text{CO} \\
|\\
\text{COO}^- \\
\text{Ketoglutarate}
\end{array}
$$

Note that the reaction is reversible and can, therefore, be used to synthesize glutamate as well as to oxidatively deaminate it. Glutamate is the best substrate for the enzyme but a lower dehydrogenase activity can be demonstrated for some monocarboxylic amino acids such as alanine. The enzyme can exist as rod-like polymers of indefinite length formed by the aggregation of monomers (mol. wt. about 312,000) and which are themselves composed of six subunits of mol. wt. about 52,000. The aggregated form has a powerful glutamate dehydrogenase activity and is protected from disaggregation by ADP and certain amino acids. Various compounds, however, such as guanosine triphosphate, NADH and some steroid hormones can lead to extensive disaggregation and this is accompanied by inhibition of the glutamate dehydrogenase activity but an increase in alanine dehydrogenase activity. These changes in activity and specificity suggest that the enzyme can exist in various forms depending on the presence of certain small molecules, i.e., it is an allosteric molecule acting as a regulatory enzyme in amino acid metabolism.

7.6. Transamination

Most amino acids can participate in reactions of this type which can be summarized:

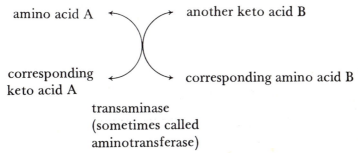

amino acid A $\rightleftharpoons$ another keto acid B

corresponding keto acid A $\rightleftharpoons$ corresponding amino acid B

transaminase
(sometimes called
aminotransferase)

Thus alanine-glutamate transaminase will catalyze the specific reaction:

$$
\begin{array}{ll}
\text{CH}_3 & \text{COOH} \\
| & | \\
\text{CH}-\text{NH}_3^+ + \text{CH}_2 \\
| & | \\
\text{COO}^- & \text{CH}_2 \\
& | \\
& \text{CO} \\
& | \\
& \text{COO}^-
\end{array}
\rightleftharpoons
\begin{array}{ll}
\text{CH}_3 & \text{COOH} \\
| & | \\
\text{CO} & + \text{CH}_2 \\
| & | \\
\text{COO}^- & \text{CH}_2 \\
& | \\
& \text{CH}-\text{NH}_3^+ \\
& | \\
& \text{COO}^-
\end{array}
$$

Alanine   Ketoglutarate   Pyruvate   Glutamate

Various transaminases occur and two other important ones are glutamate transaminase and alanine transaminase which can convert a range of different α-amino acids into their corresponding keto acids using α-ketoglutarate and pyruvate respectively as amino group acceptors, i.e.,

amino acid + α-ketoglutarate $\rightleftharpoons$ keto acid + glutamate

amino acid + pyruvate $\rightleftharpoons$ keto acid + alanine

The action of transaminases is to transfer the α-amino group from one acid to another but there is no net deamination since no ammonia is formed. However, by coupling with glutamate dehydrogenase, the amino group can be liberated, e.g.,

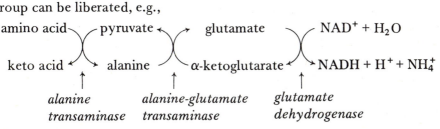

amino acid $\rightleftharpoons$ pyruvate $\rightleftharpoons$ glutamate $\rightleftharpoons$ NAD$^+$ + H$_2$O

keto acid $\rightleftharpoons$ alanine $\rightleftharpoons$ α-ketoglutarate $\rightleftharpoons$ NADH + H$^+$ + NH$_4^+$

alanine   *alanine-glutamate*   *glutamate*
transaminase   *transaminase*   *dehydrogenase*

or alternatively,

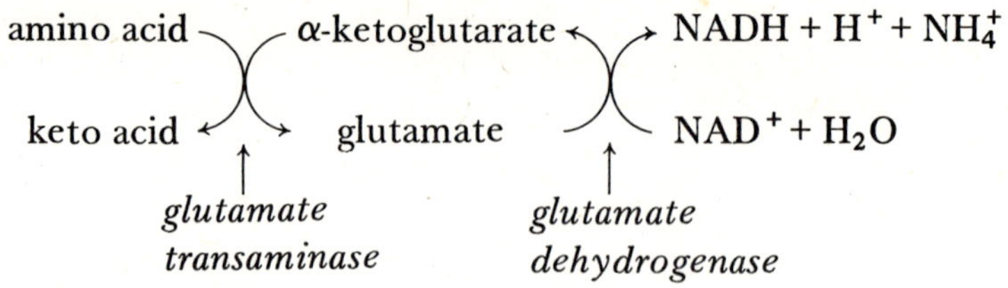

amino acid $\diagdown$ $\diagup$ α-ketoglutarate $\diagup$ $\diagdown$ NADH + H$^+$ + NH$_4^+$

keto acid $\diagup$ $\diagdown$ glutamate $\diagdown$ $\diagup$ NAD$^+$ + H$_2$O

glutamate   *glutamate*
transaminase   *dehydrogenase*

In either case, the overall reaction is:

$$R \cdot CH \cdot NH_3 \cdot COO^- + NAD^+ + H_2O = R \cdot CO \cdot COO^- + NADH + H^+ + NH_4^+$$
Amino acid   α Keto acid

The reactions are all freely reversible and can be used if necessary for the biosynthesis of amino acids if the corresponding α-keto acid is available, e.g., aspartate can be made from oxaloacetate. The amino acids which the body can make for itself are called *non-essential*, meaning that it is not essential for them to be supplied in the diet (but they are, of course, essential in the sense that protein biosynthesis cannot occur without them). Those amino acids which the body cannot synthesize must be supplied ready made and are called the *essential* ones (i.e., essential in the diet). The number of these varies with the species but in man about nine amino acids are considered to be essential. Higher animals appear to have lost the ability to make certain of the keto acids which by transamination would have yielded the "essential" amino acids. Plants and many microorganisms can make these keto acids without difficulty.

All transaminases require pyridoxal phosphate as an essential co-enzyme. This is the phosphorylated form of Vitamin B$_6$ and has the structure shown but for convenience may be regarded as X-CHO since it is

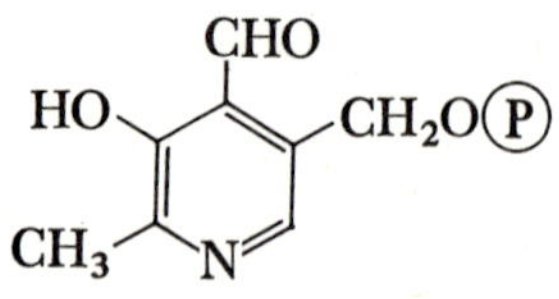

CHO
HO $\diagdown$ $\diagup$ CH$_2$O(P)
CH$_3$ $\diagup$ N

Pyridoxal
phosphate

the aldehyde group which plays a key role in facilitating the transfer of the amino group. By condensation with the amino acid A to form a Schiff's base, rearrangement and hydrolysis, a keto acid A is formed and the coenzyme is converted into pyridoxamine phosphate (Fig. 7.5). This then condenses with a different keto acid B to form another Schiff's base which on rearrangement and hydrolysis forms a new amino acid B and regenerates the original coenzyme.

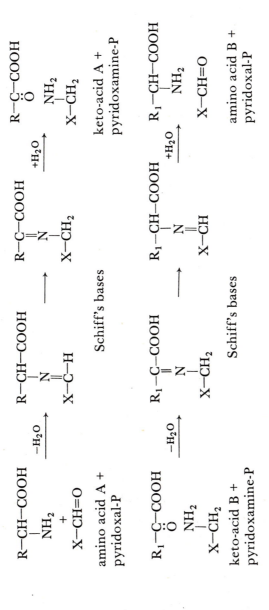

Fig. 7.5 Role of pyridoxal phosphate in transamination.

7.7. Non-oxidative Deaminations

In addition to the enzymes already mentioned which remove amino groups from amino acids by oxidation or transamination (and which are usually active with a variety of amino acids), there are also a few more specific enzymes which effect deamination by non-oxidative processes.

Three enzymes which catalyze analogous reactions are:

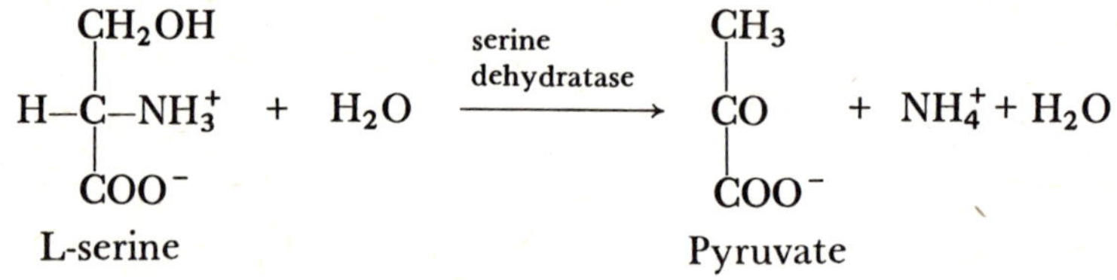

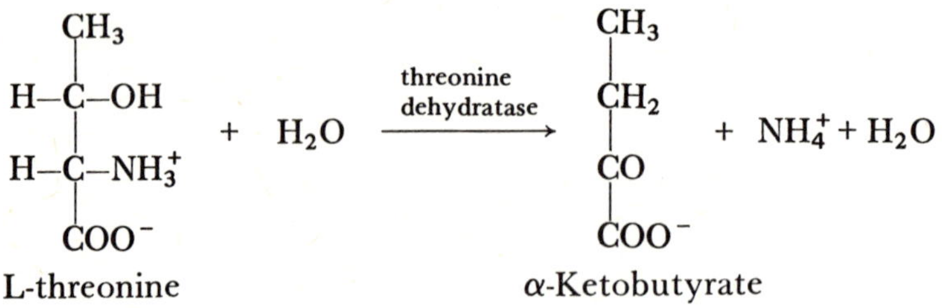

$$
\begin{array}{c}
CH_2OH \\
| \\
H-C-NH_3^+ \\
| \\
COO^-
\end{array}
\quad + \quad H_2O
\quad \xrightarrow{\text{serine dehydratase}} \quad
\begin{array}{c}
CH_3 \\
| \\
CO \\
| \\
COO^-
\end{array}
\quad + \quad NH_4^+ + H_2O
$$

L-serine   Pyruvate

$$
\begin{array}{c}
CH_3 \\
| \\
H-C-OH \\
| \\
H-C-NH_3^+ \\
| \\
COO^-
\end{array}
\quad + \quad H_2O
\quad \xrightarrow{\text{threonine dehydratase}} \quad
\begin{array}{c}
CH_3 \\
| \\
CH_2 \\
| \\
CO \\
| \\
COO^-
\end{array}
\quad + \quad NH_4^+ + H_2O
$$

L-threonine   α-Ketobutyrate

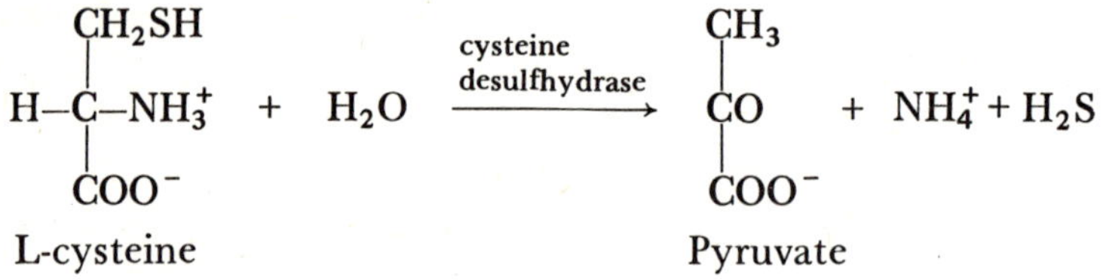

$$
\begin{array}{c}
CH_2SH \\
| \\
H-C-NH_3^+ \\
| \\
COO^-
\end{array}
\quad + \quad H_2O
\quad \xrightarrow{\text{cysteine desulfhydrase}} \quad
\begin{array}{c}
CH_3 \\
| \\
CO \\
| \\
COO^-
\end{array}
\quad + \quad NH_4^+ + H_2S
$$

L-cysteine   Pyruvate

These three enzymes require pyridoxal phosphate as a coenzyme and a Schiff's base is believed to be an intermediate. Threonine dehydratase is of particular interest in providing a good example of an enzyme that is inducible in mammalian systems—examples are much more common in bacterial systems. If rats are fed intensively with a casein hydrolysate, the threonine dehydratase activity in their livers may increase 300-fold. This induction is repressed by dietary glucose, and may be considered, therefore, as one of the responses of the liver to a need for gluconeogenesis since α-ketobutyrate can be metabolized further to pyruvate and hence to glucose. The increase in activity is due to new enzyme being made since the effect is prevented by administration of an inhibitor of protein biosynthesis, such as puromycin.

Some bacteria possess *aspartase* which catalyzes the reaction:

$$
\begin{array}{ccc}
\begin{array}{c}
\text{COOH} \\
| \\
\text{CH}_2 \\
| \\
\text{H--C--NH}_2 \\
| \\
\text{COOH}
\end{array}
&
\xrightleftharpoons{\text{aspartase}}
&
\begin{array}{c}
\text{COOH} \\
| \\
\text{CH} \\
\| \\
\text{CH} \\
| \\
\text{COOH}
\end{array}
\quad + \quad \text{NH}_3
\end{array}
$$

L-aspartic acid   Fumaric acid

This enzyme does not occur in higher animals who use an alternative route for the degradation of aspartate to a compound on the tricarboxylic acid cycle—transamination giving oxaloacetate. However, aspartate is sometimes converted (through an intermediate) into fumarate in mammalian systems when it is acting as an amino group donor. An example occurs later in the urea cycle in the conversion of citrulline into arginine.

7.8. Fate of the Ammonia Produced by Deamination of Amino Acids

Since about 70 g of amino acids may be deaminated daily, quite a large amount of ammonia (as ammonium ions) is produced but its concentration must be kept very low at all times since ammonia is surprisingly toxic. The blood level is normally 0·1 to 0·2 mg/100 ml (i.e., only about one thousandth that of glucose) and in the rabbit, a blood level of 5 mg/100 ml is fatal. The body has to remove, therefore, ammonia as fast as it is produced. Some fish excrete free ammonia but they have an unlimited supply of water to facilitate this. Land animals detoxify ammonia by converting it to some harmless and easily excreted compound. Mammals use urea for this purpose, a very soluble and non-toxic compound but its elimination is accompanied by an appreciable volume of water since the ability of the kidney to concentrate urine is limited. Animals such as birds, reptiles, and some flies in which a low body weight or water conservation is important, do not produce urea but uric acid instead since this is a rather insoluble compound and they are able to excrete a solid urine without appreciable loss of water.

Before discussing the conversion of ammonia to urea, the formation of glutamine will be briefly mentioned. Glutamine synthetase is an enzyme present in most tissues promoting the reaction:

$$
\begin{array}{ccc}
\begin{array}{c}
\text{COOH} \\
| \\
\text{CH}_2 \\
| \\
\text{CH}_2 \\
| \\
\text{CH--NH}_3^+ \\
| \\
\text{COO}^-
\end{array}
\quad + \ \text{NH}_3 + \text{ATP}
&
\longrightarrow
&
\begin{array}{c}
\text{CONH}_2 \\
| \\
\text{CH}_2 \\
| \\
\text{CH}_2 \\
| \\
\text{CH--NH}_3^+ \\
| \\
\text{COO}^-
\end{array}
\quad + \ \text{ADP} + \text{P}_i
\end{array}
$$

Glutamate   Glutamine

(Note that glutamine is a bad name, but accepted because of long usage, since the compound is not an amine but an amide.) Tissues also contain glutaminase which will hydrolyze the amide linkage:

$$\text{glutamine} + H_2O \rightarrow \text{glutamate} + NH_3$$

Starting with α-ketoglutarate, we now have a system for storing two ammonia molecules in a safe form:

$$\alpha\text{-ketoglutarate} \underset{\substack{-NH_3 \\ \text{glutamate} \\ \text{dehydrogenase}}}{\overset{+NH_3}{\rightleftharpoons}} \text{glutamate} \underset{\substack{-NH_3 \\ \text{glutaminase}}}{\overset{\substack{+NH_3 \\ \text{glutamine} \\ \text{synthetase}}}{\rightleftharpoons}} \text{glutamine}$$

Glutamine can serve three useful purposes:

(a) Ammonia is produced by deamination in most tissues but is converted to urea only in the liver and, therefore, has to be taken there by the blood. The concentration of ammonia in the blood can be kept low and toxic effects avoided by incorporation of the ammonia into glutamine, a safe form for transport.

(b) Glutamine provides a small but useful store of amino nitrogen for biosynthesis. It is the most plentiful free amino acid in the blood and accounts for about a quarter of the free amino nitrogen there.

(c) When it is necessary to neutralize acidic components in the urine (e.g., sulphuric and phosphoric acids from a high protein diet), the kidney uses some ammonium ions for this purpose, obtaining them by the hydrolysis of glutamine, and thus is able to conserve plasma Na^+ which otherwise might have had to be used in order to bring the pH of the urine to a physiologically acceptable level.

7.9. Nitrogen Balance and the Urea Cycle

Three situations are possible:

(i) Nitrogen intake is greater than output = positive N balance
    e.g., a growing child, a pregnant woman
(ii) Nitrogen intake less than output = negative N balance
    e.g., a patient losing weight, a girl slimming, a famine victim
(iii) Nitrogen intake exactly equals output = true N balance
    e.g., an adult whose body weight is constant.

The third situation can be considered the normal one in which there is no net gain or loss of body protein, in that the 15 g or so of nitrogen ingested daily must be balanced by the excretion of the same amount.

This occurs largely in the urine in several different forms, the approximate distribution being:

85% as urea
3% as ammonium ion
5% as creatinine
1% as uric acid
6% in other forms

It is seen that urea represents the main end product of nitrogen metabolism and its formation must now be considered.

Urea is made in the liver which is capable of producing up to 100 g a day should the diet be very rich in protein. The pathway of biosynthesis was first elucidated by Krebs and Henseleit in 1931 who measured the amount of urea produced by liver slices suspended in a suitable medium to which various amino acids and ammonium ions could be added. They found that exceptionally high rates of urea formation were observed if ornithine and ammonium ions were present together. One molecule of ornithine (an amino acid) led to the production of over 20 molecules of urea, the nitrogen being derived from the ammonium ions, so that the role of ornithine was a catalytic one. Since liver was known to contain a highly active arginase, the enzyme catalyzing the reaction:

$$\text{arginine} \xrightarrow{\text{arginase}} \text{ornithine} + \text{urea}$$

it occurred to Krebs and Henseleit that the catalytic role of ornithine could be accounted for if there was some mechanism for the regeneration of arginine from ornithine, i.e.,

$$\text{ornithine} + CO_2 + 2\,NH_3 \rightarrow \text{arginine} + 2\,H_2O$$

This led to a search for possible intermediates between ornithine and arginine, and a candidate which suggested itself was the amino acid citrulline which had previously been isolated from the Water Melon, *Citrullus*. When tested in the liver slice system, citrulline showed a similar active catalytic effect on urea synthesis from ammonium ions, and this led to the suggestion of the Urea Cycle which is represented in outline in Fig. 7.6.

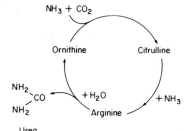

Fig. 7.6 The urea cycle in a simplified form.

This was the first example of a metabolic cycle to be discovered and antedated the well-known tricarboxylic acid by about five years.

The overall reaction is:

$$2 NH_3 + CO_2 = NH_2-CO-NH_2 + H_2O \quad \Delta G^0 = +60 \text{ kJ } (+14 \text{ kcal})$$

Subsequent work has confirmed that the cyclic mechanism is essentially correct in principle and has elucidated the finer details. In the overall reaction given in Fig. 7.6, there is an increase in free energy (ΔG^0 positive) and from the thermodynamic point of view one would not expect the cycle to turn spontaneously in the direction of urea synthesis unless the cycle was driven by coupling to some energy yielding reactions. The reverse reaction, the hydrolysis of urea to carbon dioxide and ammonia, takes place with a decrease in free energy and thus urea will break down spontaneously in the presence of the appropriate enzyme, urease.

The details of the biosynthetic cycle will now be described in greater detail. The first step is the formation of carbamyl phosphate from carbon dioxide and ammonia. This compound is energy-rich and its formation requires the participation of two molecules of ATP:

$$CO_2 + NH_3 + 2ATP + H_2O \xrightarrow{\text{carbamyl phosphate synthetase}} {}^-O-\overset{\overset{\displaystyle O^-}{|}}{\underset{\underset{\displaystyle O}{\|}}{P}}-O-CO-NH_2 + 2ADP + P_i$$

carbamyl phosphate

The enzyme contains biotin as its prosthetic group and initially carbon dioxide and one ATP react to give an "active-CO_2-enzyme" complex which subsequently reacts with ammonia in a reaction involving a second molecule of ATP as a phosphate donor (cf., acetyl CoA synthetase, another enzyme containing biotin which was described in Chapter 5).

In the next stage the carbamyl phosphate reacts with ornithine to give citrulline:

$$
\begin{array}{l}
\text{H} \\
| \\
\text{N(H} + \text{O}-\overset{\overset{\displaystyle O^-}{|}}{\underset{\underset{\displaystyle O}{\|}}{P}}-\text{O}-\text{CO}-\text{NH}_2 \\
| \\
\text{CH}_2 \\
| \\
\text{CH}_2 \\
| \\
\text{CH}_2 \\
| \\
\text{CH}-\text{NH}_3^+ \\
| \\
\text{COO}^- \\
\text{ornithine}
\end{array}
\xrightarrow[\text{transcarbamylase}]{\text{ornithine}}
\begin{array}{l}
\text{NH}-\text{CO}-\text{NH}_2 \\
| \\
\text{CH}_2 \\
| \\
\text{CH}_2 \\
| \\
\text{CH}_2 \qquad + \text{ P}_i \\
| \\
\text{CH}-\text{NH}_3^+ \\
| \\
\text{COO}^- \\
\text{citrulline}
\end{array}
$$

The citrulline is now converted into arginine in two stages. It first reacts with aspartate to give arginosuccinate, a condensation which requires the participation of one molecule of ATP. The arginosuccinate is then split into arginine and fumarate.

```
NH₂                    COOH
|                      |
CO          H₂N—CH
|                      |
NH                    CH₂        arginosuccinate
|                      |          synthetase
CH₂        +       COOH        ──────────────→
|                      Aspartate      + ATP
CH₂
|
CH₂
|
CH—NH₃⁺
|
COO⁻
Citrulline        aspartic acid
```

```
NH            COOH                          NH           COOH
||            |                             ||           |
C——NH——CH                            C—NH₂        CH
|             |               argino        |            ||
NH           CH₂            succinase      NH           CH
|             |           ──────────→      |            |
CH₂        COOH                        CH₂    +     COOH
|                                            |            Fumaric
CH₂      + AMP + PPᵢ                   CH₂          acid
|                                            |
CH₂                                        CH₂
|                                            |
CH—NH₃⁺                                CH—NH₃⁺
|                                            |
COO⁻                                       COO⁻
Arginosuccinic                         Arginine
acid
```

The true overall reaction is now seen to be

$$NH_3 + CO_2 + 3\ ATP + 2\ H_2O + aspartate$$
$$= urea + 2\ ADP + AMP + fumarate + 2P_i + PP_i$$

$$\Delta G^0 \text{ about } -40 \text{ kJ } (-10 \text{ kcal})$$

and since there is a decrease in free energy, the process will proceed in the direction of urea synthesis.

It has previously been pointed out that the deamination of amino acids by transamination leads to the formation of glutamate from which free ammonia can be liberated by the NAD-dependent glutamate dehydrogenase. This is probably the main source of the ammonia used in the synthesis of carbamyl phosphate. The second nitrogen in urea was seen to be derived from the amino group of aspartate and not from free ammonia directly. However, the fumarate produced can be converted back into aspartate by a scheme in which glutamate again supplied the amino group.

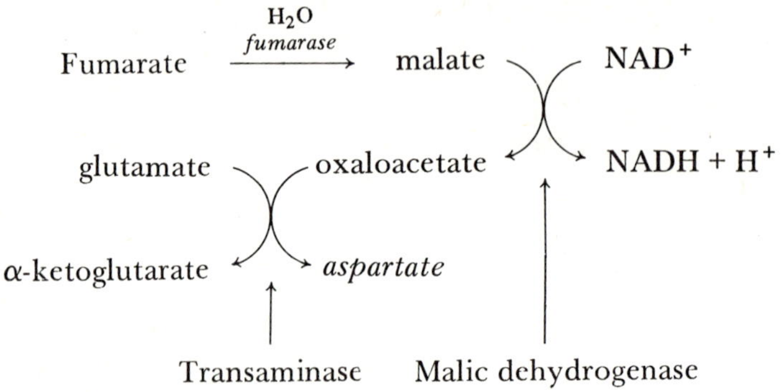

$$
\text{Fumarate} \xrightarrow[\text{fumarase}]{H_2O} \text{malate} \qquad NAD^+
$$

glutamate ⟍ ⟋ oxaloacetate ⟍ ⟋ $NADH + H^+$

α-ketoglutarate ⟋ ⟍ *aspartate*

Transaminase   Malic dehydrogenase

7.10. Amino Acid Decarboxylases

These enzymes occur in many tissues and catalyze reactions of the general type:

$$
\underset{\underset{COO^-}{|}}{R-CH-NH_3^+} \longrightarrow R-CH_2NH_2 + CO_2
$$

The products are primary amines and frequently possess powerful physiological properties. For instance, histidine is decarboxylated to give histamine which strongly stimulates smooth muscle, dilates capillaries, increases the flow of gastric secretion, etc. Other examples are:

glutamic acid $\xrightarrow{-CO_2}$ γ-amino-butyrate (active in the C.N.S.)

5-hydroxy-tryptophan $\xrightarrow{-CO_2}$ serotonin (a vasoconstrictor and synaptic transmitter substance)

3 : 4-di-hydroxyphenylalanine $\xrightarrow{-CO_2}$ dopamine (precursor of adrenaline)

tyrosine $\xrightarrow{-CO_2}$ tyramine (a hypertensive agent)

The production in the body of such active materials necessitates the provision of some mechanism for their conversion into inactive substances (detoxification). Mono- and di-amine oxidases are widely distributed in tissues and oxidize the amines to aldehydes, e.g.,

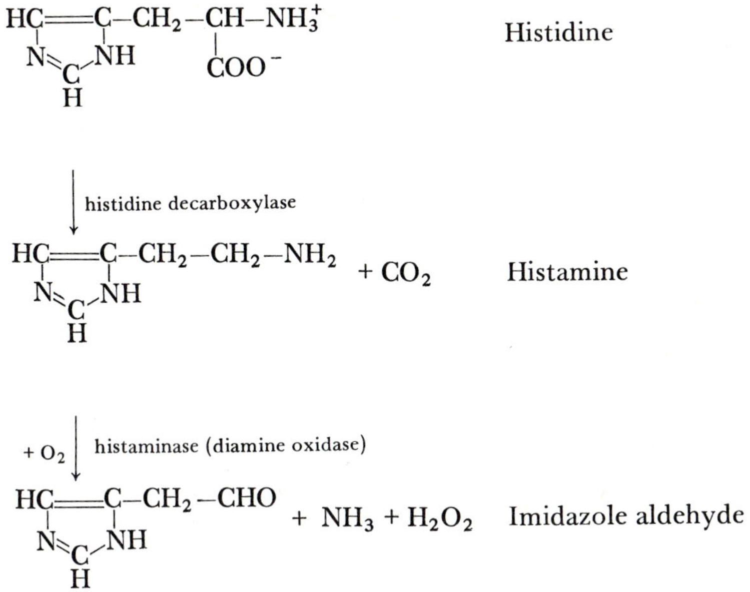

$$HC = C-CH_2-CH-NH_3^+ \qquad \text{Histidine}$$

histidine decarboxylase

$$HC = C-CH_2-CH_2-NH_2 \; + CO_2 \qquad \text{Histamine}$$

$+ O_2$ | histaminase (diamine oxidase)

$$HC = C-CH_2-CHO \; + NH_3 + H_2O_2 \qquad \text{Imidazole aldehyde}$$

Patients suffering from psychiatric depression may be treated with drugs which are specific inhibitors of mono-amine oxidases and their therapeutic action may be related to a blocking of the normal oxidation of naturally-occurring amines. Toxic effects have resulted from the eating of certain cheeses by patients while being treated with these drugs. This bizarre effect is due to the cheeses containing sufficient tyramine (formed by microbiological decarboxylation of tyrosine) to produce a marked rise in blood pressure and other cardiovascular changes when the usual detoxifying enzyme is being inhibited.

Before considering the origin of other nitrogenous constituents in the urine, it is convenient to discuss briefly some aspects of the metabolism of a few of the amino acids.

7.11. Metabolism of Some Specific Amino Acids

(a) Glycine and serine

Glycine is the unique α-amino acid which contains no asymmetrical carbon atom and is not a substrate for either D- or L-amino acid oxidases. There is, however, a special glycine oxidase (also a flavoprotein) which will catalyze the reaction:

$$CH_2-NH_3^+ \qquad \qquad \qquad \qquad CHO$$
$$\qquad \qquad + O_2 + H_2O \xrightarrow{\text{glycine oxidase}} \qquad \qquad + H_2O_2 + NH_4^+$$
$$COO^- \qquad \qquad \qquad \qquad \qquad COO^-$$

Glycine   Glyoxalate

Alternatively, glyoxalate can be produced from glycine by a trans-amination reaction. Glyoxalate is rapidly oxidized by liver to formate and carbon dioxide. The formate can be used as a source of single carbon units in biosynthesis. These occur as derivatives of tetrahydrofolic acid (THFA) which can exist in various interchangeable forms at different levels of oxidation and can act as donors of $-CHO$, $-CH_2OH$ and $-CH_3$ groups.

Experiments with isotopically-labelled compounds have shown that glycine and serine are readily interconvertible and that both the α and β carbon atoms of serine can be derived from the α carbon of glycine.

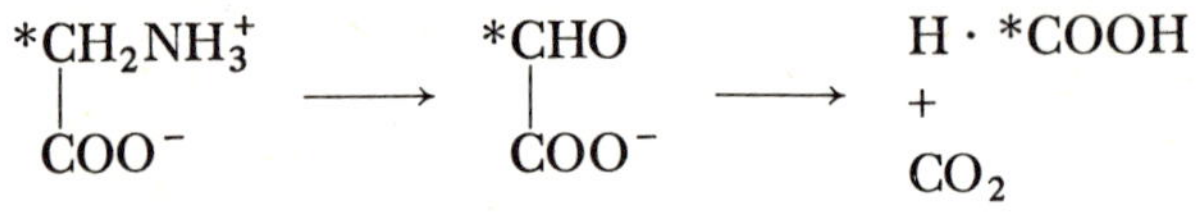

$$
\begin{array}{ccccc}
*CH_2NH_3^+ & & *CHO & & H\cdot *COOH \\
| & \longrightarrow & | & \longrightarrow & + \\
COO^- & & COO^- & & CO_2
\end{array}
$$

and then

$$
\begin{array}{ccccc}
 & & *CH_2{-}NH_3^+ & & *CH_2OH \\
H\cdot *COOH & + & | & \xrightarrow{+\,2H} & *CH{-}NH_3^+ \\
 & & COO^- & & | \\
 & & & & COO^-
\end{array}
$$

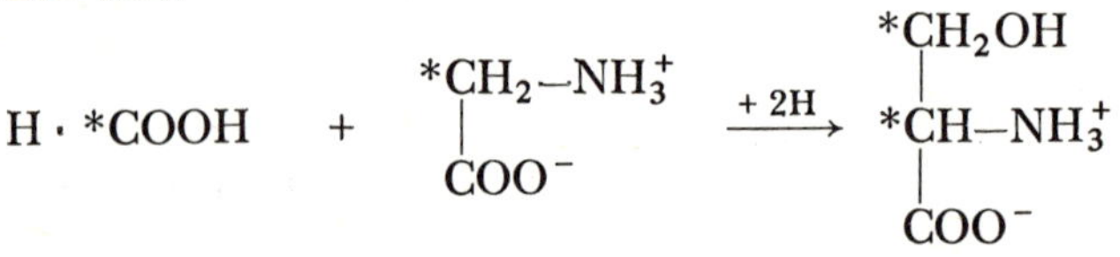

Glycine   Serine

The formate in the first reaction can yield a formyl derivative of THFA and this can be reduced by NADPH to another THFA derivative in which the C_1 is at the oxidation level of $-CH_2OH$ (viz., N^5,N^{10}-methylene THFA) and which can act as a donor of a $-CH_2OH$ group, and using this compound, the cell can readily interconvert glycine and serine:

$$
\begin{array}{ccc}
 & & CH_2OH \\
CH_2{-}NH_3^+ & & | \\
| \quad + N^5,N^{10}\text{-methylene THFA} \rightleftharpoons & & CH{-}NH_3^+ + THFA \\
COO^- & & | \\
 & & COO^-
\end{array}
$$

Glycine   Serine

Since serine can be converted to pyruvate by serine dehydratase

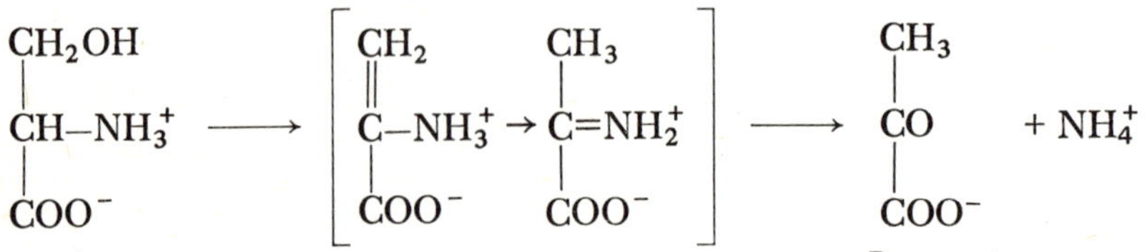

$$
\begin{array}{ccc}
CH_2OH & \left[\begin{array}{cc} CH_2 & CH_3 \\ \| & \\ C{-}NH_3^+ \rightarrow C{=}NH_2^+ \\ | & | \\ COO^- & COO^- \end{array} \right] & CH_3 \\
CH{-}NH_3^+ \longrightarrow & & CO \qquad + NH_4^+ \\
| & & | \\
COO^- & & COO^- \\
\text{Serine} & & \text{Pyruvate}
\end{array}
$$

it is possible for glycine also to be metabolized by conversion to pyruvate (via serine) and so join the main pathway of carbohydrate metabolism. A route for the biosynthesis of glycine and serine by liver that has been proposed is:

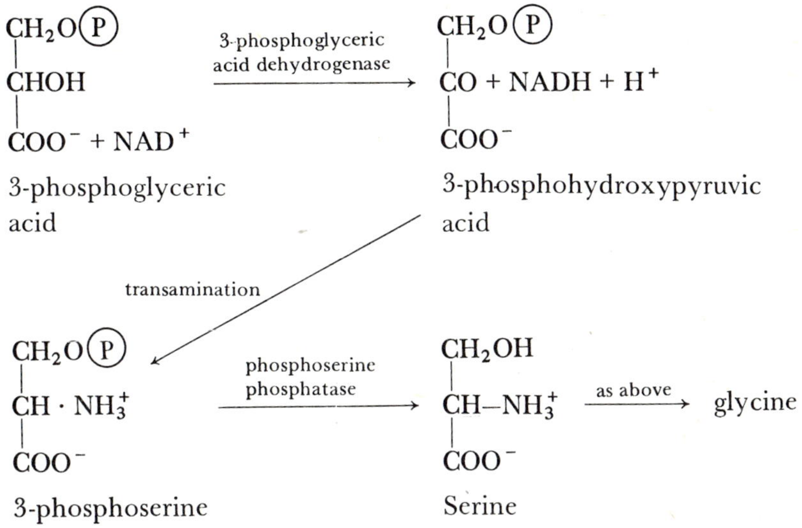

$$
\begin{array}{ccc}
\underset{|}{CH_2O\,\textcircled{P}} & \xrightarrow[\text{acid dehydrogenase}]{\text{3-phosphoglyceric}} & \underset{|}{CH_2O\,\textcircled{P}} \\
\underset{|}{CHOH} & & CO + NADH + H^+ \\
COO^- + NAD^+ & & COO^- \\
\text{3-phosphoglyceric} & & \text{3-phosphohydroxypyruvic} \\
\text{acid} & & \text{acid}
\end{array}
$$

transamination

$$
\begin{array}{ccc}
\underset{|}{CH_2O\,\textcircled{P}} & \xrightarrow[\text{phosphatase}]{\text{phosphoserine}} & \underset{|}{CH_2OH} \\
\underset{|}{CH\cdot NH_3^+} & & CH-NH_3^+ \xrightarrow{\text{as above}} \text{glycine} \\
COO^- & & COO^- \\
\text{3-phosphoserine} & & \text{Serine}
\end{array}
$$

Serine can give rise to ethanolamine by decarboxylation which can be methylated to choline, two compounds which form part of the structures of the two most abundant phosphoglycerides, while acetylcholine is important because of its key role in the transmission of nerve impulses.

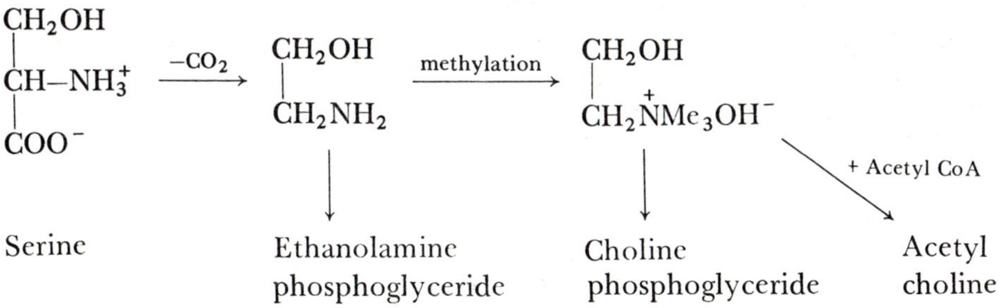

$$
\begin{array}{cccc}
\underset{|}{CH_2OH} & \xrightarrow{-CO_2} & \underset{|}{CH_2OH} & \xrightarrow{\text{methylation}} & \underset{|}{CH_2OH} \\
\underset{|}{CH-NH_3^+} & & CH_2NH_2 & & CH_2\overset{+}{N}Me_3OH^- \\
COO^- & & & & \\
& & & & \searrow\ \text{+ Acetyl CoA}
\end{array}
$$

| Serine | Ethanolamine phosphoglyceride | Choline phosphoglyceride | Acetyl choline |

Glycine is a useful and versatile molecule which can be used by the cell for a variety of purposes, as illustrated by three examples.

(i) BIOSYNTHESIS OF HIPPURIC ACID

Benzoic acid is derived from the diet and as it cannot be metabolized further, it is excreted in the urine after having been conjugated with glycine to form hippuric acid:

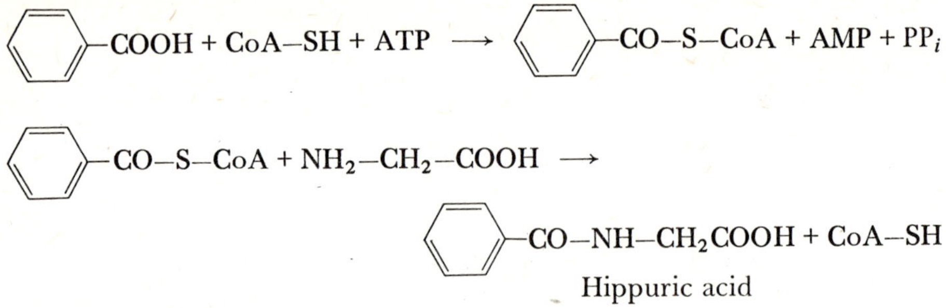

$$\text{⟨⟩—COOH} + \text{CoA–SH} + \text{ATP} \longrightarrow \text{⟨⟩—CO–S–CoA} + \text{AMP} + \text{PP}_i$$

$$\text{⟨⟩—CO–S–CoA} + \text{NH}_2\text{–CH}_2\text{–COOH} \longrightarrow$$

$$\text{⟨⟩—CO–NH–CH}_2\text{COOH} + \text{CoA–SH}$$

Hippuric acid

Detoxification often involves the conversion of an unwanted compound into a more polar and less lipid-soluble one in order to facilitate its excretion by the kidneys. The synthesis of hippuric acid takes place in the liver and can be used as the basis of a clinical test of the efficiency of liver function. A normal healthy liver should be able to synthesize glycine and conjugate it with benzoate at such a rate that at least 3 g of hippuric acid can be synthesized and excreted in the urine in the 4 h following an oral dose of 6 g of sodium benzoate in solution.

(ii) PORPHYRIN SYNTHESIS

At the end of their life of about 120 days, red blood cells are broken down and the haem of haemoglobin degraded and excreted as bile pigments at the rate of about 300 mg a day. New haem has to be synthesized at this rate to supply the new red cells. Haem is a porphyrin, a complex structure whose carbon and nitrogen skeleton only is shown in Fig. 7.7 (see Fig. 9.8 for complete structure). It can be synthesized from two simple molecules—succinate and glycine. The 8 carbon atoms and 4 nitrogen atoms shown in heavy print are derived from the 8 molecules of

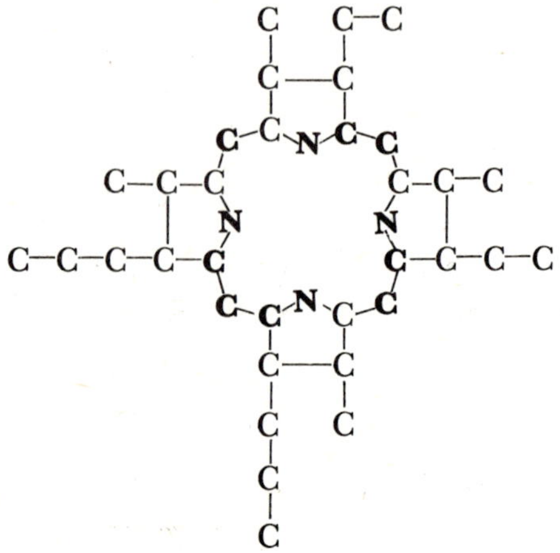

Fig. 7.7 The skeleton of the porphyrin ring, as in haem.

glycine used in the biosynthesis of each haem molecule. The porphyrin ring is a structure that you should be able to recognize (but not necessarily remember in detail) since it occurs in compounds which are involved in some of the most fundamental processes in living systems, e.g., chlorophyll in photosynthesis, cytochromes in cell respiration, haemoglobin and myoglobin in oxygen transport and in the prosthetic group of some enzymes.

(iii) PURINE BIOSYNTHESIS

Adenine and guanine are two of the most important purine bases since they form a part of RNA and DNA while adenine occurs in ATP, NAD, CoA, FAD and other key compounds, so it is advantageous that the organism should be able to synthesize them as required. The purines are derivatives of the heterocyclic ring system shown and work with isotopically labelled compounds proved that the heart of the molecule, shown in heavier type, was derived from a molecule of glycine. The biosynthetic pathway is known and is somewhat complex since nucleotides are synthesized initially rather than the free bases and the starting point is ribose phosphate upon which the purine ring is built up stepwise.

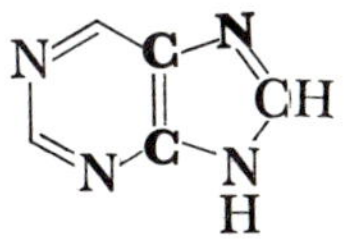

The purine ring

(b) Amino acids containing sulphur (cysteine and methionine)

Cysteine is an analogue of serine in which one oxygen atom has been replaced by a sulphur atom. The thiol group, $-SH$ is reactive and a pair of them can be readily oxidized to give a disulphide:

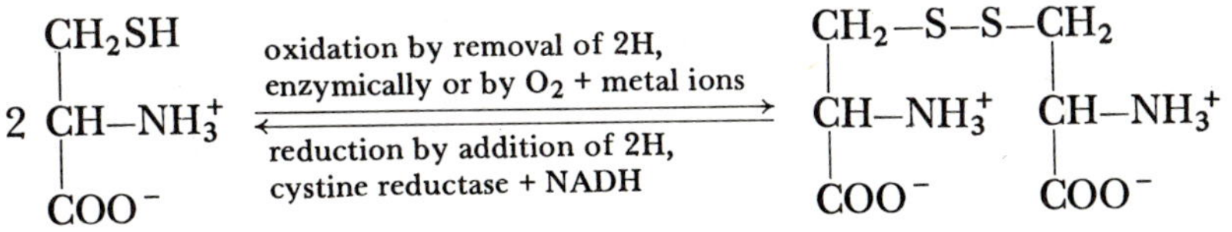

When disulphide formation occurs between two cysteine residues incorporated into polypeptide chains, then inter-chain bridges or loops are formed (Chapter 4.5).

Cysteine is a constituent of *glutathione*, a tripeptide that is widely distributed in tissues and has the structure shown overleaf. It can be represented in shorthand as G—SH since the thiol group is the main functional point of interest, and like cysteine, G–SH can form a disulphide

$$2\,\text{G–SH} \underset{+2\text{H}}{\overset{-2\text{H}}{\rightleftharpoons}} \text{G–S–S–G}$$

reduced glutathione — oxidized glutathione

Some enzymes have an essential –SH grouping at their active centres (a cysteine residue) and are inactivated if this gets oxidized. It has been suggested that one important role of glutathione is to maintain these enzymes in the active reduced form. If glutathione gets oxidized, it can be reduced again enzymically by glutathione reductase when NADPH acts as the hydrogen donor. It may also play some part in reduction reactions, as for example in the inactivation of insulin by glutathione-insulin transhydrogenase when glutathione donates hydrogen atoms which break the disulphide bridges by reduction and allow the two chains of insulin to fall apart. Glutathione also acts as a coenzyme with a few enzymes.

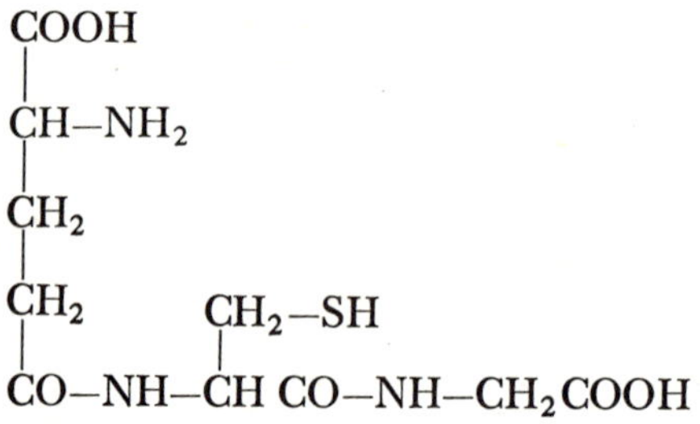

COOH
CH–NH2
CH2
CH2
CH2–SH
CO–NH–CH CO–NH–CH2COOH

Glutathione (γ-glutamyl-cysteinyl-glycine)

Methionine is an interesting amino acid because it has a methyl group on its sulphur atom which can be used for biological methylations after the amino acid has been converted into "active methionine". This is achieved by reaction with ATP:

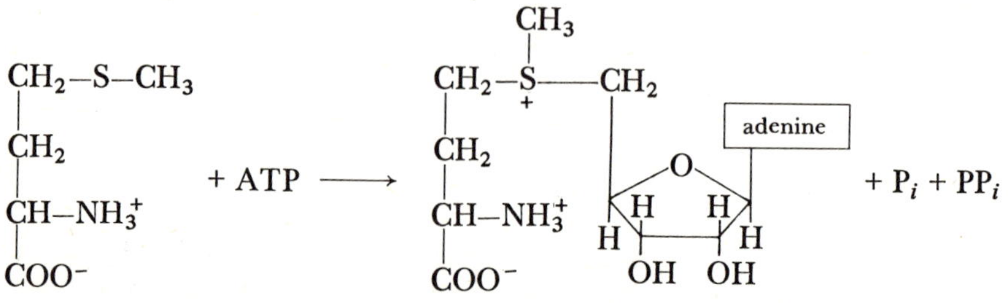

CH2–S–CH3
CH2
CH–NH3+
COO−
+ ATP
CH3
CH2–S–CH2
+
adenine
CH2
O
CH–NH3+
H H
H H
COO−
OH OH
+ Pi + PPi

Methionine S-adenosylmethionine ("active" methionine)

The methyl group can be transferred from this derivative to various acceptors, e.g., to guanidoacetic acid in the final stage of the synthesis of creatine:

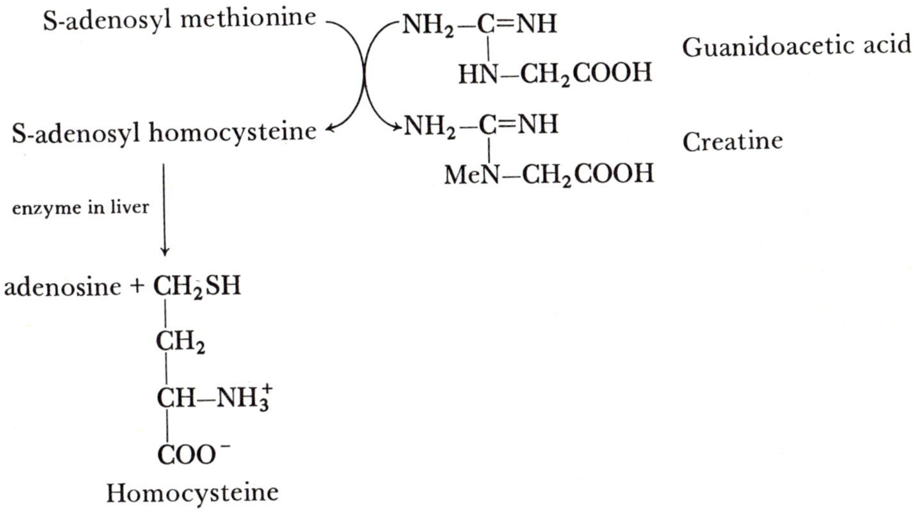

S-adenosyl methionine $\diagdown$ $\diagup$ $\;$ NH$_2$–C=NH

$\qquad\qquad\qquad\qquad\qquad$ |

$\qquad\qquad\qquad\qquad\qquad$ HN–CH$_2$COOH $\qquad$ Guanidoacetic acid

S-adenosyl homocysteine $\leftarrow$ $\;$ NH$_2$–C=NH

$\qquad\qquad\qquad\qquad\qquad$ |

$\qquad\qquad\qquad\qquad\qquad$ MeN–CH$_2$COOH $\qquad$ Creatine

enzyme in liver

adenosine + CH$_2$SH

$\qquad\qquad$ |

$\qquad\qquad$ CH$_2$

$\qquad\qquad$ |

$\qquad\qquad$ CH–NH$_3^+$

$\qquad\qquad$ |

$\qquad\qquad$ COO$^-$

$\qquad$ Homocysteine

Methylated purine bases, N-Me-nicotinamide and adrenaline are other examples of compounds formed by transmethylation in this manner. The demethylated adenosyl derivative formed is broken down in the liver to adenosine and *homocysteine* (so named because it is the next acid up the homologous series with one more CH$_2$ group than cysteine), and this is a useful compound since it can be used for the synthesis of cysteine. It is not possible for the cell to do this directly by removal of the extra CH$_2$ group, but a reaction with serine is utilized so that the sulphur atom is transferred.

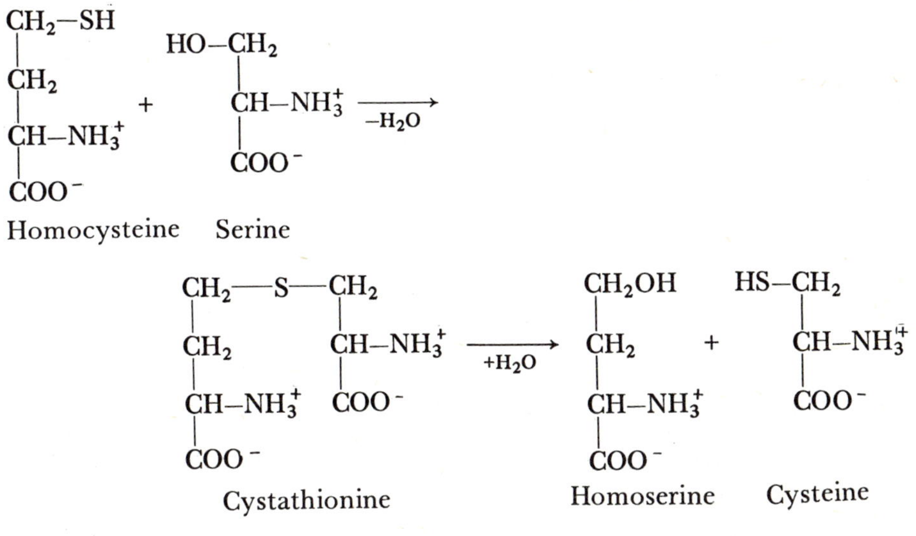

CH$_2$–SH

|

CH$_2$ $\qquad$ HO–CH$_2$

| $\qquad\qquad$ |

CH–NH$_3^+$ $\quad$ + $\quad$ CH–NH$_3^+$ $\;\xrightarrow{-H_2O}$

| $\qquad\qquad\qquad$ |

COO$^-$ $\qquad\qquad$ COO$^-$

Homocysteine $\quad$ Serine

CH$_2$—S—CH$_2$ $\qquad\qquad$ CH$_2$OH $\quad$ HS–CH$_2$

| $\qquad\qquad$ | $\qquad\qquad\qquad$ | $\qquad\qquad$ |

CH$_2$ $\qquad$ CH–NH$_3^+$ $\;\xrightarrow{+H_2O}$ CH$_2$ $\quad$ + $\quad$ CH–NH$_3^+$

| $\qquad\qquad$ | $\qquad\qquad\qquad$ | $\qquad\qquad$ |

CH–NH$_3^+$ $\quad$ COO$^-$ $\qquad\qquad$ CH–NH$_3^+$ $\qquad$ COO$^-$

| $\qquad\qquad\qquad\qquad\qquad\qquad$ |

COO$^-$ $\qquad\qquad\qquad\qquad\qquad$ COO$^-$

$\qquad$ Cystathionine $\qquad\qquad\qquad$ Homoserine $\quad$ Cysteine

The homoserine (note reason for name) now undergoes transamination to give α-ketobutyric acid which can be metabolized completely, yielding useful energy. Homocysteine can also be used for the regeneration of methionine since it can be methylated by a methyl derivative of tetrahydrofolic acid.

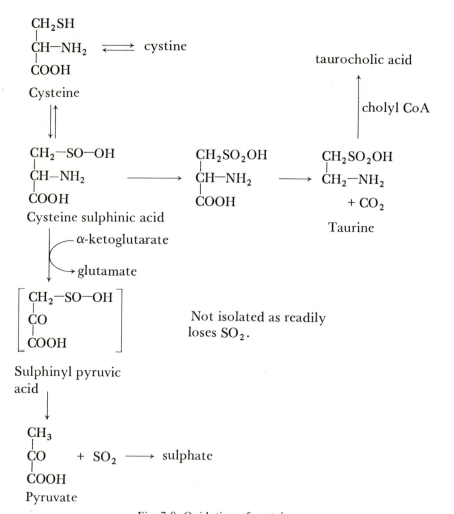

Fig. 7.8 Oxidation of cysteine.

Methionine is an essential amino acid for man and if there is an adequate supply of this, then cysteine is not essential since it can be synthesized as indicated above from methionine.

The main pathway of metabolism of excess cysteine leads to pyruvate and inorganic sulphate (Fig. 7.8) but taurine can be formed as a side product when required for the synthesis of taurocholic acid. This is one of the bile salts and results from the condensation of taurine with the CoA derivative of cholic acid, a steroid acid. The $-SO_3H$ grouping in taurocholic acid produces a highly polar location in an otherwise non-polar molecule and thus yields an efficient emulsifying agent.

(c) Phenylalanine and tyrosine

The body is unable to synthesize the phenyl ring in phenylalanine so that this amino acid must be supplied in the diet, but the other aromatic amino acid, tyrosine, is non-essential if there is an adequate supply of phenyl-alanine since it can be made from this by a hydroxylating enzyme in the liver. The reaction is not reversible and the phenolic group once introduced cannot be removed again, so tyrosine cannot replace phenyl-alanine. The degradation of the aromatic ring can be achieved as indicated in Fig. 7.9. This pathway is of medical interest because various genetic defects are known in which different enzymes are missing (inborn errors of metabolism) so that various clinical conditions arise, usually from the accumulation of intermediates whose further normal metabolism is blocked, e.g.,

> Enzyme A missing—*phenylketonuria* produced. Blockage of the main pathway diverts phenylalanine to phenylpyruvic acid and derivatives whose accumulation leads to mental deficiency.
>
> Enzyme B missing—*tyrosinosis* produced. A very rare condition in which p-OH-phenylpyruvic acid and tyrosine are excreted in the urine.
>
> Enzyme C missing—*alcaptonuria* produced. Homogentisic acid is excreted in the urine which may turn black on standing.
>
> Enzyme D missing—*albinism* produced. The natural melanin pigments of skin, hair and eyes not formed.

Tyrosine is used by the body as the starting point for the synthesis of various small molecules which contain an aromatic ring, e.g., adrenaline, noradrenaline, thyroxine and tyramine. In the presence of oxygen and oxidases, tyrosine can be converted into a reactive quinone containing an indole ring system which polymerizes to give melanin pigments. These contain long chains of conjugated double bonds and can absorb light of most wavelengths in the visible spectrum and thus appear black or very dark in colour.

(d) Tryptophan

This is an essential amino acid although not a great deal is required and a safe daily intake has been assessed at 0·5 g. It can be used to make nicotin-amide by a complex series of reactions which involve an ingenious opening of the aromatic ring and its closing to give a pyridine ring instead, with a heterocyclic N atom. This ability of the body to synthesize a vitamin may seem to be a contradiction in terms as a vitamin is usually defined as a factor which must be supplied in the diet. The diet which is lacking in

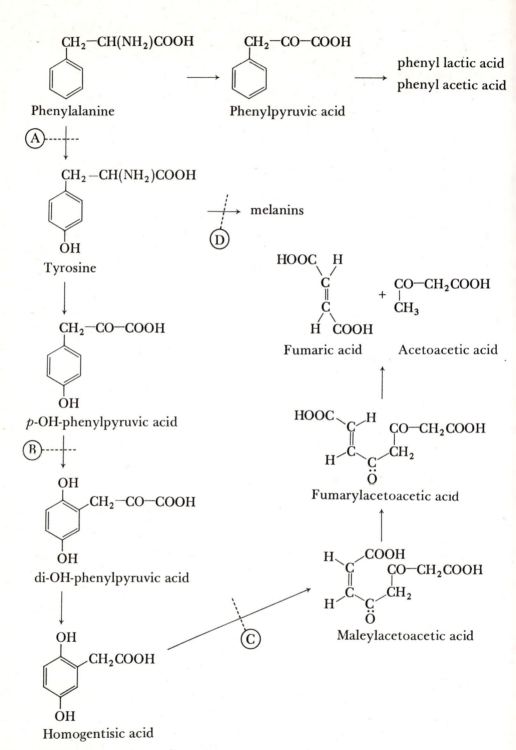

Fig. 7.9 Degradation of phenylalanine and tyrosine and the location of missing enzymes due to genetic defects.

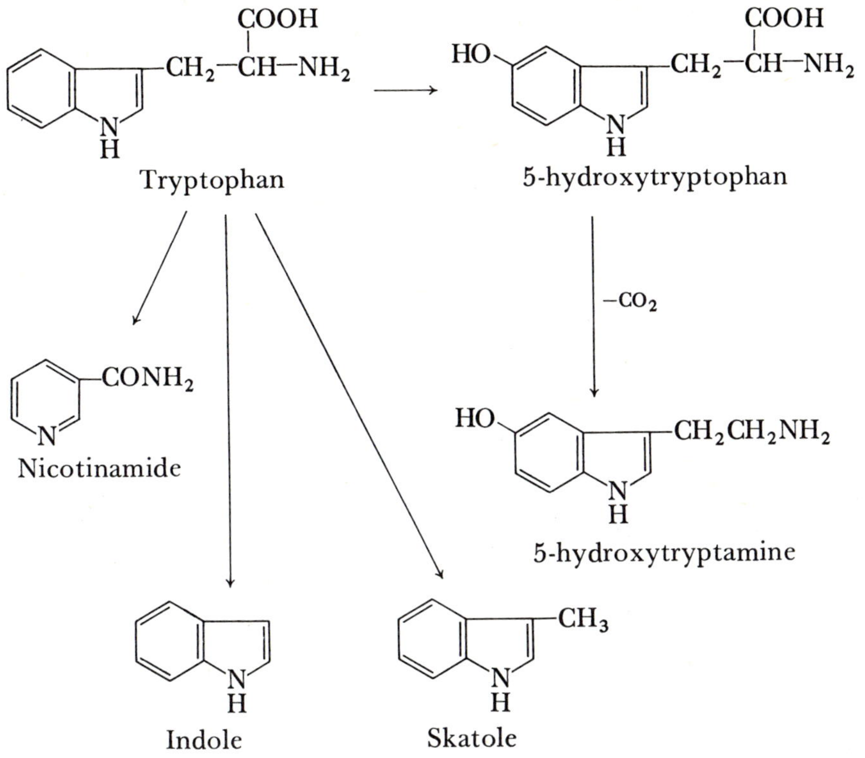

Fig. 7.10 Metabolism of tryptophan.

nicotinamide and gives rise to the disease of pellagra is unfortunately usually also deficient in tryptophan.

A normal person uses about 1% of his daily intake of tryptophan to make an important derivative, 5-hydroxy-tryptamine or serotonin which is a powerful vasoconstrictor. It is released from platelets during blood clotting and also occurs in gastro-intestinal and nervous tissue.

Bacteria in the large intestine can convert some unabsorbed tryptophan into two bases, indole and skatole, which contribute to the unpleasant smell of faeces.

7.12 The Origin of the Creatinine in the Urine

After urea, creatinine is the next most plentiful nitrogenous constituent in the urine, excretion normally being in the range of 1-3 g/day. Its origin is largely endogenous, that is, it originates in the body rather than being derived from the diet. The total amount of creatine plus creatine phosphate in an individual remains fairly constant and is proportional to his muscular development. A fraction—about 2%—of this creatine and creatine phosphate is converted each day into creatinine, so that the excretion of this is also fairly constant. The conversion may perhaps be regarded as a spontaneous side reaction which the body is unable to prevent, and once cyclization to creatinine has occurred, the molecule is of no further use and is excreted.

The daily loss of creatine is made good by synthesis in a process in which three different amino acids take part as shown in Fig. 7.11. In the first stage, the amidine group of arginine is transferred to glycine to give guanidoacetic acid, which is then methylated by "active methionine" to yield creatine. The first reaction occurs in the kidneys and the second in the liver. The creatine kinase in muscle can then convert creatine into creatine phosphate which serves as a store of energy phosphate which can be used to regenerate ATP from ADP.

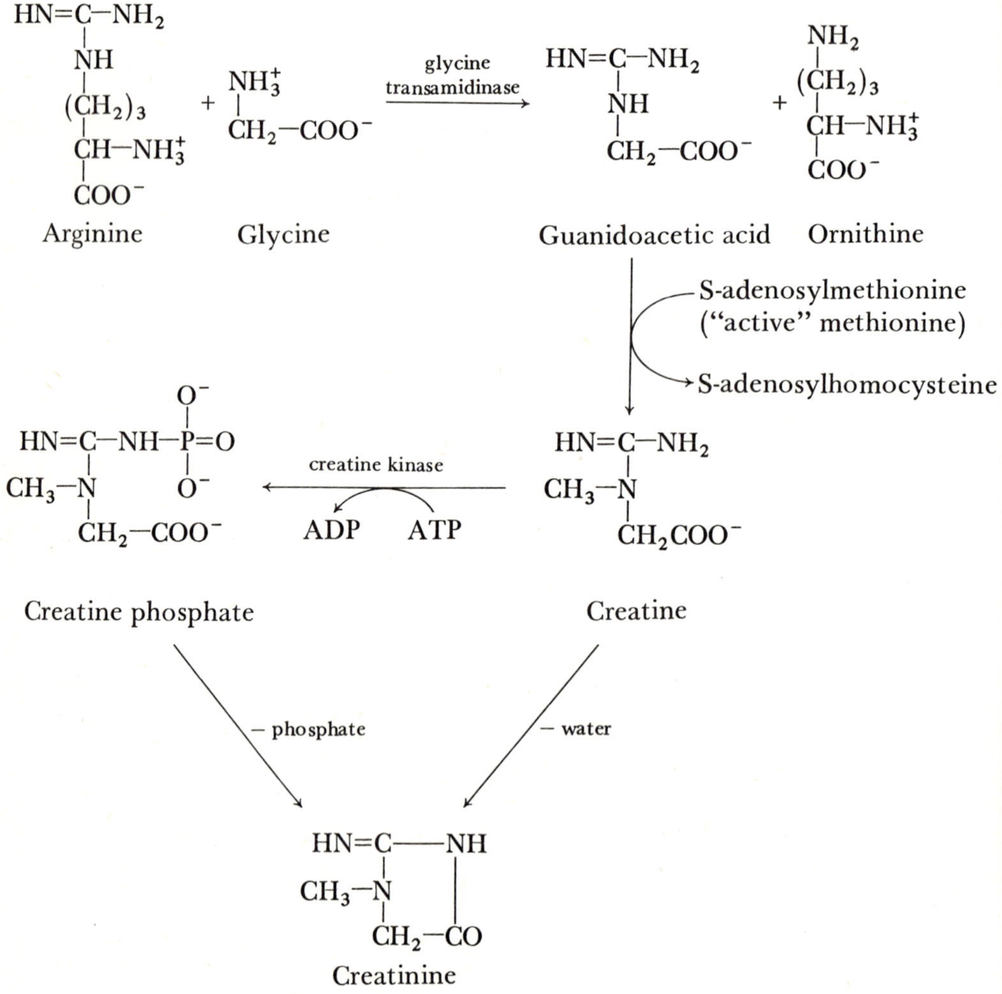

Fig. 7.11 Biosynthesis of creatine phosphate and formation of creatinine.

7.13. Purine and Pyrimidine Metabolism

The purine bases, adenine and guanine occur in the diet, largely as components of nucleic acids and are also synthesized *de novo* by the tissues. That which is surplus to requirement is converted by oxidation to uric acid, which forms the end product of purine metabolism in man, and is excreted in the urine. In most mammals other than man and the higher apes, an additional enzyme, uricase occurs which degrades uric acid further to allantoin which is a more soluble compound. The penalty that man has to pay for "losing" uricase during his evolution is that he is susceptible to gout, a painful condition due to the separation of sharp uric acid crystals in his joints. The development of allopurinol for the treatment of gout and other hyperuricaemic conditions is a good example of biochemical knowledge being used to relieve a metabolic disorder. It will be seen from Fig. 7.12 that if xanthine oxidase could be inhibited, the

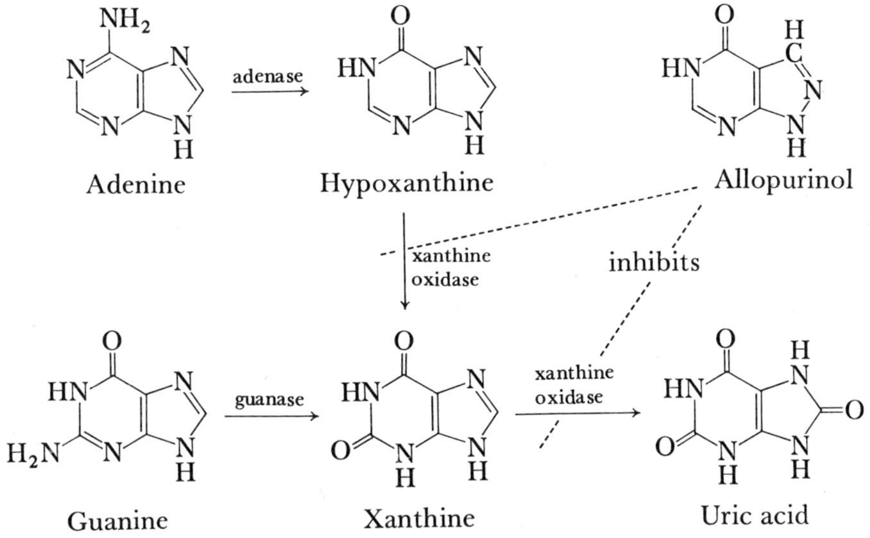

Fig. 7.12 Outline of purine base metabolism.

final stages in uric acid biosynthesis would be blocked and hypoxanthine would become one end-product of purine metabolism, which would be advantageous since hypoxanthine is more soluble and more easily excreted. You will remember that malonate is an inhibitor of succinic dehydrogenase because it is a structural analogue of succinate, the true substrate, and will also form a complex at the active centre but cannot react. It was logical, therefore, to seek an inhibitor of xanthine oxidase among structural analogues of hypoxanthine, and of the "tailor-made" compounds which were examined, the one now called allopurinol was found to be the most effective inhibitor of xanthine oxidase. The close resemblance of its structure to that of hypoxanthine should be noted— merely a —CH: and a —N: being interchanged in position. The compound was found to be relatively non-toxic and is a useful drug for reducing uric acid production *in vivo*.

The pyrimidine rings in cytosine, uracil and thymine are metabolized by reduction, ring opening and hydrolysis and complete metabolism of the aliphatic compounds produced (e.g., uracil is degraded to β-alanine, $NH_2CH_2CH_2COOH$ which is then degraded to acetyl CoA). The nitrogen atoms of the pyrimidines eventually form urea but there is no other recognizable end-product of pyrimidine metabolism.

8

The Structure and Metabolism of Lipids*

8.1. Structure of the Various Classes of Lipids

Lipids are compounds of dissimilar function and chemical composition which are related to each other only in their solubility properties; that is, they are relatively insoluble in water and soluble in many organic solvents. Many lipids contain esterified long-chain fatty acids in their molecular structure.

The principal classes of lipids are as follows:

(a) Free fatty acids (FFA)

These fall into three groups:

(i) saturated, e.g., palmitic $CH_3 \cdot (CH_2)_{14} COOH$ and stearic $CH_3 \cdot (CH_2)_{16} COOH$,

(ii) mono-unsaturated, e.g., oleic $CH_3 \cdot (CH_2)_7 CH{=}CH \cdot (CH_2)_7 COOH$,

(iii) polyunsaturated, e.g., linoleic $CH_3 \cdot (CH_2)_4 CH{=}CH \cdot CH_2 \cdot CH {=} CH \cdot (CH_2)_7 COOH$.

(b) The glycerides

The alcoholic hydroxyl groups of the trihydric alcohol, glycerol, $CH_2 OH \cdot CHOH \cdot CH_2 OH$ are involved in ester linkages with fatty acids. When each of the three hydroxyl groups forms an ester with a fatty acid, a triglyceride (neutral fat) is obtained. This class of lipid is quantitatively the most important in a normal dietary intake and is the form in which lipids

---

* Because of the extensive use of formulae in this chapter the latter are written for clarity in a rather more extended form than elsewhere in the book.

are mainly found in the subcutaneous and intraperitoneal adipose tissue. Triglycerides have the general structure

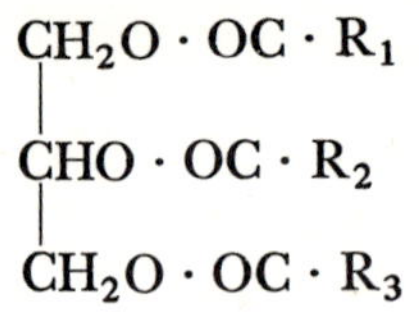

$$CH_2O \cdot OC \cdot R_1$$
$$CHO \cdot OC \cdot R_2$$
$$CH_2O \cdot OC \cdot R_3$$

where the fatty acid chain residues, R_1, R_2 and R_3 may be the same or different.

Mono- and di-glycerides in which one or two of the hydroxyl groups of glycerol are respectively involved in ester linkages with fatty acids are also known to exist in tissues in small amounts and play a physiological role particularly in the intestinal absorption of fats.

Glycerides are found throughout the plant and animal kingdoms and are essential to living matter. They vary in composition from species to species and appear to be a biological solution to the problem of storing, transporting and utilizing the fatty acids required for metabolic processes. Glycerides are used normally for the economic storage of biological energy yielding approximately 38 kJ (9 kcal) per g compared with 17 kJ (4 kcal) for carbohydrate and 23 kJ (5·5 kcal) for protein.

(c) Phospholipids

The existence of well-defined groups of phosphorus-containing lipids (phospholipids or phosphatides) such as the lecithins, cephalins and sphingomyelins has been recognized for a long time but our knowledge of their chemical structures came comparatively late because suitable methods for their purification were not initially available.

Phospholipids may be conveniently classified into two groups according to their chemical structure:

(i) GLYCEROPHOSPHOLIPIDS

Such as the lecithins, phosphatidyl ethanolamines, plasmalogens and inositides whose chemical structure is derived from glycerol. In the following structure R_1, R_2 etc. are fatty acid chain residues:

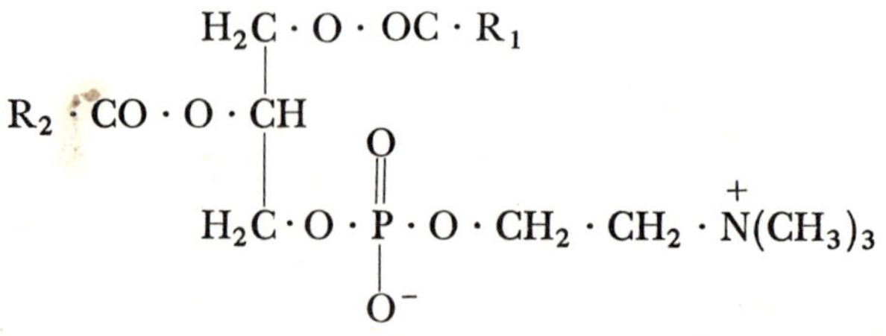

$$H_2C \cdot O \cdot OC \cdot R_1$$
$$R_2 \cdot CO \cdot O \cdot CH$$
$$H_2C \cdot O \cdot \overset{O}{\underset{O^-}{\overset{\|}{P}}} \cdot O \cdot CH_2 \cdot CH_2 \cdot \overset{+}{N}(CH_3)_3$$

Lecithins (phosphatidyl cholines)

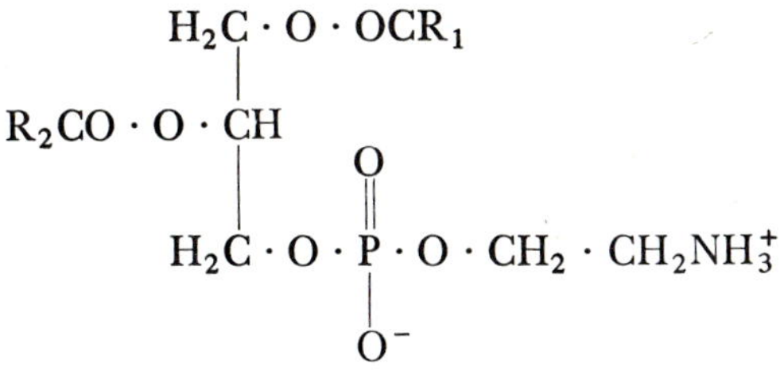

$$H_2C \cdot O \cdot OCR_1$$

$$R_2CO \cdot O \cdot CH$$

$$H_2C \cdot O \cdot \overset{\overset{\displaystyle O}{\parallel}}{\underset{\underset{\displaystyle O^-}{|}}{P}} \cdot O \cdot CH_2 \cdot CH_2NH_3^+$$

Phosphatidyl ethanolamines

The lecithins differ from the phosphatidyl ethanolamines and phosphatidyl serines in the nature of the nitrogenous compound attached to phosphoric acid. In lecithins it is choline whereas in the latter it is either ethanolamine, as shown, or the amino acid serine, $HOCH_2 \cdot CH(NH_2) \cdot COOH$.

The fatty acid residues, R_1, R_2 may be alike or different and may be saturated or unsaturated. Native lecithins generally contain both saturated and unsaturated fatty acid residues but wholly saturated lecithins have been obtained from certain mammalian tissues.

Plasmalogens (acetal phosphatides) These differ from lecithins and phosphatidyl ethanolamines in that an unsaturated ether grouping on the C-1 atom of glycerol replaces the fatty acid ester linkage

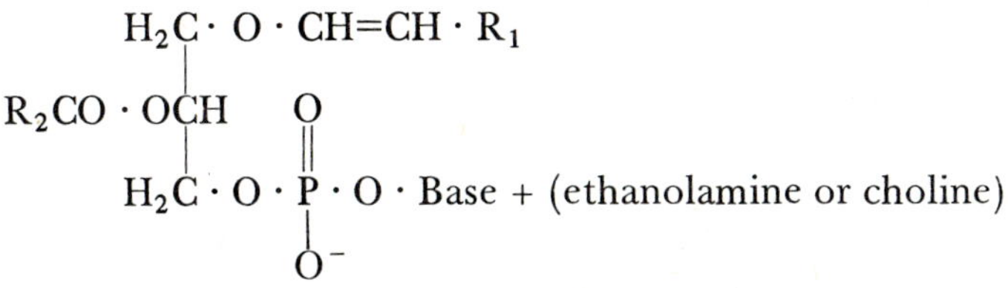

$$H_2C \cdot O \cdot CH{=}CH \cdot R_1$$

$$R_2CO \cdot OCH$$

$$H_2C \cdot O \cdot \overset{\overset{\displaystyle O}{\parallel}}{\underset{\underset{\displaystyle O^-}{|}}{P}} \cdot O \cdot Base + (\text{ethanolamine or choline})$$

General structure of plasmalogens

Plasmalogens constitute an appreciable proportion of the total phospholipids of brain, nerve and muscle.

Inositol phospholipids These exist in three forms; monophosphoinositides, diphosphoinositides and triphosphoinositides based on inositol mono-, di- and triphosphates respectively. The compounds are analogous to the phosphatidyl cholines and phosphatidylethanolamines with inositol replacing the nitrogenous bases, e.g., monophosphoinositides. They have the structure shown

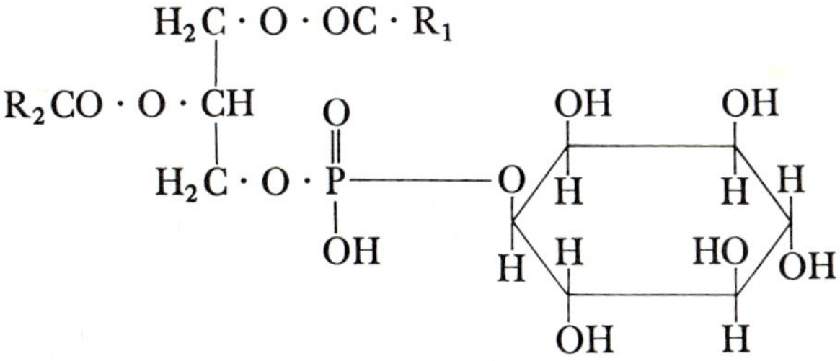

General structure of monophosphoinositides

(ii) SPHINGOLIPIDS (SPHINGOMYELINS AND CEREBROSIDES)

These compounds contain the base, sphingosine, which is a long-chain unsaturated amino alcohol of chemical structure $CH_3 \cdot (CH_2)_{12} \cdot CH=CH \cdot CH(OH) \cdot CH(NH_2) \cdot CH_2OH$.

Sphingomyelins differ from the lecithins in that sphingosine replaces glycerol. A fatty acid is linked to the amino group of the base and phosphorylcholine is attached to the terminal hydroxyl group.

$$CH_3 \cdot (CH_2)_{12} \cdot CH=CH \cdot CH(OH) \cdot \underset{\underset{R \cdot CO \cdot NH}{|}}{CH} \cdot CH_2O \cdot \overset{\overset{O}{\parallel}}{\underset{\underset{O^-}{|}}{P}} \cdot O \cdot CH_2CH_2 \cdot \overset{+}{N}(CH_3)_3$$

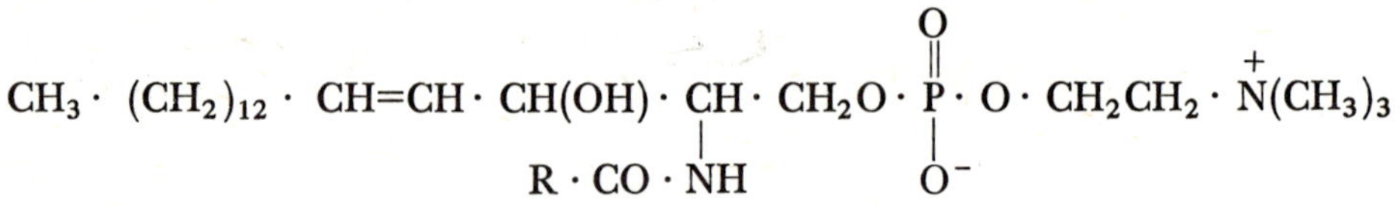

General structure of sphingomyelins

Sphingomyelins are the most common sphingosine derivatives containing phosphorus but corresponding ethanolamine derivatives (Sphingoethanolamines) have been found in egg yolk. Sphingomyelins differ from each other in the nature of their constituent fatty acids. Very long-chain acids, including α-hydroxy acids, predominate. These lipids occur particularly in brain but are also present in kidney, liver and other organs.

Cerebrosides contain no phosphoric acid or choline. These groups are replaced by a hexose sugar, generally galactose although glucose-containing cerebrosides are known. The hexose is attached to the terminal hydroxyl group of sphingosine by an ether linkage. The acid present is one of a number of unsaturated or hydroxy acids containing 22 or more carbon atoms.

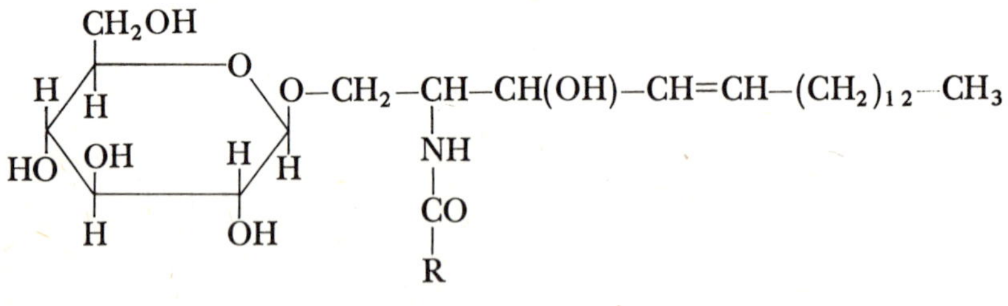

General structure of cerebrosides

The cerebrosides are found in most tissues but, as their name implies, are more abundant in brain and nerves. Another group of more complex representatives of the sphingoglycolipids, known as the gangliosides, are also found particularly in nervous tissue; these are characterized by possessing one or more sialic acid residues as part of their molecular structure.

(d) Sterols

The sterols are secondary alcohols which are found in the free form and as esters of fatty acids. They occur in association with lipids and have similar solubility properties. For this reason they may be regarded as derived lipids.

The sterols are examples of a large and important group of compounds known as the steroids, which have a common polycyclic ring structure. The naturally-occurring steroids which include the sterols, the bile acids, the adreno-cortical steroids and the sex hormones, take part in a wide variety of biological processes.

The basic carbon ring structure characteristic of the steroids is shown:

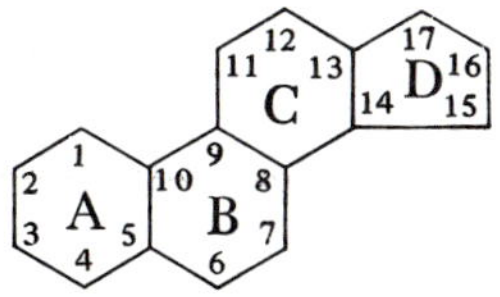

General basic structure of a steroid

Four substituent groups are usually attached to the ring structure at carbon atoms 3, 10, 13 and 17. In some steroids, groups are also present in other positions. Methyl groups are usually the substituents at positions 10 and 13. Position 3 may be occupied by a hydroxyl group or a keto group while at position 17 there may be one of a variety of groups.

Most steroids occur only in trace amounts in cells but the sterols are extremely abundant. Sterols contain an alcoholic hydroxyl group at C-3 and a branched aliphatic chain of 8 or more carbon atoms at C-17; they occur either as free alcohols or as long-chain fatty acid esters of the C-3 hydroxyl group.

Cholesterol is the most abundant sterol in animal tissues and occurs in the free and combined forms.

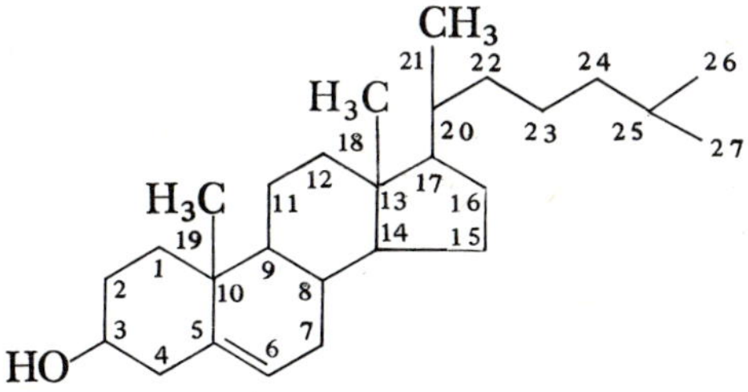

Cholesterol

Cholesterol is a white, crystalline, optically active solid—insoluble in water but readily extractable from cells by organic solvents. It is found in abundance in the plasma membranes of many animal cells and, in much smaller amounts, in the membranes of mitochondria and the endoplasmic reticulum. Cholesterol is not found in plants, fungi or yeasts which contain other sterols including ergosterol which is converted to vitamin D_2 (ergocalciferol) on irradiation by sunlight. No sterols are found in bacteria.

(e) Prostaglandins

For some 40 years it has been known that extracts of semen will cause contraction of the gut and also a drop in blood pressure when administered to rabbits. Mainly due to Bergstrom and his group in Sweden, the structure of these lipid compounds has now been elucidated and they are known under the general name of prostaglandins. They are C_{20} compounds and the structure of prostaglandins is shown.

Structure of a typical prostaglandin (PGE_1)

They are very easily dehydrated and then lose biological activity. There are six different prostaglandins of importance which differ in the number and position of —OH and =O groups and in the number of double bonds. The prostaglandins are found in many different tissue extracts, e.g., eye, kidney, lung and normal human serum. They are also found in a coral in the Gulf of Mexico and this provides the commercial source. As might be expected, the fatty acids are the biological precursors, so that if arachidonic acid is incubated with sheep prostate gland, the E_2 type of prostaglandin is synthesized (the intracellular site of synthesis is the endoplasmic reticulum). The E type compounds reduce the blood pressure while the F type compounds lack this effect but have other activities.

It seems that a particular tissue synthesizes a specific prostaglandin and this has a local activity. Once the prostaglandin passes into the general circulation, it is degraded and loses activity.

8.2. Importance of Fatty Acids as a Source of Energy

Except during the few hours after the ingestion of carbohydrates, the oxidation of long-chain fatty acids supplies more than half the total energy needs of the body. The normal quantity of fatty acid stored in the human body, largely in the intraperitoneal and subcutaneous tissues, will provide the energy needs for several weeks. A 70 kg man, for example, may contain approximately 6 kg of triglycerides in adipose tissue alone although, as observation will readily confirm, the size of the fat stores is a very variable factor. If we assume that a sedentary worker requires about 8400 kJ (2000 kcal)* per day to provide his energy needs, then, since each g of fat would supply 38 kJ of energy, this would be obtained by the oxidation of only 220 g of fat. Hence the fat stores in adipose tissue

* Under the S.I. system a kcal is not defined but it is equal to 4·2 kJ.

can supply the energy needs for over 27 days. The importance of fat metabolism to the total economy of the body can thus be readily appreciated.

Whilst it has been recognized by stockbreeders for over a century that carbohydrates can be converted by the body to fat, it is important to note that net synthesis of carbohydrate from fat cannot take place in man. The reason for this is that glucose can be readily converted by glycolysis to pyruvic acid and hence to acetyl CoA which takes part in fatty acid biosynthesis; the irreversible nature of the pyruvate → acetyl CoA reaction prevents the reverse process.

The term "fat" is used for the triglycerides which are the most labile class of lipids. The structural lipids, comprising cholesterol and its esters, the phospholipids and the glycolipids, are also important since they are found in all cells; even when the triglyceride reserves are completely exhausted following a prolonged fast, the structural lipids do not disappear. The precise function of the structural lipids, apart from being a part of cellular material, is not yet fully known (but see Chapter 11.2).

8.3. The Digestion and Absorption of Dietary Fat

The bulk of dietary fat is in the form of triglycerides. No lipase is present in saliva or gastric juice although regurgitated pancreatic lipase has sometimes been found in the stomach. Hence, although ingested lipids are warmed to body temperature in the stomach, thus softening them and assisting in their subsequent emulsification, no digestive action on the triglycerides occurs until the stomach contents reach the duodenum when pancreatic juice and bile are secreted. The presence of the bile salts and the bile phospholipids in the duodenum assist in the production and stabilization of an emulsion of the triglycerides 500-1000 nm in diameter. Emulsification exposes a large surface area of triglyceride to the action of pancreatic lipase which accumulates at the oil-water interface. The enzyme specifically acts upon the 1 and 3 ester bonds of the triglyceride producing a mixture of 1,2-diglyceride and 2-monoglyceride. The secondary alcoholic 2-ester linkage is hydrolyzed only slowly, migration of the fatty-acyl residue from the C-2 to C-1 position being necessary before hydrolysis. The chemical processes occurring in the lumen of the small intestine are represented schematically in Fig. 8.1.

The main products of hydrolysis are free fatty acids and 2-monoglycerides which, together with the bile salts and bile phospholipids, are dispersed into micelles of diameter 3-8 nm which enter the cells of the intestinal mucosa. Within the cells, most of the free fatty acids and monoglycerides from the micelles are resynthesized to triglycerides in association with the endoplasmic reticulum. The triglycerides are then released into the bloodstream with phospholipid and protein as droplets which are essentially chylomicrons (lipoproteins). The bile salts from the micelles are released again into the intestinal lumen and absorbed lower down the intestine.

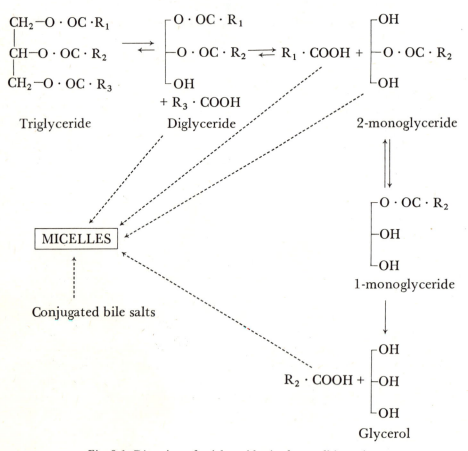

Fig. 8.1 Digestion of triglycerides in the small intestine.

Biosynthesis of triglycerides in the intestinal cells   This can be achieved using as substrate either monoglycerides or α-glycerophosphate (Fig. 8.2). In each case, free fatty acid has first to be converted into an acyl CoA derivative:

$$R \cdot COOH + CoA—SH + ATP \xrightarrow{\text{thiokinase}} R \cdot CO—SCoA + AMP + PP_i$$

Fatty acid   Acyl CoA

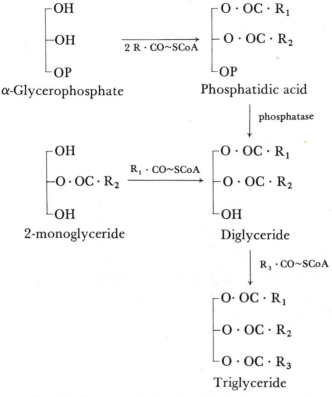

α-Glycerophosphate   Phosphatidic acid

2-monoglyceride   Diglyceride

Triglyceride

Fig. 8.2 Biosynthesis of triglycerides in intestinal cells.

The α-glycerophosphate (i.e., L-glycerol 3-phosphate) may be formed either from glycerol or from glucose:

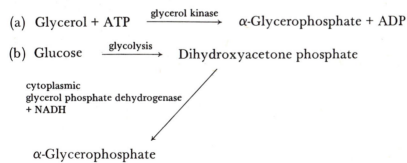

(a) Glycerol + ATP $\xrightarrow{\text{glycerol kinase}}$ α-Glycerophosphate + ADP

(b) Glucose $\xrightarrow{\text{glycolysis}}$ Dihydroxyacetone phosphate

cytoplasmic
glycerol phosphate dehydrogenase
+ NADH

α-Glycerophosphate

Formation of chylomicrons The resynthesized triglyceride is formed into droplets of 100-500 nm in diameter known as chylomicrons which also contain about 2% protein and 7% phospholipid. Since inhibitors of protein synthesis such as puromycin reduce fat absorption, it is likely that the synthesis of protein is required in the formation of chylomicrons and that they are representative of a class of lipid-protein complexes called lipoproteins. Chylomicrons then pass through the base of the mucosal cells into the extracellular fluid and hence into the lymphatics. They are only present in the blood plasma of normal subjects on a normal diet during the absorptive state.

Most of the long-chain fatty acids are absorbed into the lymph as chylomicrons but a small proportion is not reesterified in the mucosa and enters the portal blood where it is bound to albumin.

The fat-soluble vitamins, A, D, E and K and, to a limited extent, cholesterol probably enter the intestinal cells with phospholipid in a different type of micelle from that in which fatty acids are absorbed. Fine dispersion into micelles in the small intestine seems to be a necessary pre-requisite for their absorption which is greatly reduced when micelle formation is depressed. Thus obstructive jaundice, when the bile salt concentration in the gut lumen is decreased, is frequently accompanied by deficiencies of vitamins, A, D and particularly K, which are stored in the body only to a very limited extent. Some absorption of triglyceride still occurs in the absence of bile but negligible absorption of the fat-soluble vitamins takes place under this circumstance. In contrast, pancreatic steatorrhoea due to lipase deficiency and the consequential impairment of triglyceride hydrolysis, results in a failure to absorb triglyceride but not the fat-soluble vitamins.

8.4. Plasma Lipids

Blood plasma, in the post-absorptive state, normally contains the following lipids in a concentration (mg/100 ml) within the limits stated:

Triglycerides (10-150); non-esterified fatty acids (10-17); free and esterified cholesterol (150-250); bile acids (0·2-3); bile salts (5-12); total phospholipids (150-250); total plasma lipids therefore vary between 320 and 700 mg/100 ml.

The chylomicrons are normally a small and transient contributor to the plasma lipids but in subjects on high fat diets or suffering from uncontrolled diabetes they may be present in very large amounts. With the exception of the liver and the brain, most tissues effect the hydrolysis of the triglycerides of chylomicrons and absorb the resulting free fatty acids (FFA). This hydrolysis is effected by a lipase which is located at the luminal surface of the capillary endothelial cells of the tissues concerned. This lipase is called "clearing factor lipase" or "lipoprotein lipase". The most active tissues in this respect are muscle and adipose tissue. In muscle, the fatty acids are mainly oxidized for energy, while in adipose tissue, they are reconverted to triglyceride and stored as such. The extent of uptake by adipose tissue is high in the fed state and low in the fasting, whereas the uptake by muscle is high in the fasting state and low in the fed. These changes in the extent of uptake are achieved by variations in the activity of clearing factor lipase in these two tissues brought about through hormonal controls (see Chapter 13.4).

Triglycerides are also carried in the plasma in other lipoproteins called very low density (VLD) lipoproteins. These are smaller than the chylomicrons and they are released from the liver rather than the intestine. However, the triglycerides carried by the VLD-lipoproteins are dealt with in a similar fashion to chylomicron triglyceride, being taken up by muscle or adipose tissue according to the nutritional state.

Other lipoproteins (α or high density, β or low density) are also present in plasma. The high and low density lipoproteins are probably produced from the very low density lipoproteins as the triglyceride in the latter is removed, although high density lipoproteins may also have independent functions. Most of the plasma cholesterol is in the lipoproteins of low density whereas the high density lipoproteins are rich in phospholipids.

The FFA in plasma which are bound to serum albumin are derived almost entirely from adipose tissue. This tissue contains another lipase, called "mobilizing lipase" (formerly called hormone-sensitive lipase). The rate of release of the FFA from the fat depots depends on the activity of the mobilizing lipase which is controlled by hormones (see Chapter 13.4). The metabolic turnover of FFA is very high (biological half-life of 2-3 min.); despite their low concentration in plasma, this rate of turnover is sufficient to account for all the fat oxidized by a fasting animal. The factors affecting the concentration of fatty acids in plasma are shown diagrammatically in Fig. 8.3.

To summarize, therefore, fatty acids are transported in the blood in three main forms:

  (i) as FFA attached to plasma albumin
 (ii) as triglyceride fatty acids carried in chylomicrons, released from the intestine into the thoracic duct thence into the plasma
(iii) as triglyceride fatty acids in very low density lipoproteins released from the liver into the plasma.

During fasting, the FFA are mobilized from adipose tissue and pass to the extra-hepatic tissues where they are oxidized to CO_2 and water or to the liver where they are either oxidized to CO_2 and ketone bodies or are reesterified to triglycerides. Approximately 30% of the FFA enter the liver, the uptake being determined by the plasma concentration and the blood flow. Any re-formed triglycerides are re-transported from the liver as VLD lipoproteins. They are then acted upon by clearing factor lipase and the resulting FFA are oxidized by the extra-hepatic tissues.

During conditions when the energy intake exceeds the energy utilization, triglyceride fatty acids are present in plasma as chylomicrons and as very low density lipoproteins (which have come from the liver after transformation of dietary glucose into fatty acids and then into tri-

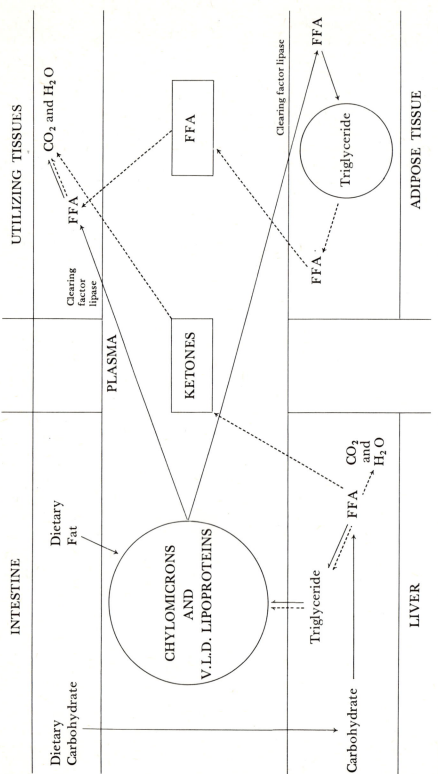

Fig. 8.3 Predominant patterns of fat transport in energy surplus (——) and energy deficient (- - -) situations.

glycerides). Both forms of triglyceride are acted upon by clearing factor lipase in the adipose and other utilizing tissues yielding FFA which are resynthesized to triglyceride or oxidized to CO_2 in these tissues.

8.5. The Fatty Acid Oxidation Cycle

(a) Initial activation

The first step is the activation of the fatty acid by its condensation with CoA according to the equation:

$$R \cdot COOH + HS-CoA + ATP \xrightarrow{\text{Thiokinase}} R \cdot CO-S \cdot CoA + AMP + PP_i$$

Fatty acid $\qquad\qquad\qquad\qquad\qquad\qquad$ Acyl CoA

This reaction, which occurs in the cytoplasm of the cells, is catalyzed by fatty acid thiokinase. At least three different fatty acid thiokinases with almost identical properties have been isolated, each specific for a fatty acid chain of a given length. The most important is the long-chain fatty acid thiokinase which catalyzes the activation of C_{12}-C_{22} acids.

The pyrophosphate formed in the reaction is hydrolyzed by pyrophosphatase giving two molecules of inorganic phosphate. The net effect of pyrophosphatase action is to prevent the reverse reaction and thus to pull the overall equilibrium of the activation step far in the direction of fatty acyl-CoA formation. There is evidence that the hydrolysis of pyrophosphate is very important in the overall activation process since pyrophosphatase inhibitors such as fluoride inhibit oxidation of free fatty acids.

Since long-chain fatty acids possess only a very limited capacity to cross the inner membranes of mitochondria as CoA esters, a transfer of the fatty acyl group occurs from CoA to a carrier molecule, carnitine (γ-trimethylamino-β-hydroxybutyric acid) $(CH_3)_3 \overset{+}{N} \cdot CH_2 \cdot CH(OH) \cdot CH_2 \cdot COOH$. This transfer is catalyzed by fatty acyl CoA-carnitine transferase, as follows:

$$R \cdot CO \cdot SCoA + (CH_3)_3 \overset{+}{N} \cdot CH_2 CH(OH) \cdot CH_2 COOH \rightleftharpoons$$

$\qquad$ Acyl CoA $\qquad\qquad$ Carnitine

$$CoA-SH + (CH_3)_3 \overset{+}{N} \cdot CH_2 \cdot \underset{\underset{O \cdot OC \cdot R}{|}}{CH} \cdot CH_2 COOH$$

$\qquad\qquad\qquad\qquad\qquad\qquad\qquad$ Fatty-acyl carnitine ester

The free energy change in this reaction is small and so a high-energy acyl-carnitine linkage is produced. The fatty-acyl carnitine ester readily passes through the inner membrane into the mitochondrion where the fatty acyl group is transferred back from carnitine to intramitochondrial

CoA by the action of intramitochondrial carnitine fatty-acyl CoA trans-ferase.

The fatty acyl-CoA is then used as substrate by the β-oxidation enzyme complex which is located in the inner matrix compartment of the mitochondrion. The reaction proceeds as follows:

(b) First dehydrogenation step

$$R \cdot CH_2 \cdot CH_2 \cdot CH_2 \cdot CO \cdot SCoA + FAD \rightleftharpoons$$
Fatty acyl CoA

$$R \cdot CH_2 \cdot CH{=}CH \cdot CO \cdot SCoA + FAD \cdot H_2$$
Unsaturated fatty acyl CoA

The hydrogen acceptor is FAD which is the prosthetic group of the fatty acyl-CoA dehydrogenase which catalyzes this reaction. Four different FAD-containing dehydrogenases have been isolated, each specific for a given chain length of fatty acid. The $\Delta^{2:3}$ unsaturated bond produced is in the *trans* configuration. (Note: this system of nomenclature means that the double bond is between C-2 and C-3 in the chain with the carboxyl group containing C-1.)

The reduced FAD is reoxidized by a second flavoprotein, electron-transferring factor (EFT) which transfers reducing equivalents to the cytochrome c region of the hydrogen transport system and hence ultimately to atmospheric oxygen (see Chapter 9.6).

(c) Hydration step

The addition of water to the unsaturated fatty acyl-CoA ester is a stereospecific step which produces the L stereoisomer of the 3-hydroxy acyl-CoA by the action of the enzyme enoyl hydratase

$$R \cdot CH_2 CH{=}CH \cdot CO \cdot SCoA + H_2O \rightleftharpoons$$
Unsaturated fatty acyl CoA

$$R \cdot CH_2 \cdot \overset{\overset{\displaystyle OH}{|}}{CH} \cdot CH_2 \cdot CO \cdot SCoA$$
3-hydroxy acyl CoA

This enzyme cannot hydrate $\Delta^{3:4}$-unsaturated esters. Although it is able to hydrate $\Delta^{2:3}$-*cis* esters, this would give the corresponding D stereo-isomer and, since the second dehydrogenation step in fatty acid oxidation is specific for the L stereoisomer, the production of this stereoisomer would be abortive in fatty acid oxidation.

(d) Second dehydrogenation step

$$R \cdot CH_2 \cdot \overset{\overset{\displaystyle OH}{|}}{CH} \cdot CH_2 \cdot CO \cdot SCoA + NAD^+ \rightleftharpoons$$

3-hydroxy acyl CoA

$$R \cdot CH_2 \cdot \overset{\overset{\displaystyle O}{\|}}{C} \cdot CH_2 \cdot CO \cdot SCoA + NADH + H^+$$

3-keto acyl CoA

The L-3-hydroxy fatty acyl-CoA dehydrogenase which catalyzes this reaction is relatively non-specific for chain length but absolutely specific for the L-stereoisomer. The NADH is reoxidized by donating its electron equivalents to the NADH dehydrogenase of the respiratory chain.

(e) Thiolytic cleavage step

The final step in the cycle, catalyzed by β-ketothiolase, results not only in chain-shortening by two carbon atoms but also produces two thio-esters of CoA. Thus the process of oxidation can be repeated

$$R \cdot CH_2 \cdot CO \cdot CH_2 \cdot CO \cdot SCoA + H \cdot SCoA \rightleftharpoons$$

3-keto acyl CoA

$$R \cdot CH_2 \cdot CO \cdot SCoA + CH_3CO \cdot SCoA$$

Fatty acyl CoA   acetyl CoA

The final product of oxidation from most naturally-occurring fatty acids which contain an even number of carbon atoms will be two molecules of acetyl-CoA whose normal fate will be to enter the tricarboxylic acid cycle by condensing with oxaloacetate to give citrate. An acid containing an odd number of carbon atoms, however, will yield propionyl-CoA and acetyl-CoA in the final stage.

Propionyl-CoA, which is also the catabolic product of certain amino acids such as valine and isoleucine, is carboxylated to methylmalonyl-CoA as follows:

$$CH_3CH_2 \cdot CO \cdot SCoA + CO_2 \longrightarrow CH_3\overset{\overset{\displaystyle COOH}{|}}{CH} \cdot CO \cdot SCoA$$

Propionyl CoA   Methylmalonyl CoA

This reaction is catalyzed by the biotin-containing enzyme, propionyl carboxylase, and the methylmalonyl-CoA is then converted by means of an unusual reaction to succinyl-CoA which, in turn, yields succinic acid through the action of succinyl thiokinase.

$$CH_3 \cdot \overset{\overset{\displaystyle COOH}{|}}{CH} \cdot CO \cdot SCoA \longrightarrow CoAS \cdot OC \cdot CH_2 \cdot CH_2 \cdot COOH$$

Methylmalonyl CoA   Succinyl CoA

The succinic acid is, of course, an intermediate in the tricarboxylic acid cycle and these three reaction steps account for the glycogenic property of propionyl-CoA.

The conversion of methylmalonyl-CoA to succinyl-CoA is catalyzed by the enzyme, methylmalonyl-CoA isomerase which has a vitamin B_{12} coenzyme, cobamide. In the reaction, the thio-ester group, $(-CO \cdot SCoA)$ migrates from C-2 to C-3 of methylmalonyl-CoA in exchange for a hydrogen atom. Such an exchange between a hydrogen atom on one carbon atom and a substituent group of a vicinal carbon atom seems to be characteristic of all known reactions in which cobamide is the coenzyme. When the methylmalonyl-CoA isomerase reaction is deficient, such as in patients suffering with pernicious anaemia who are deficient in vitamin B_{12}, large quantities of methylmalonic and propionic acids are excreted in the urine.

Each successive cycle of β-oxidation requires a fresh molecule of CoA for the final thiolytic cleavage step. Since the CoA concentration in the cell is rather small and there is no significant accumulation of any intermediate, the acetyl CoA removed from the fatty acid being oxidized must itself be rapidly metabolized and the CoA made available for subsequent β-oxidation cycles. The regeneration of free CoA mainly occurs when acetyl CoA combines with oxaloacetate to enter the tricarboxylic acid cycle.

(f) Balance sheet for the oxidation of palmitic acid

During one turn of the β-oxidation cycle, one molecule of acetyl CoA and two pairs of hydrogen atoms are removed from palmitoyl CoA. For the complete breakdown of the C_{16} acid, therefore, seven turns of the cycle are required to produce 8 molecules of acetyl CoA. Fourteen pairs of hydrogen atoms are removed and enter the respiratory chain, seven pairs in the form of reduced FAD coenzyme of fatty acyl-CoA dehydrogenase and seven pairs in the form of the reduced NAD of β-ketoacyl dehydrogenase. The passage of electrons from $FAD \cdot H_2$ to oxygen leads to generation of two and that from NADH to O_2 to three molecules of ATP.

Hence the equation is

$$\text{Palmitoyl-CoA} + 7\text{CoA—SH} + 7O_2 + 35P_i + 35\text{ADP} \longrightarrow$$
$$8 \text{ acetyl CoA} + 35\text{ATP} + 42H_2O.$$

When a molecule of acetyl CoA enters the tricarboxylic acid cycle and is oxidized to CO_2 and water a further 12 molecules of ATP are produced by substrate-level and oxidative phosphorylation as follows:

$$\text{Acetyl CoA} + 2O_2 + 12\text{ADP} + 12P_i \longrightarrow$$
$$\text{CoA} \cdot \text{SH} + 12\text{ATP} + 2CO_2 + 13H_2O.$$

For the complete oxidation of palmitoyl-CoA to CO_2 and water, the overall equation, therefore, becomes:

Palmitoyl-CoA + $23O_2$ + $131P_i$ + $131ADP \longrightarrow$

$$CoA \cdot SH + 16CO_2 + 131ATP + 146H_2O$$

Since 1 molecule of ATP is required for the original activation of 1 molecule of free palmitic acid, the net yield of ATP is 130 molecules. 40% of the standard free energy of the oxidation is recovered as the phosphate bond energy of ATP $((130 \times 30\cdot7/9830) \times 100)$ where $\Delta G^{0\prime}$ for palmitic acid $\rightarrow 16CO_2 + 16H_2O$ is -9830 kJ $(-2340$ kcal$)$. (It may be more correct to say that the net yield is 129 ATP since in the initial activation ATP $\rightarrow$ AMP which is the equivalent of 2 ATP $\rightarrow$ 2 ADP.)

8.6. Metabolism of the ketone bodies

The mammalian liver diverts some of the acetyl-CoA produced from fatty acid oxidation into free acetoacetate, β-hydroxybutyrate and acetone which are transported in the blood to peripheral tissues where they can be oxidized by the tricarboxylic acid cycle. Acetoacetic acid, β-hydroxybutyric acid and acetone are frequently called the ketone bodies.

The parent compound of the group, acetoacetic acid, in the form of its thio-ester with CoA, arises metabolically in two ways: some from the last four carbon atoms during the oxidation of long-chain fatty acids but mainly from the β-ketothiolase catalyzed condensation of two molecules of acetyl-CoA.

$$CH_3 \cdot COSCoA + CH_3 \cdot CO \cdot SCoA \rightleftharpoons$$
Acetyl CoA   Acetyl CoA

$$CH_3 \cdot CO \cdot CH_2CO \cdot SCoA + CoA{-}SH$$
Acetoacetyl CoA

The major pathway concerned in the deacylation of acetoacetyl-CoA to free acetoacetic acid involves β-hydroxy-β-methyl glutaryl-CoA (HMG–CoA) as "intermediate".

$$CH_3 \cdot CO \cdot CH_2 \cdot COSCoA + CH_3COSCoA \rightleftharpoons$$
Acetoacetyl CoA   Acetyl CoA

$$HOOC \cdot CH_2 \cdot \overset{\overset{OH}{|}}{\underset{\underset{CH_3}{|}}{C}} \cdot CH_2 \cdot CO \cdot SCoA + CoA{-}SH$$

$$\downarrow$$

$$CH_3 \cdot CO \cdot CH_2 \cdot COOH + CH_3 CO \cdot SCoA$$
Acetoacetic acid   Acetyl CoA

The free acetoacetic acid can be converted either into acetone, a reaction which occurs spontaneously under physiological conditions, or into D-β-hydroxybutyric acid by the NAD-linked enzyme, β-hydroxybutyrate dehydrogenase which acts only on the free acid and yields the D stereoisomer exclusively.

The mixture of acetoacetic and β-hydroxybutyric acids then diffuses from the liver cells into the blood and passes to peripheral tissues. Normally the ketone bodies occur in only very low concentration in the blood but their concentration may rise to very high levels in conditions such as starvation and uncontrolled diabetes mellitus when the rate of formation of ketone bodies in the liver exceeds the capacity of peripheral tissues to oxidize them. This state, with high blood concentration of the ketone bodies (ketonaemia) and consequent high excretion of the compounds into the urine (ketonuria), is known as ketosis.

Normally, the peripheral tissues are able to activate the acetoacetic acid by transforming it to its CoA ester. Two enzymic mechanisms occur for this activation:

(i) $ATP + CH_3 \cdot CO \cdot CH_2COOH + CoA{-}SH \longrightarrow$
    Acetoacetic acid

$$CH_3 \cdot CO \cdot CH_2 \cdot COSCoA + AMP + PP_i$$
Acetoacetyl CoA

(ii) $HOOC \cdot CH_2 \cdot CH_2 \cdot COSCoA + CH_3 \cdot CO \cdot CH_2 \cdot COOH \rightleftharpoons$
    Succinyl CoA   Acetoacetic acid

$$CH_3 \cdot CO \cdot CH_2 \cdot COSCoA + HOOC \cdot CH_2 \cdot CH_2 \cdot COOH.$$
Acetoacetyl CoA   Succinic acid

The first reaction is the general thiokinase reaction and the second is catalyzed by succinyl-CoA-acetoacetate transferase.

The acetoacetyl CoA produced in either reaction then undergoes thiolytic cleavage into two molecules of acetyl CoA which enter the tricarboxylic acid cycle and are oxidized. It should be noted that the liver is by far the most important site in the body for the production of the ketone bodies but is unique among the organs and tissues in that it cannot utilize ketone bodies. The peripheral tissues make very good use of the ketone bodies and indeed in heart muscle, under some metabolic conditions, the ketone bodies are an important source of energy.

8.7. The Oxidation of Unsaturated Fatty Acids

The double bonds of the naturally-occurring unsaturated fatty acids are in the *cis* configuration and, for most of these acids, occur at such positions in the carbon chain that the successive removal of two carbon fragments from the carboxylic end of the molecule up to the position of the first double bond yields a $\Delta^{3:4}$-unsaturated fatty acyl CoA, e.g., oleoyl CoA has a *cis* 9-10 double bond,

$$CH_3(CH_2)_7CH=CH \cdot CH_2 \cdot CH_2 \mid CH_2 \cdot CH_2 \mid CH_2 \cdot CH_2 \mid CH_2 \cdot CO \cdot SCoA.$$

Oleoyl CoA

Removal of six carbon atoms by three turns of the β-oxidation cycle will, therefore, yield a $\Delta^{3:4}$-*cis* acid

$$CH_3 \cdot (CH_2)_7 \cdot CH=CH \cdot CH_2 \cdot COSCoA.$$

3:4-*cis*-dodecenoyl CoA

Since enoylhydratase cannot hydrate $\Delta^{3:4}$-unsaturated fatty acyl-CoA esters and would, in any case, produce a D stereoisomer of the 3-hydroxyl acyl-CoA from the *cis* double bond (whereas fatty acyl-CoA dehydrogenase is absolutely specific for the L-stereoisomer), it is not possible for the oxidation of unsaturated fatty acids to follow an identical pathway to that for saturated acids.

In fact, the anomaly has been resolved by the demonstration of an auxiliary enzyme, $\Delta^{3:4}$ *cis*-$\Delta^{2:3}$ *trans* enoyl-CoA isomerase, which catalyzes the conversion of a $\Delta^{3:4}$ *cis* to a $\Delta^{2:3}$ *trans* double bond. Hence, with this extra enzyme step operating, unsaturated fatty acids can be oxidized by the same oxidation cycle enzymes as the saturated acids.

The oxidation of polyunsaturated essential fatty acids such as linoleic acid, requires a second auxiliary enzyme since they contain two or more *cis* double bonds so that, even when the first $\Delta^{3:4}$ *cis* bond is converted to the $\Delta^{2:3}$ *trans* bond by $\Delta^{3:4}$ *cis*-$\Delta^{2:3}$ *trans* enoyl-CoA isomerase, the removal of the next molecule of acetyl-CoA would produce an 8-carbon thio-ester with one remaining double bond, admittedly in the $\Delta^{2:3}$ position, but still with a *cis* configuration. Hence a second auxiliary enzyme, 3-hydroxy fatty acyl-CoA epimerase, is required to change the configuration of the D stereoisomer of 3-hydroxy-acyl-CoA to L by catalyzing the epimerization of C-3. When this has been accomplished, the oxidation cycle can then proceed as for saturated fatty acids until 4 molecules of acetyl CoA are produced from the 8-carbon thio-ester.

$$\underset{\text{Linoleyl-CoA}}{CH_3 \cdot (CH_2)_4 \cdot \overset{cis}{CH=CH} \cdot CH_2 \cdot \overset{cis}{CH=CH} \cdot (CH_2)_7 \cdot CO \cdot S \cdot CoA}$$

$\downarrow$ *β-oxidation*

$$3\ CH_3\ COSCoA + CH_3 \cdot (CH_2)_4 \cdot \overset{cis}{CH=CH} \cdot CH_2 \cdot \overset{cis}{CH=CH} \cdot CH_2 \cdot CO \cdot SCoA$$

$\downarrow$ $\Delta^{3:4}$cis-$\Delta^{2:3}$trans *enoyl-CoA isomerase*

$$CH_3 \cdot (CH_2)_4 \cdot \overset{cis}{CH=CH} \cdot CH_2 \cdot CH_2 \cdot \overset{trans}{CH=CH} \cdot CO \cdot SCoA$$

$\downarrow$ *β-oxidation*

$$\underset{\text{Acetyl CoA}}{2\ CH_3\ COSCoA} + CH_3 \cdot (CH_2)_4 \cdot \overset{cis}{CH=CH} \cdot CO \cdot SCoA$$

$\downarrow$ *enoyl hydratase*

$$\underset{\text{D-3-hydroxyoctanoyl-CoA}}{CH_3 \cdot (CH_2)_4 \cdot \underset{OH}{\overset{H}{C}} \cdot CH_2 \cdot CO \cdot SCoA}$$

$\searrow$ *3-hydroxy fatty acyl-CoA epimerase*

$$\underset{\substack{\text{4 CH}_3\text{CO SCoA} \\ \text{Acetyl CoA}}}{} \xleftarrow{\text{β-oxidation}} \underset{\text{L-3-hydroxy octanoyl-CoA}}{CH_3 \cdot (CH_2)_4 \cdot \underset{H}{\overset{OH}{C}} \cdot CH_2 \cdot CO \cdot SCoA}$$

8.8. The Biosynthesis of Saturated Fatty Acids

Plants, moulds, yeasts and bacteria synthesize both glycerides and their component fatty acids; animals can synthesize much, but not all, of their glyceride and fatty acid requirements. They also ingest plant glycerides and modify them.

It has long been known that ingested carbohydrate can be converted in the mammal into fat. Since the capacity of higher animals to store glycogen is limited, any carbohydrate ingested in excess of the immediate energy needs is converted into fatty acids which may be stored in large amounts as triglycerides in adipose tissues.

Following the discovery of fatty acid oxidation in the mitochondrion and the role played in this oxidation by CoA it is not surprising that it was widely anticipated that the biosynthesis of fatty acids would be found to occur by a reversal of the same enzymic steps involved in their oxidation. However, it soon became clear that certain observations were inconsistent with this view.

(i) The weak capacity of isolated liver mitochondrial preparations to incorporate labelled acetate into long-chain fatty acids whereas the cytoplasmic supernatant from the same preparations effected this at a high rate.

(ii) Citrate was required for maximum rates of incorporation of acetate into long-chain fatty acids, yet the citrate was not itself incorporated into the product nor was it utilized as a reductant since the effect was not obtained by the addition of NADH or NADPH instead of citrate.

(iii) Carbon dioxide or bicarbonate was essential for fatty acid synthesis by liver extracts although labelled carbon dioxide was not itself incorporated into the product.

The role of carbon dioxide did, in fact, provide an important clue to the mechanism of fatty acid biosynthesis. Wakil and his associates demonstrated that the CO_2 was incorporated into malonyl CoA ($COOH \cdot CH_2 \cdot CO \cdot SCoA$) as one of the carboxyl carbon atoms, and that it was malonyl CoA which was the immediate precursor of the two-carbon fragments entering into fatty acid synthesis.

We now have an understanding of the intracellular location and the enzymic steps of fatty acid synthesis. Complete synthesis of long-chain saturated fatty acids from acetyl CoA occurs only in the soluble fraction of the cytoplasm. It is a process which is catalyzed by a complex of seven proteins called the fatty acid synthetase complex. The intermediates are not the thio-esters of CoA as in fatty acid oxidation but of a low-molecular weight protein known as acyl carrier protein (ACP). ACP isolated from *E. coli* is a heat-stable protein of molecular weight 10,000 whose primary structure has been established. The single sulphydryl group with which the acyl intermediates are esterified is contributed by 4′-phosphopantetheine as in CoA.

Acyl carrier protein forms a complex with one or more of the six other enzyme proteins needed for the complete synthesis of palmitic acid. The complex in yeast consists of seven, tightly associated proteins which can be dissociated into individual peptide chains but the separate sub-units are not enzymically active. In pigeon liver the complex can be dissociated into two major components without loss of activity.

The function of ACP in fatty acid synthesis is analogous to that of CoA in oxidation in that each serves as an anchor to which acyl intermediates are esterified during the reactions by which aliphatic chains are built up or broken down.

The synthesis of palmitic acid   The source of all the carbon atoms of newly synthesized palmitic acid is cytoplasmic acetyl CoA. Acetyl CoA arises inside the mitochondrion from the oxidative decarboxylation of pyruvic acid. Since, as we have previously seen, acetyl CoA cannot readily pass through the mitochondrial membrane into the cytoplasm, it is condensed with oxaloacetate to form citrate which passes through the membrane to the cytoplasm where it undergoes cleavage by the following reaction:

$$\text{Citrate} + \text{ATP} + \text{CoA} - \text{SH} \rightarrow \text{CH}_3\text{CO} \cdot \text{SCoA} + \text{ADP} + P_i + \text{oxaloacetate}.$$

Another mechanism for the transport of acetyl-CoA from the mito-chondrion to the cytoplasm involves the transference of the acetyl group to carnitine and the passage of acetylcarnitine through the mitochondrial membrane to the cytoplasm, where the acetyl group is re-transferred to cytoplasmic CoA. This process is similar to, but in the opposite direction to, that discussed for the transport of fatty acyl-CoA into the mito-chondrion for β-oxidation.

Malonyl-CoA is formed from cytoplasmic acetyl-CoA and CO_2 through the action of the biotin-containing enzyme, acetyl-CoA carboxylase. Biotin, the prosthetic group of the enzyme, serves as an intermediate carrier of CO_2 in the following two-step reaction (see also Chapter 5.4):

$$CO_2 + \text{ATP} + \text{biotin-enzyme} \rightleftharpoons \text{carboxybiotin-enzyme} + \text{ADP} + P_i$$

$$\Big\downarrow + \text{acetyl CoA}$$

$$\text{malonyl CoA} + \text{biotin-enzyme}$$

Acetyl CoA carboxylase is a relatively specific enzyme and, although it will catalyze the carboxylation of propionyl-CoA to methyl-malonyl-CoA, the latter reaction takes place at a much lower rate. The production of malonyl-CoA is the rate-limiting step for the synthesis of fatty acids and hence acetyl-CoA carboxylase is the regulatory enzyme; its activity is greatly increased by the positive allosteric modulators, citrate and isocitrate, and this accounts for the earlier observation of the stimulatory effect of citrate on fatty acid synthesis.

Acetyl CoA is the primer of palmitic acid synthesis and its methyl and carboxyl carbon atoms respectively become the C-16 and C-15 of palmitic acid, numbering from the carboxyl group as C-1. This fact indicates that chain growth begins at the carboxyl group of acetyl CoA and proceeds by the successive additions of acetyl residues at the carboxyl end of the growing chain. The acetyl residue successively added is derived from the two carbon atoms in malonyl CoA nearest the CoA group*,

$$*COSCoA$$
$$*CH_2 \qquad\qquad \text{malonyl CoA}$$
$$COOH$$

the third carbon of the unesterified carboxyl group is lost as CO_2.

Thus the free carboxyl group of malonyl-CoA is displaced by the carboxyl carbon of the acyl-CoA which is undergoing lengthening to form CO_2 and a β-ketoacyl derivative. The latter is next reduced to a β-hydroxyacyl derivative at the expense of NADPH, followed by dehydration to $\Delta^{2:3}$-unsaturated acyl derivative which is then reduced to a saturated acyl derivative by a second molecule of NADPH. The cycle is then repeated a further six times before the C_{16}-palmitic acid stage is reached.

The original condensing materials, acetyl-CoA and malonyl-CoA, are first transferred to ACP by specific enzymes and subsequent reactions of the fatty acid synthetase cycle occur in the form of the ACP derivatives as shown in Fig. 8.4.

The overall reaction for the synthesis of palmitic acid is

$$CH_3 \cdot CO \cdot SCoA + 7HOOC \cdot CH_2 \cdot CO \cdot SCoA + 14NADPH + 14H^+ \longrightarrow$$
$$\text{Acetyl CoA} \qquad\qquad \text{Malonyl CoA}$$

$$CH_3,(CH_2)_{14}COOH + 7CO_2 + 8CoA \cdot SH + 14NADP^+ + 6H_2O$$
$$\text{Palmitic acid}$$

or, since 1 molecule of ATP and 1 molecule of carbon dioxide is utilized for each molecule of malonyl-CoA produced from acetyl-CoA, the equation can be formulated as

$$8CH_3CO \cdot SCoA + 7ATP + 14NADPH + 14H^+ \longrightarrow$$
$$\text{Acetyl CoA}$$

$$CH_3(CH_2)_{14}COOH + 14NADP^+ + 8CoA \cdot SH + 7ADP + 7P_i + 6H_2O.$$
$$\text{Palmitic acid}$$

The NADPH arises principally from the NADP-dependent oxidation of glucose 6-phosphate in the cytoplasm (the pentose phosphate pathway, see Chapter 6.13). Since 14 molecules are required to form 1 molecule of palmitic acid, 7 molecules of glucose 6-phosphate must be oxidized to ribulose 5-phosphate and carbon dioxide. The pentose phosphate cycle is very active in those tissues, such as liver and adipose tissue, in which the rate of fatty acid synthesis is high. Another source of NADPH in the liver is the oxidation by malic enzyme of malate to pyruvate and CO_2.

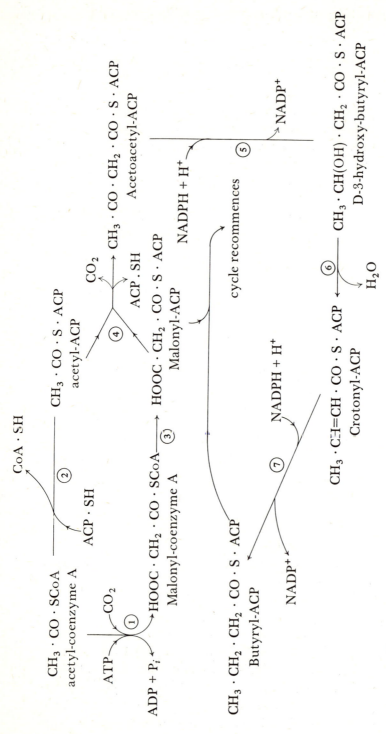

Fig. 8.4 Pathway and enzymes for the biosynthesis of fatty acids from acetylcoenzyme A.

Enzymes. ① acetylcoenzyme A carboxylase
② acetylcoenzyme A–ACP-transacylase
③ malonylcoenzyme A–ACP-transacylase
④ acetyl-malonyl-ACP condensing enzyme (β-ketoacyl-ACP synthetase)
⑤ β-ketoacyl-ACP reductase
⑥ enoyl-ACP-dehydratase
⑦ enoyl-ACP-reductase

The differences between the enzyme reactions in synthesis and oxidation of palmitic acid may be summarized as follows:

1. Intracellular location: synthesis occurring in the cytoplasm and oxidation in the mitochondrion.
2. The form in which two-carbon fragments are added or removed: as malonyl CoA in synthesis and as acetyl CoA in oxidation.
3. The stereoisomeric configuration of the β-hydroxyacyl intermediate: being the D form in synthesis and the L form in oxidation.
4. The identity of the acyl carrier: acyl carrier protein in synthesis, CoA in oxidation.
5. The pyridine nucleotide specificity of the β-ketoacyl → β-hydroxy acyl reaction: NADP in synthesis and NAD in oxidation.
6. The electron donor-acceptor system for the crotonyl-butyryl step: FAD in oxidation, NADP in synthesis.

These differences illustrate the way in which chemically-opposing processes are segregated from each other within the cell.

The lengthening of pre-existing fatty acid chains may be brought about by mitochondrial and endoplasmic reticulum enzyme systems which are unable to carry out the *de novo* synthesis of palmitic acid. The mitochondrial system lengthens both unsaturated and saturated fatty acyl-CoA derivatives containing C_{12-16} by successive additions of acetyl-CoA. Malonyl-CoA cannot replace acetyl-CoA in this system which is the reverse of the fatty acid oxidation cycle except that the reduction of the $\Delta^{2:3}$-double bond takes place at the expense of NADPH not FAD·H_2. The system located on the endoplasmic reticulum adds on malonyl-CoA to fatty acyl-CoA esters; NADPH is again the reductant of the $\Delta^{2:3}$-double bond but ACP is *not* involved as carrier as in *de novo* cytoplasmic synthesis.

8.9. Synthesis of Unsaturated Mono- and Poly-Enoic Acids

There are four types of unsaturated acids found in mammals;

(a) *Palmitoleic family* containing $CH_3 \cdot (CH_2)_5 \cdot CH=CH-$ exemplified by palmitoleic acid itself $CH_3(CH_2)_5 CH=CH(CH_2)_7 COOH$.

(b) *Oleic acid family* containing $CH_3 \cdot (CH_2)_7 \cdot CH=CH-$ e.g., oleic acid $CH_3(CH_2)_7 CH=CH \cdot (CH_2)_7 \cdot COOH$.

(c) *Linoleic acid family* containing $CH_3(CH_2)_4 \cdot CH=CH-$ e.g., linoleic acid $CH_3(CH_2)_4 \cdot CH=CH \cdot CH_2 CH=CH \cdot (CH_2)_7 COOH$.

(d) *Linolenic acid family* containing $CH_3 \cdot CH_2 \cdot CH=CH-$ as in linolenic acid itself $CH_3 \cdot CH_2 \cdot CH=CH \cdot CH_2 CH=CH \cdot CH_2 CH=CH(CH_2)_7 COOH$.

The two most common monoenoic acids, oleic and palmitoleic acids which each possess a $\Delta^{9:10}$-*cis* double bond are synthesized in the mammal from stearic and palmitic acids respectively by a specific oxygenase associated with the endoplasmic reticulum particularly of cells of liver and adipose tissue. In this reaction, 1 molecule of molecular oxygen accepts two pairs of electrons, one pair from the fatty acyl-CoA substrate and the other from NADPH, e.g.,

$$C_{15}H_{31} \cdot CO \cdot SCoA + NADPH + H^+ + O_2 \longrightarrow$$

Palmitoyl CoA

$$CH_3 \cdot (CH_2)_5 \cdot CH{=}CH \cdot (CH_2)_7 CO \cdot SCoA + NADP^+ + 2H_2O.$$

Palmitoloeyl CoA

The four types of polyenoic acids listed act as precursors of all mammalian polyenoic acids and are formed from them by further elongation and/or desaturation. The elongation is catalyzed by either the mitochondrial or the endoplasmic reticulum enzyme systems and the desaturation involves oxygenases in the presence of NADPH. Linoleic and linolenic acids cannot be synthesized by mammals at an adequate rate and must, therefore, be obtained from plant sources in the diet.

The tetra-enoic acid, arachidonic acid, which is the most abundant polyenoic acid, is synthesized from linoleic acid as follows:

Linoleic acid $(9, 12\text{-}C_{18:2})$ $\xrightarrow[-2H]{}$ γ-linolenic acid $(6,9,12\text{-}C_{18:3})$ $\xrightarrow[+C_2]{}$

$8,11,14\text{-}C_{20:3}$ $\xrightarrow[-2H]{}$ arachidonic acid $(5,8,11,14\text{-}C_{20:4})$

Linoleic, linolenic and arachidonic acids are known as the essential fatty acids since they cannot be synthesized adequately in mammals. Weanling rats placed on a fat-free diet (that is, a diet deficient in essential fatty acids) grow only slowly and develop, after 4-6 months, a scaly dermatitis which can be relieved by the administration of linoleic, linolenic or arachidonic acids. It is very difficult to feed adult rats a diet such that they become deficient with respect to fatty acids and such deficiency has never been established in man. The specific function of the essential fatty acids is not known but there has been much discussion and research into whether the relative proportions of animal and vegetable fats in the diet are factors in the genesis of coronary artery disease.

8.10. Biosynthesis of Triglycerides, Phosphoglycerides and Sphingolipids

(a) The biosynthesis of triglycerides

The rate of biosynthesis of fatty acids appears to be well integrated with the rate of formation of triglycerides and phosphoglycerides because FFA occur only in very small amounts in plasma and cells: they do not normally accumulate.

The depot fats are actively synthesized in liver and adipose cells of mammals from L-glycerol-3 phosphate and fatty acyl-CoA by the phosphatidic acid pathway as previously described for intestinal mucosa cells.

Unlike the intestinal mucosa cells of higher animals, which can acylate the monoglycerides to di- and tri-glycerides, the phosphatidic acid pathway is the sole mechanism of triglyceride biosynthesis in liver and adipose cells.

The reactions involved occur preferentially with C_{16} and C_{18} acyl-CoA derivatives. The phosphatidic acids occur in only trace amounts but are common intermediates for the synthesis of both triglycerides and phosphoglycerides.

In the synthesis of triglycerides, the phosphate group of L-phosphatidic acid is first hydrolyzed by a phosphatase and the resulting diglyceride is converted to triglyceride by a reaction with a third molecule of fatty acyl-CoA.

The biosynthesis of phosphoglycerides from L-phosphatidic acid is the next subject for discussion.

(b) Biosynthesis of phosphoglycerides

The synthesis takes place largely in the endoplasmic reticulum and cytidine triphosphate (CTP) first reacts with phosphatidic acid to give cytidine diphosphate-diglyceride (CDP-diglyceride).

$$\text{L-phosphatidic acid} + \text{CTP} \rightarrow \text{CDP-diglyceride} + \text{PP}_i$$

The cytidine monophosphate (CMP) moiety thus becomes the carrier of the phosphatidic acid and it is displaced from CDP-diglyceride by one of three alcohols; serine, inositol or glycerol phosphate yielding phosphatidyl serine, phosphatidyl inositol and 3-phosphatidyl glycerol-1'-phosphate respectively.

For example, with serine:

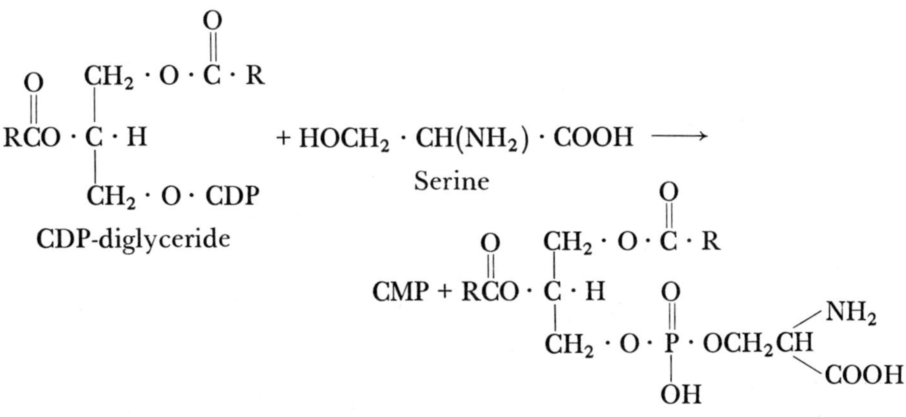

Phosphatidyl serine

The phosphatidyl serine can be decarboxylated to phosphatidylethanol-amine by an enzyme which has pyridoxal phosphate as its prosthetic group. By the consecutive transfer of three methyl groups from three molecules of S-adenosylmethionine to the amino group of ethanolamine, phosphatidylcholine can then be obtained.

Phosphatidylinositol is the precursor of two derivatives, phosphatidyl inositol monophosphate (diphosphoinositide) and phosphatidyl inositol diphosphate (triphosphoinositide) which are formed by two successive phosphorylations by ATP of free hydroxyl groups in the inositol residue. The phosphoinositides may be involved in membrane transport since they are rapidly metabolized in mitochondrial membranes and in brain.

3-Phosphatidyl glycerol-1-phosphate is dephosphorylated yielding 3-phosphatidyl glycerol which is present in many bacterial cell membranes. 3-Phosphatidyl glycerol by reaction with a second molecule of CDP-diglyceride gives diphosphatidyl glycerol (cardiolipin) which is also a major component of many bacterial cell membranes and comprises over 10% of the lipids present in the mitochondrial membrane. Cardiolipin is thought to play a role in electron transport and oxidative phosphoryla-tion.

Plasmalogens are synthesized by the reaction of CDP-choline or CDP-ethanolamine with monoacyl derivatives of an alkenyl glycerol ether, e.g.,

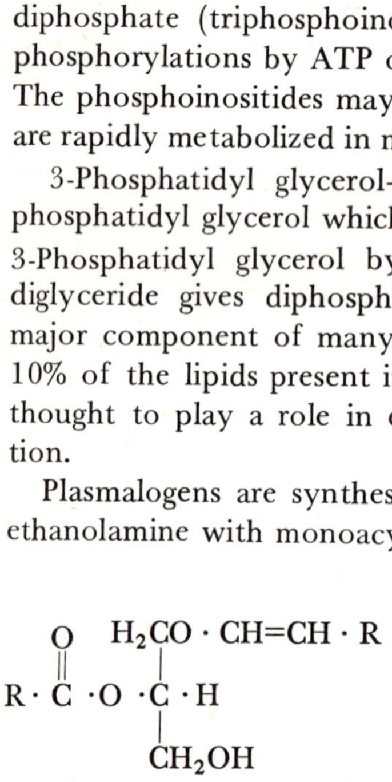

$$\begin{array}{c} \quad\quad O \quad H_2CO \cdot CH{=}CH \cdot R \\ \quad\quad \| \quad\quad\quad | \\ R \cdot C \cdot O \cdot C \cdot H \\ \quad\quad\quad\quad | \\ \quad\quad\quad CH_2OH \end{array}$$

mono acyl derivative of
alkenyl glycerol ether

$$+ \text{CDP-choline} \longrightarrow \begin{array}{c} O \quad H_2C \cdot O \cdot CH{=}CH \cdot R \\ \| \quad\quad | \\ R \cdot CO \cdot C \cdot H \cdot \quad O \\ \quad\quad | \quad\quad \| \\ \quad\quad CH_2O \cdot P \cdot O \cdot CH_2 \cdot CH_2 \cdot \overset{+}{N}(CH_3)_3 \\ \quad\quad\quad | \\ \quad\quad\quad O^- \end{array} \quad \longrightarrow \text{CMP}$$

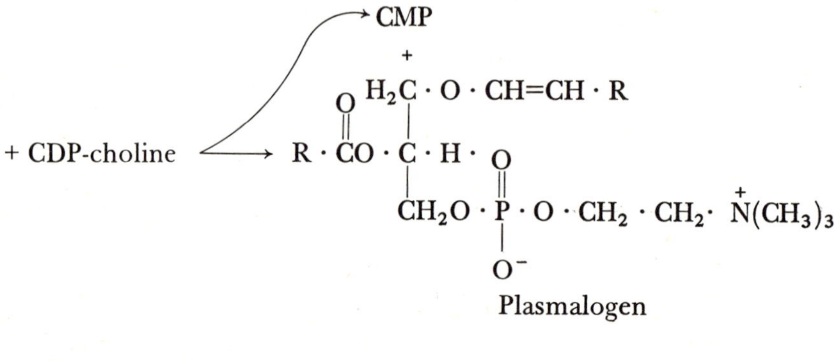

Plasmalogen

(c) Biosynthesis of sphingolipids

The building blocks of the sphingolipids are the long-chain aliphatic amino-alcohols sphingosine and dihydrosphingosine. These amines are formed from palmitoyl-CoA by the following reaction sequence:

$$CH_3 \cdot (CH_2)_{14} \cdot CO \cdot SCoA + NADPH + H^+ \longrightarrow$$
palmitoyl CoA

$$CH_3 \cdot (CH_2)_{14} \cdot CHO + NADP^+ + CoA \cdot SH$$
Palmitaldehyde

$$+ HO \cdot CH_2 \cdot CH(NH_2) \cdot COOH$$
Serine

$\downarrow$

$$CO_2 + CH_3 \cdot (CH_2)_{14} \cdot CH(OH) \cdot CH(NH_2) \cdot CH_2OH$$
Dihydrosphingosine

$\downarrow$ FAD

$$FAD \cdot H_2 + CH_3 \cdot (CH_2)_{12} \cdot CH=CH \cdot CH(OH) \cdot CH(NH_2) \cdot CH_2OH$$
Sphingosine

The amino group of sphingosine in all the sphingolipids is acylated by a long-chain fatty acyl-CoA to yield N-acyl-sphingosines, a group of compounds called ceramides.

Sphingomyelin is obtained by the reaction of a ceramide with CDP-choline:

$$\overset{\displaystyle NH \cdot OCR}{\overset{|}{CH_3 \cdot (CH_2)_{12} \cdot CH=CH \cdot CH(OH)CH \cdot CH_2OH}} + CDP\text{-choline} \longrightarrow CMP +$$
N-acyl sphingosine (ceramide)

$$CH_3 \cdot (CH_2)_{12} \cdot CH=CH \cdot CH(OH) \cdot \overset{\overset{\displaystyle NH \cdot OCR}{|}}{CH} \cdot CH_2O\overset{\overset{\displaystyle O}{\|}}{-P}-O^-$$
$$\underset{O \cdot CH_2 \cdot CH_2 \overset{+}{N}(CH_3)_3}{|}$$

Sphingomyelin

The cerebrosides, another type of sphingolipid which are hexose derivatives of ceramides, are synthesized by a reaction of a ceramide with uridine diphosphate glucose (UDPG) or UDP-galactose.

8.11. The Biosynthesis of Cholesterol

This is given in outline only and indicates the initial common pathway of biosynthesis of the ketone bodies and cholesterol.

Cholesterol is synthesized from acetyl-CoA by six enzymic steps. The first two steps in which β-hydroxy-β-methyl glutaryl-CoA is formed, have already been discussed (see Section 8.6). The third step involves the formation of mevalonic acid by the two-stage reduction of the —COSCoA group of HMG-CoA by NADPH.

$$HOOCH \cdot CH_2 \cdot \underset{\underset{CH_3}{|}}{\overset{\overset{OH}{|}}{C}} \cdot CH_2 \cdot COSCoA + 2NADPH + 2H^+ \longrightarrow$$

HMG-CoA

$$HOOC \cdot CH_2 \cdot \underset{\underset{CH_3}{|}}{\overset{\overset{OH}{|}}{C}} \cdot CH_2 \cdot CH_2OH$$

Mevalonic acid

$$+ CoA \cdot SH + 2NADP^+$$

This, or one of the first two reactions, appears to be the rate-limiting step in cholesterol synthesis since mevalonic acid is very rapidly utilized by tissues which actively synthesize cholesterol.

The fourth reaction is the formation of mevalonic acid pyrophosphate by two molecules of ATP and the conversion of the pyrophosphate to isopentenyl pyrophosphate by an ATP-requiring reaction which splits off CO_2 and water.

$$HOOC \cdot CH_2 \cdot \underset{\underset{CH_3}{|}}{\overset{\overset{OH}{|}}{C}} \cdot CH_2 \cdot CH_2 \cdot O \cdot P \cdot P \longrightarrow$$

Mevalonic acid pyrophosphate

$$\overset{CH_2}{\underset{H_3C}{>}} C \cdot CH_2CH_2O \cdot P \cdot P + H_2O + CO_2$$

Isopentenyl pyrophosphate

Isopentenyl pyrophosphate is tautomeric with dimethylallylpyrophosphate $(CH_3)_2C=CH \cdot CH_2 \cdot O \cdot P \cdot P$. These tautomers are the so-called "active isoprene units" which are key intermediates in the synthesis of cholesterol and ubiquinone in mammals and carotenoids, essential oils and rubber latex in plants.

The fifth stage results in the formation of the 15-carbon farnesyl pyrophosphate by the two-stage condensation of three isopentenyl and dimethylallyl residues.

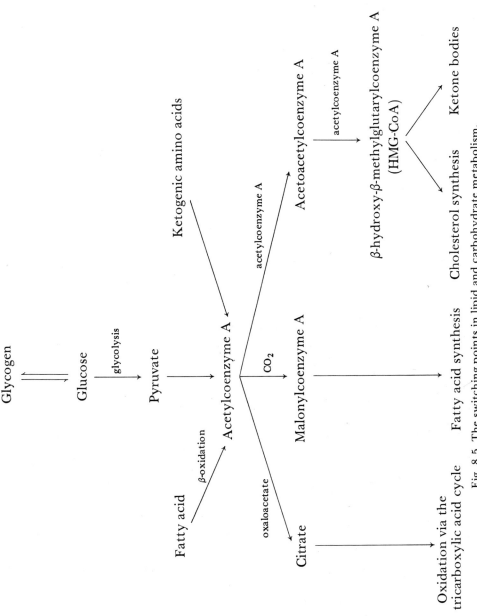

Fig. 8.5 The switching points in lipid and carbohydrate metabolism.

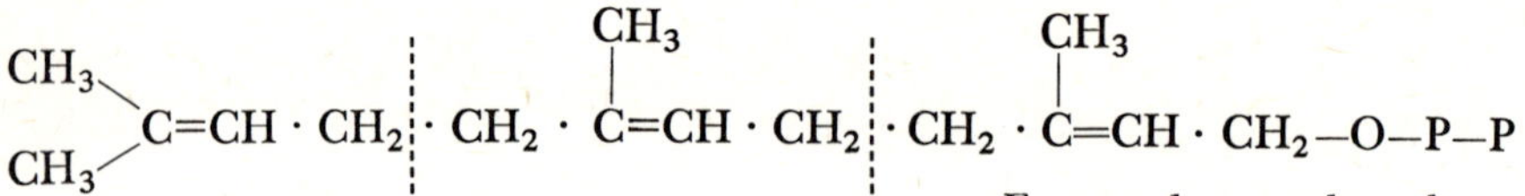

$$CH_3\diagdown \atop CH_3\diagup C{=}CH \cdot CH_2 \cdot CH_2 \cdot \overset{\overset{\displaystyle CH_3}{|}}{C}{=}CH \cdot CH_2 \cdot CH_2 \cdot \overset{\overset{\displaystyle CH_3}{|}}{C}{=}CH \cdot CH_2{-}O{-}P{-}P$$

Farnesyl pyrophosphate

In the final step, two farnesyl chains condense to give squalene which contains 30 carbon atoms. This compound is folded as shown below and simultaneous rearrangement of bonds occurs to form the four rings A, B, C and D of a C_{30} precursor of cholesterol

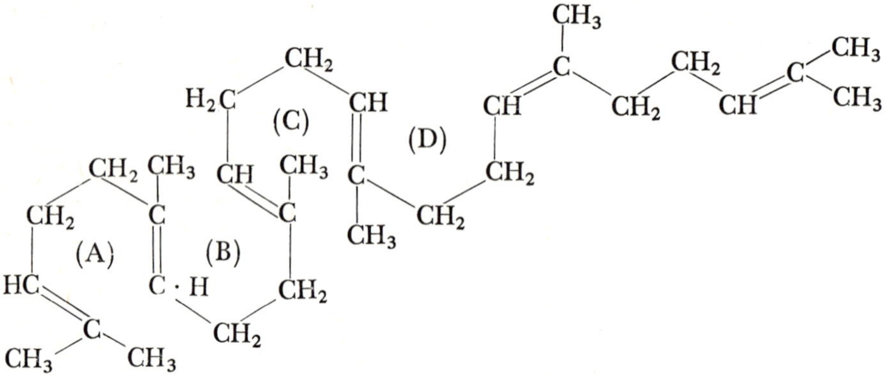

Squalene

Before the rings close, an oxygen atom is attached across the 3-4 double bond. This atom remains as an $-OH$ or $=O$ group at C_3 in all steroids. The C_{27} sterol cholesterol is then formed from the C_{30} steroid.

8.12. Integration of Lipid and Carbohydrate Metabolism

The importance of acetyl CoA and HMG-CoA as switching points in the integration of metabolic pathways is illustrated schematically in Fig. 8.5.

In fasting and uncontrolled diabetes, more acetyl CoA is produced in the liver by the oxidation of fatty acids than can be used in the tricarboxylic acid cycle. This excess acetyl CoA therefore is diverted into the HMG-CoA pathway leading to increased formation of cholesterol and the ketone bodies. The increase in cholesterol formation is very limited, so most of the HMG-CoA is converted to the ketone bodies whose blood and urine concentration may become very high in diabetes.

9

Bioenergetics of Mitochondria

In 1910, one of the pioneers of modern biochemistry, Otto Warburg, observed that the oxidation of many organic acids took place in a "large granule" fraction of tissue extracts. Yet more than 30 years passed before the development of good methods of subcellular fractionation enabled Kennedy and Lehninger to show that fatty acid oxidation was a mitochondrial function. In the last 20 years, the mitochondrion has been investigated with such remarkable thoroughness that we now have a very detailed, though as yet incomplete, understanding of its structure and function. This chapter, concerned with biological oxidation, will try to give a simplified account of the mitochondrion, its structure and biogenesis, its internal organization of enzymes and respiratory chain carriers, its relationship with the cytoplasm and its physiological function in conserving the energy derived from oxidation by regenerating ATP. When this account of the mitochondrion has been completed there will remain to be considered only a few other oxidation systems taking place elsewhere in the cell—in the endoplasmic reticulum and in the microbodies or peroxisomes; these are described in Chapter 10.

9.1. The Ultrastructure of Mitochondria

Mitochondria are a ubiquitous feature of almost all eukaryotic cells —anaerobic fungi and the mammalian erythrocyte being exceptional in this respect. If we examine electron micrographs of various tissues we are likely to be impressed by the diversity in size, shape, internal structure

(b) Muscle

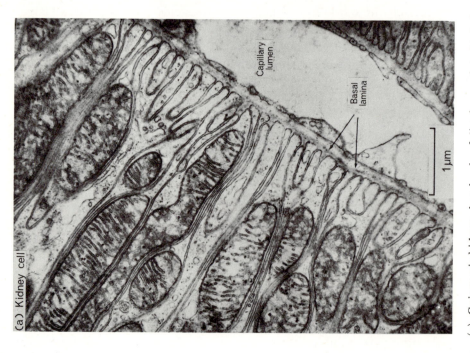

(a) Kidney cell

(a) Guinea pig kidney—basal pole of distal convoluted tubule.   (b) Cat-heart—papillary muscle from right ventricle.

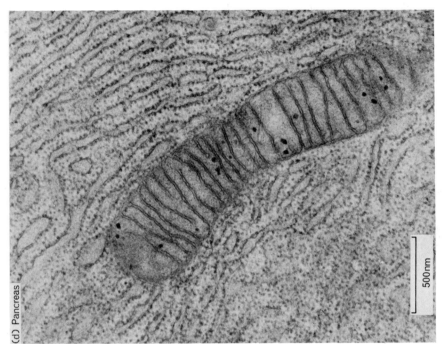

(d) Pancreas

500nm

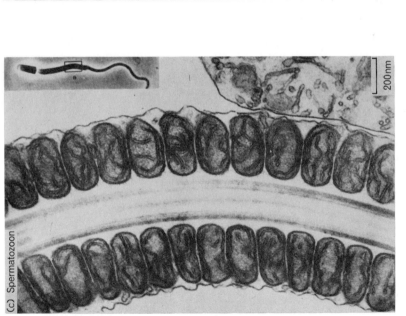

(c) Spermatozoon

200nm

(c) Spermatozoon from bat epididymis.   (d) Rat pancreas—part of exocrine cell.

(a, b and c from Fawcett.)

Fig. 9.1 Mitochondria from various tissues.

and intracellular location of mitochondria. Even so, certain logical patterns seem to emerge. Tissues with high rates of oxygen consumption are richer in mitochondria than those with low metabolic rates. Within cells, mitochondria are often located in close proximity to structures which demand a major and/or rapid supply of ATP. Some examples are shown in Fig. 9.1. In kidney tubule cells, ATP is provided by the many mitochondria packed between the infoldings of the plasma membrane as a substrate for the ATPase at the basal pole of the cell. In muscle cells they are located close to myofibrils. In a spermatozoon they provide the energy for the motive power of the tail. In the pancreas, where energy is needed for the biosynthesis of secreted proteins, they are sandwiched between layers of rough endoplasmic reticulum.

The presence of an internal membrane system, the *cristae mitochondrialis*, within the mitochondrion is common to all types of mitochondria, though again there is much diversity in its appearance. In liver mitochondria, the cristae are few compared with those in the kidney. The density with which the cristae are packed is related to the energy demands of the tissue—being very densely packed, for example, in the mitochondria of the cricothyroid muscle of the bat (Fig. 9.2). This is the muscle which generates the continuous ultrasonic squeak of the bat in flight. The surface area of the cristal membranes from 1 ml of such mitochondria amounts to 50 m^2, equivalent to the floor area of two good sized rooms. One might suppose that the specialized functions of a mitochondrion could be served by simply enclosing the appropriate enzymes within a limiting membrane. But the arrangement of the functional components in a membrane system offers several advantages. First, higher local concentrations of reactants may be achieved than if all were distributed uniformly. Secondly, it is possible to determine a particular sequence of reactions by the spatial order of the enzymes in the membrane. Thirdly, *vectorial* reactions are possible. This term implies that components in a membrane may be arranged asymmetrically in relation to the two surfaces—an arrangement that confers a directional quality on the movement of reactants and products to one or other side of the membrane.

Although it is not apparent in a single thin section of a mitochondrion, it is known that the cristae are derived from one continuous membrane which encloses an inner space. This scheme (Fig. 9.3) subdivides the mitochondrion into two membranes—inner and outer—and two spaces—the matrix and the intermembrane space. Of these subdivisions it is the complex inner membrane which is the key to the special properties of mitochondria.

9.2. Biogenesis of Mitochondria

In size, mitochondria are generally about 2 or 3 μm long and 1 μm in diameter—near the limit of resolution with the light microscope. They are

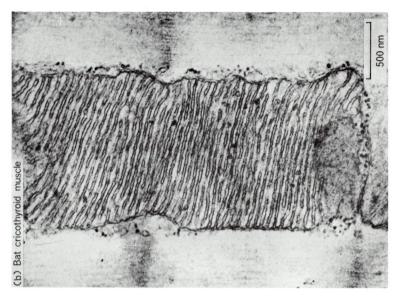

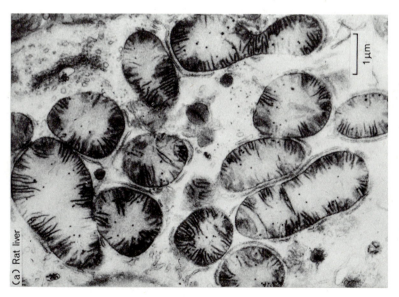

Fig. 9.2 The structure of mitochondria from bat liver and cricothyroid muscle.
(a) Bat liver. The cristae are short and do not span the whole width of the mitochondrion. The matrix occupies much of the space of the organelle.
(b) Cricothyroid muscle. In contrast to (a) the cristae are densely packed leaving much less space for the matrix. (a and b from Fawcett.)

comparable in size with many bacteria and indeed the resemblance between the two structures extends to many aspects. Like bacteria, mitochondria multiply by binary fission—a property which gives them some degree of independence within the cell. They contain DNA and RNA and are capable of effecting protein biosynthesis. The mitochondrial chromosome is a circular double strand of DNA. The mitochondrial ribosomes are smaller than those in the cytoplasm and protein biosynthesis in mitochondria is susceptible to inhibition by the antibiotic Chloram-

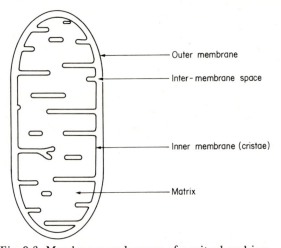

Fig. 9.3 Membranes and spaces of a mitochondrion.

phenicol (see Chapter 12.33). In each of these respects the mitochondrion resembles bacteria and differs from the analogous system in the rest of the mammalian cell. From these points, and other evidence, it is widely held that mitochondria have evolved from symbiotic bacteria—an example of endosymbiosis. In their present form their independence from the rest of the cell is very limited. Only a minority of the proteins, no more than 30 in number, are coded for and synthesized within the mitochondrion. The rest of the proteins together with the lipids are synthesized elsewhere in the cell. Indeed, the outer membrane is similar in form to the endoplasmic reticulum and probably arose from this structure, while the inner membrane is probably evolved from the symbiont.

9.3. Localization of Enzymes within the Mitochondrion

The activities of a mitochondrion are determined by the complement of enzymes and other factors within it and by their structural arrangement in the two membranes and two spaces of the mitochondrion. Table 9.1 gives the details of this functional organization. The matrix contains "soluble" enzymes, though in some cases these are complex multienzyme systems. It effects most of the TCA cycle reactions, some steps in gluconeogenesis, β-oxidation of fatty acids and amino acid metabolism leading to the synthesis of urea. The inner membrane contains some important dehydrogenases, all the cytochromes and other factors concerned with the generation of ATP. In comparison, the outer membrane and intermembrane space have rather limited roles—the activation of fatty acids on the membrane and certain nucleotide reactions in the space.

TABLE 9.1. The location of enzymes in mitochondria

Site	Functions	Enzymes
Outer Membrane	activation of fatty acids	fatty acid CoA synthetase
		monoamine oxidase
Intermembrane space	nucleotide interactions	myokinase (ATP + AMP ⇌ 2 ADP) nucleotide diphosphokinase (e.g. ATP + UDP ⇌ ADP + UTP)
Inner Membrane	H and electron transport	cytochromes, succinic DH β-hydroxybutyrate DH α-glycerophosphate DH NADH DH
	generation of ATP	"coupling" factors
	fatty acid translocation	carnitine fatty acyl CoA transferase
	Other transport systems	
Matrix	further dehydrogenations & TCA cycle reactions	citrate synthetase isocitrate DH fumarase malate DH aconitase
	fatty acid oxidation	fatty acid β-oxidation enzymes
	deamination and transamination urea synthesis	glutamic DH transaminases urea cycle enzymes
	initial gluconeogenesis reactions	pyruvate carboxylase PEP carboxykinase

9.4. The Mitochondrion in its Intracellular Environment

It follows from consideration of this specialized range of activities (described above) that the mitochondrion needs to receive and to deliver many different metabolites. This flux of materials is shown in Fig. 9.4.

Some of these—O_2, CO_2, urea, pyruvate, acetoacetate, β-hydroxy-butyrate penetrate the membranes and spaces without hindrance. The other metabolites, the nucleotides, di- and tricarboxylic acids, fatty acids and cations are not freely permeable and their entry and exit through the inner membrane is regulated by several mechanisms, which will be described after we have considered the oxidative and energy conserving activities in more detail.

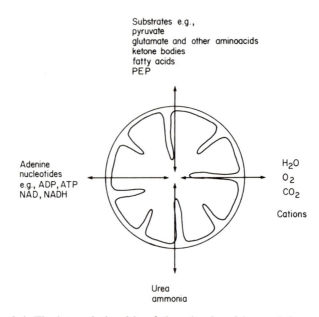

Fig. 9.4 The interrelationship of the mitochondrion and the cytosol.

9.5. Mitochondrial and Submitochondrial Preparations

Biochemists have a strong inclination to take complex systems apart to simplify their investigation, hoping ultimately to isolate each enzymic step. The mitochondrion has been subjected to the same approach. One objective has been to obtain inner and outer membranes in undamaged forms. The other objective has been to fragment the inner membrane: to dissect, if possible, the different functional units that, together, comprise the respiratory chain and the ATP generating systems. In either case the starting point is a preparation of "pure" mitochondria obtained by a technique such as that shown in Fig. 9.5 (see also Fig. 12.9).

Inner and outer membranes can be separated by a mild procedure—mitochondria swell in hypotonic phosphate solution—the outer membrane bursts while the inner membrane distends but remains intact. Centrifugation enables the two fractions to be separated (Fig. 9.6). The outer one is freely permeable to many substances, charged or uncharged, even up to 10,000 mol. wt. The inner membrane is osmotically active—it swells reversibly and may well contain a contractile protein. It is highly selective in its permeability, which is regulated by several specific permeases, (i.e., enzymes capable of effecting transport in and out of the mitochondrion). The two membranes are clearly different when viewed in the electron microscope. The outer one is thin and smooth surfaced: the inner is thicker and in negatively stained preparations shows a covering of particles, usually stalked and about 8 nm diameter, on the inner surface. They are concerned with ATP generation.

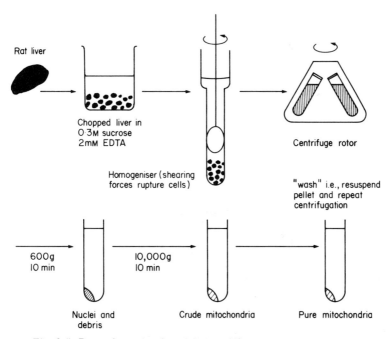

Fig. 9.5 Preparing mitochondria by differential centrifugation.

The other type of submitochondrial particle arises from the fragmentation of the inner membrane. This process requires more drastic treatments—prolonged sonication, for example, of swollen mitochondria produces fragments possessing most of the inner membrane activities. Further fragmentation involves the use of detergents, lipid solvents or enzymes to separate some of the individual functions of the inner membrane. One procedure produces four complexes each containing different enzymes and carriers and each representing a different part of the respiratory chain. To understand the roles of these enzymes and carriers and how the Complexes I-IV may be reorganized into a functioning whole it is first necessary to describe each of the carriers in a little more detail.

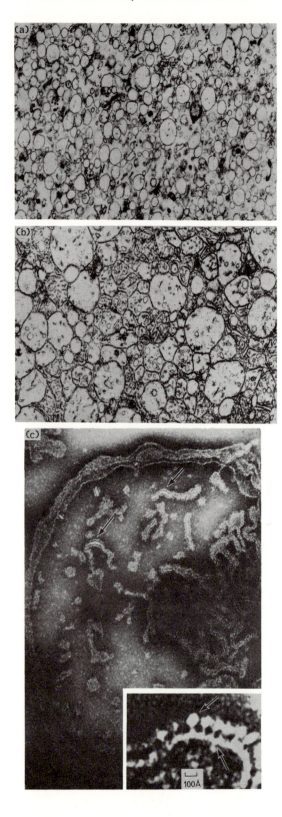

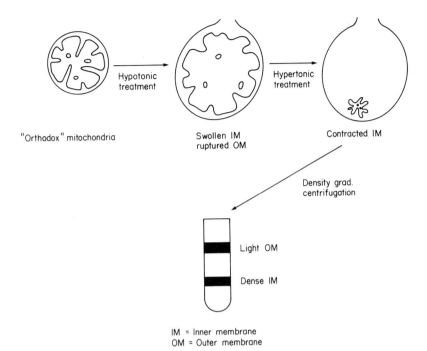

"Orthodox" mitochondria   Swollen IM   Contracted IM
                           ruptured OM

Density grad.
centrifugation

Light OM

Dense IM

IM = Inner membrane
OM = Outer membrane

Fig. 9.6 Preparation of inner and outer membranes. Electron micrographs of
(a) Outer membrane. (b) Inner membrane (thin sections). (c) Inner membrane (negative
staining). Both types of membrane tend to form closed vesicles. (a) and (b) are shown
at the same magnification, the outer membrane is thinner than the inner. When the
membrane is visualized by negative staining, i.e. when it is seen against the electron-
dense background of the stain, the external surface is coated with particles or knobs
about 80 Å (8 nm) diameter. In the intact organelle these structures project into the
matrix (see also Fig. 9.27). (a and b from Racker, c from Fernandez-Moran.)

9.6. The Chemistry of the Components of the Respiratory Chain

(a) The cytochromes

David Keilin discovered in 1925 that all sorts of cells from mammals,
bacteria, yeasts, etc., contained coloured substances with spectra characterized by three absorption bands in the visible light (Fig. 9.7). The bands
are more intense if the cell extract or suspension is kept in the reduced
form by removal of O_2 or, more easily, by the addition of a pinch of
dithionite. Each band could be resolved into more than one component
and Keilin concluded that the observations were best explained by the
presence of three substances which he named cytochromes a, b and c,
each with a slightly different absorption spectrum. Keilin's discovery is
undoubtedly one of the fundamental advances in biochemistry and it is
astonishing to realize that he was anticipated in this discovery (in 1886)
by a general practitioner in Wolverhampton, named MacMunn. He, too,
had examined many living tissues with a hand spectroscope, had seen the

absorption bands and realized that the reversible changes brought about by oxidation and reduction were explained by their role in cellular respiration. He called the pigments "histohaematins". MacMunn's discovery was initially ignored by the scientific world and then later his histohaematins were unfairly dismissed by the leading biochemist of the day, Hoppe-Seyler, as being no more than degradation products of haemoglobin. In this way progress in a major field was delayed for forty years.

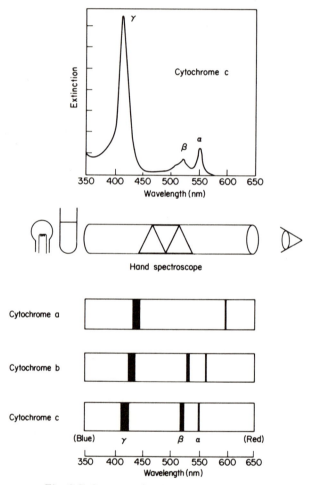

Fig. 9.7 Spectra of reduced cytochromes.

Ferroporphyrins The spectrum of the cytochromes arises from the prosthetic group which is bound to the protein. This group is a ferroporphyrin comprising four substituted pyrroles, linked by —CH= bridges to form a flat ring. The iron atom is held at the centre of this ring. This structure is similar to haem in haemoglobin and, indeed in cyto-

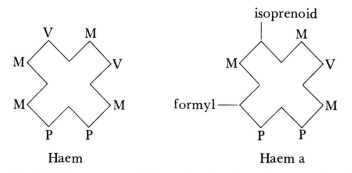

Haem porphyrin

This porphyrin is found in cytochromes b, c & c_1 as well as in all haem proteins, e.g. haemoglobin.

Haem a (Formyl porphyrin)

Found in cytochrome aa_3, it contains two variations—a formyl group—CHO replaces one methyl group and an isoprenoid chain replaces one vinyl group.

These structures can be represented in a shorthand form $(M = -CH_3 \ V = -CH=CH_2$ and $P = -CH_2 CH_2 COOH)$:

Fig. 9.8 The structure of the porphyrins of cytochromes a, b and c.

chromes b and c the ferroporphyrin is identical to that in haemoglobin, differing only in the nature of the attachment to the apo-protein (Fig. 9.8). One point of attachment involves the Fe atom, 4 valencies of which are in the (horizontal) plane of the pyrrole rings, the two other (vertical) valencies involve amino acid residues on the protein. In haemoglobin one of these latter valencies is available to combine with oxygen. Haemoglobin is, of course, ineffective as an oxygen carrier if the Fe is oxidized to the ferric form, but the essential role of the cytochromes as electron carriers depends on the reversible change of ferrous to ferric, $Fe^{2+} \underset{+e}{\overset{-e}{\rightleftharpoons}} Fe^{3+}$. The three classes of cytochromes (a, b and c) are thus defined in terms of the structural details of the ferroporphyrin groups. But the protein confers special properties too. At least 30 cytochromes have been described, of which only five, cytochromes a, a_3, b, c_1, c, are located in the inner mitochondrial membrane (Table 9.2).

TABLE 9.2. Mitochondrial Cytochromes

Cytochrome	Mol. wt.	Prosthetic groups	Properties
aa_3	240,000	Haem a (formylporphyrin) Cu	(= cytochrome oxidase) oxidized by O_2 inhibited by CO, CN^-, N_3^-, S^{2-}.
b	30,000	haem porphyrin	
c	13,000	haem porphyrin	not oxidized by O_2
c_1	37,000	haem porphyrin	

Cytochrome aa_3: *cytochrome oxidase* Cytochrome oxidase is a component of the inner mitochondrial membrane and its purification requires the use of a detergent, such as deoxycholate, in the extraction process. The purified cytochrome, which is green in colour, has a molecular weight of about 240,000 including some lipid. The smallest active preparation of cytochrome oxidase has a molecular weight of about 130,000 and contains 2 ferroporphyrin groups, referred to as haem a or formylporphyrin and two atoms of copper. The reason for designating it as "aa_3" implying the existence of two distinct cytochromes stems from an old observation by Keilin that when a crude tissue extract was treated with either carbon monoxide or cyanide the characteristic α and γ absorption bands of cytochrome oxidase became split—an additional band appearing alongside each of the original pair. This was interpreted as showing that only one component, designated cytochrome a_3, reacted with the inhibitors. However, all attempts to separate a from a_3 have ended in failure. It is now accepted that the two components form part of an integral complex. It is possible to dissociate purified cytochrome oxidase by electrophoresis in a gel containing a detergent (sodium dodecylsulphate). In this system seven subunits, now inactive, with molecular weights in the range 7000-40,000 have been resolved. It is probable that the three largest subunits are synthesized on mitochondrial ribosomes. At present it is not possible to designate some of the subunits as a and others as a_3 nor to be sure which of them bind Cu and which haem a. It is however clear that cytochrome a_3 includes the subunit(s) that react with the inhibitors of cytochrome oxidase—CN^-, N_3^- (azide) S^{2-} and CO and which also possess the unique ability (among the mitochondrial cytochromes) to combine with oxygen.

Cytochrome b   This is a heat-labile cytochrome that is rather firmly bound to the inner membrane and for long resisted attempts to purify it. It now appears to have a subunit mol. wt. of 30,000 containing one molecule of haemporphyrin. After many years of dispute, its place in the main respiratory chain is now generally accepted.

Cytochromes c *and* c_1 Cytochrome c is the only cytochrome that is easily extracted from mitochondria by salt solutions. It is a small, very basic, protein, mol. wt. 13,000, and therefore, presented a comparatively simple task to protein chemists to elucidate its amino acid sequence. We now know the primary structure of cytochrome c from a large number of sources, ranging from bacteria and fungi to vertebrates. Although by no

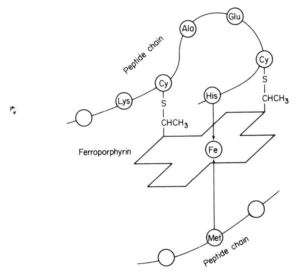

Fig. 9.9 Mode of attachment of ferroporphyrin to peptide chain of cytochrome c.

means identical, the similarities in sequence are very marked: relatively few mutations in the cytochrome c gene have been perpetuated during evolution — only one mutation per 2.6×10^7 years. This suggests that the molecule is so well designed for its function that any potential for improvement is limited. Cytochrome c has the same porphyrin as cytochrome b but it is differently linked to the protein. In addition to bonds holding the Fe atom to histidyl and methionyl residues the two vinyl chains of the porphyrin are covalently linked to cysteinyl side chains as shown in Fig. 9.9.

Cytochrome c_1 shows only slight differences in the absorption spectrum compared with cytochrome c and it is difficult to resolve the two components in intact mitochondria. However, if mitochondria are washed extensively, cytochrome c is leached out leaving cytochrome c_1, which remains firmly bound to the inner membrane.

(b) Flavoproteins

There are a number of distinct dehydrogenases in mitochondria, each employing the same isoalloxazine hydrogen carrier but differing as regards substrate specificity. Isoalloxazine is present in the vitamin riboflavine and in its active form it exists as flavine-adenine dinucleotide (FAD): isoalloxazine-ribitol-phosphate-phosphate-ribose-adenine (see Chapters 3 and 5 for details of structure). As a prosthetic group in flavoproteins it serves as a hydrogen acceptor.

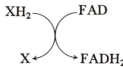

This reaction is generally shown as a two-equivalent transfer, although electron spin resonance studies confirm that the transfer goes in two steps, with a free radical semiquinone as the intermediate. Some flavoproteins contain a metal, typically iron. This is bound in a way that is quite unlike that in haemproteins and, therefore, is referred to as "non-haem iron". It appears to be bound to sulphur atoms of cysteinyl residues of the proteins. Treatment with acid releases both iron and H_2S from such protein. Probably the iron is bound to a subunit of the enzyme which is easily separated from the flavoprotein. In spite of this difference from the cytochromes the iron serves the same function as an electron acceptor, cycling between the oxidized and reduced states.

The mitochondrial flavoproteins are not able to react directly with oxygen. In mitochondria they depend on another carrier, ubiquinone (UQ), to achieve the reoxidation of $FADH_2 - FAD$. In isolated systems, certain dyes, e.g., phenazine methosulphate, can be employed to link the flavoprotein to oxygen. Five mitochondrial flavoproteins are listed in Table 9.3—they serve to oxidize NADH, succinate and fatty acyl CoA (for which there are several enzymes specific for different fatty acid chain lengths). Lipoyl dehydrogenase is part of each of the complexes

TABLE 9.3. Mitochrondrial flavoproteins

Name	Functional role	Prosthetic group	Metal
NADH Dehydrogenase	NADH → UQ	FAD	Fe (non haem iron)
Succinate Dehydrogenase	Succinate → UQ	FAD	Fe (non haem iron)
Fatty acyl CoA Dehydrogenase	Fatty acyl CoA → ETF	FAD	
Lipoyl Dehydrogenase	reduced lipoate → NAD	FAD	
Electron transporting Flavoprotein (ETF)	Fatty acyl CoA dehydrogenase → UQ ? other dehydrogenases → UQ		

concerned with pyruvate and α-ketoglutarate oxidation. It has the unusual ability to transfer electrons to NAD so that the electron pathway is complex for these α-ketoacids:

$$\text{Pyruvate} \rightarrow \text{Lipoate} \rightarrow \text{FAD} \rightarrow \text{NAD} \rightarrow \text{FAD} \rightarrow \text{UQ}$$

Electron transporting flavoprotein (ETF) is concerned primarily with accepting electrons from the $FADH_2$ of fatty acyl CoA DH, which is unable to link directly with ubiquinone:

$$FA\ CoA \rightarrow FAD \rightarrow (ETF)FAD \rightarrow UQ$$

ETF and fatty acyl CoA dehydrogenase together constitute 80% of the mitochondrial flavoproteins.

Before leaving this topic, it is worth remembering that other flavo-proteins are located in other parts of the cell. They too are de-hydrogenases or oxidases, but they are not linked with the mitochondrial respiratory chain. They have the ability to be reoxidized directly by oxygen, with H_2O_2 as the product. Some of these—D amino acid oxidase and urate oxidase, for example, are located in organelles called micro-bodies or peroxisomes and will be discussed in Chapter 10.

(c) Ubiquinone (also known as CoQ)

Solvents, such as acetone, extract various quinones from mitochondria. Such compounds can readily undergo oxidation and reduction:

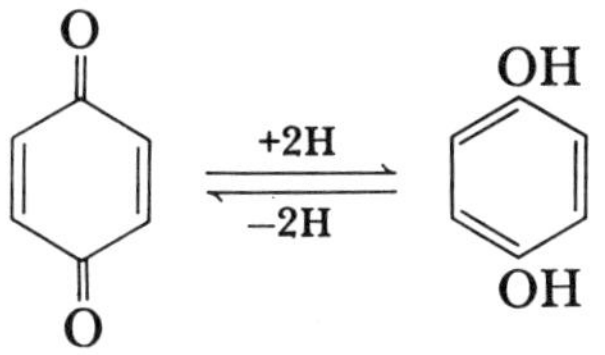

and their place in the respiratory chain became clear from the observation that succinate oxidation ceased after acetone extraction of mitochondria and could be restored by the addition of certain quinones. The particular quinone which serves this function in mitochondrial respiration is known as Ubiquinone 50 (UQ50).

Ubiqinone 50

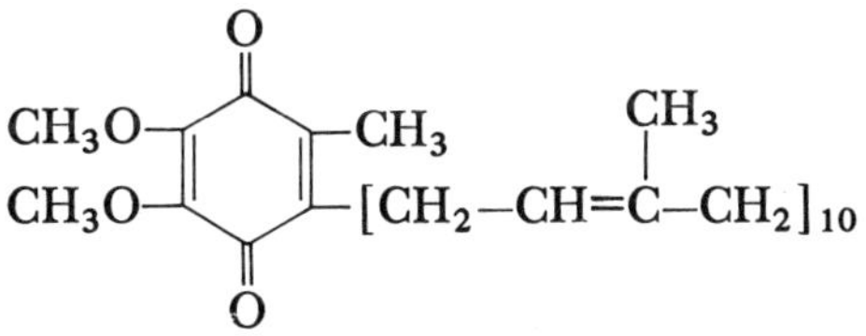

This is one of a family of compounds, so named because of their ubiquitous distribution, which differ only in the length of the long hydrocarbon "tail", UQ50 having a 50 C side chain.

9.7. The Organization of the Respiratory Chain

(a) Complexes I-IV

The organization of the components of the respiratory chain in the inner mitochondrial membrane has been elucidated in various ways. One approach follows from the further fragmentation of the sub-mitochondrial particles into four complexes (see 9.5 above). Their functions are summarized in Fig. 9.10. Complex I is the flavoprotein which reoxidizes NADH at the expense of UQ. Complex II has an analogous role in connection with the oxidation of succinate. Complex III oxidizes UQH$_2$ at the expense of cytochrome c and Complex IV—cytochrome oxidase— oxidizes reduced cytochrome c at the expense of oxygen.

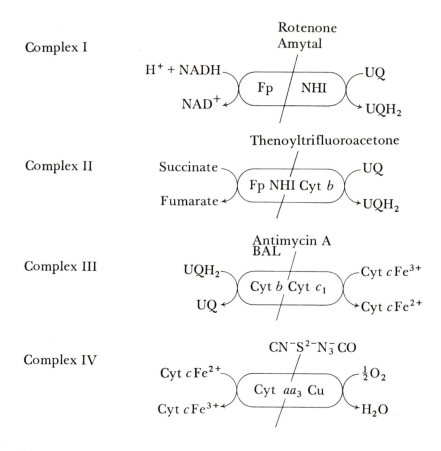

See Table 9.4 concerning the actions of the inhibitors

Fig. 9.10 The function of complexes I, II, III and IV. (Fp: flavoprotein, NHI: non-haem iron, BAL: British anti-Lewisite, 2,3-dimercaptopropanol).

The assembly of those complexes in the inner membrane is a matter for speculation. One possible arrangement is shown in Fig. 9.11. It illustrates the roles of the two small molecules of the chain—UQ and cytochrome c—in linking the larger complexes together. The four complexes account for the oxidation of only NADH and succinate. Other flavoproteins are present, in particular electron transferring flavoprotein (ETF) and fatty acyl CoA DH. Since they represent most of the flavoproteins in the mitochondrion their relationship to the rest of the chain is of major importance. It is probable that ETF links some matrix enzymes—fatty acyl CoA DH, glycerol phosphate DH and possibly other flavoproteins to ubiquinone.

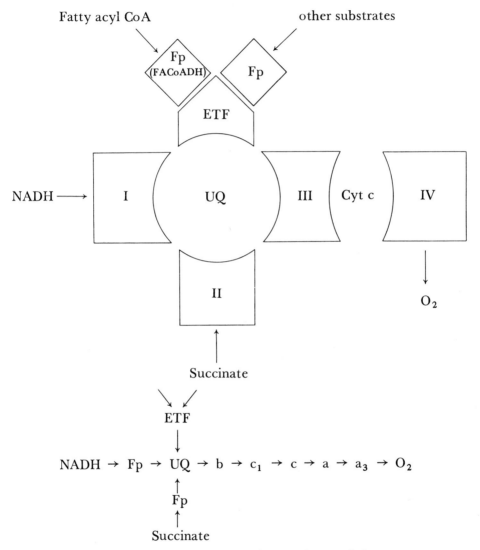

Fig. 9.11 Possible assembly of the respiratory chain.

Another component which needs to be mentioned is the so-called "structural" protein, a hypothetical cement of a hydrophobic nature which holds the components in the right orientation in the membrane. Certainly, the extraction of the cytochromes is hampered by their contamination with a very insoluble protein which can only be separated

by treatments with detergents and organic solvents. However, there is
evidence that some preparations of "structural protein" have been derived
from denatured ATPase of the inner membrane.

(b) Specific inhibitors of the respiratory chain

Inhibitors of different regions of the respiratory chain have proved very
valuable in confirming the functional arrangement. Some inhibitors of
complexes I-IV are shown in Fig. 9.10 and their effects on isolated intact
mitochondria lend support to the scheme. Cyanide blocks the oxidation

TABLE 9.4. Inhibitors and Uncouplers

Reagent	Site	Effect
Cyanide Azide Sulphide Carbon monoxide	Cyt a_3 (Complex IV cytochrome oxidase)	blocks all respiratory chain oxidation
Antimycin A British anti-Lewisite (BAL)	Cyt b (Complex III)	blocks oxidation of all substrates
Thenoyltrifluoroacetone Malonate	Succinic DH (Complex II)	blocks succinate oxidation
Rotenone Barbiturates (e.g., Amytal)	NADH DH (Complex I)	blocks oxidation of NADH linked substrates
Oligomycin	Coupling factor	blocks phosphorylation (blocks oxidation only if coupled)
Dinitrophenol	? Makes membrane permeable to H^+ ? hydrolyzes X $\sim$ I	uncouples phosphorylation and activates ATP ase
Valinomycin Gramicidin	Make membrane permeable to K^+ and Na^+	uncouple phosphorylation by diverting energy to ion transport
Atractyloside	ATP/ADP translocase	blocks phosphorylation of external ADP (block oxidation only if coupled)

of all substrates. So, too, does antimycin A (an antibiotic extracted from a
Streptomyces). Rotenone (a powerful insecticide) or Amytal (a barbiturate
sedative) blocks the oxidation of the NAD-linked substrates, without
affecting succinate oxidation. Thenoyl trifluoroacetone blocks the latter
but not the NAD-linked systems. Table 9.4 summarizes the effects of
inhibitors and uncoupling reagents on mitochondria (see also Section 9.8).

(c) The steady state of the respiratory chain

A third approach to the organization of the chain derives from the study of the oxidation-reduction states of each of the component carriers. This can be done spectroscopically in intact mitochondria or even in whole cell suspensions, because each carrier has a characteristic absorption spectrum which changes as it passes from the reduced to the oxidized state. The principle is simple but the instrumentation necessary to record these changes rapidly tends to be extremely complicated. The mitochondrion normally functions in what are "steady state" conditions. From the cytosol it receives a steady supply of substrates, processes them and discharges the products. The respiratory chain, working optimally, is like a continuous production line. Each step in the process operates at a constant rate—the molecules of each carrier are alternatively reduced and

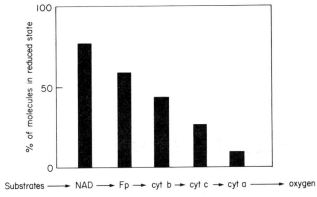

Fig. 9.12 Steady state conditions of the respiratory chain.

oxidized as the flow of electrons is maintained. The process is a dynamic one, yet at any one instant the proportion of any one carrier in the oxidized and reduced forms, will be constant. At the start of the respiratory chain the environment is strongly reducing—electron-donating. At the finish conditions are strongly oxidizing—with oxygen the final electron acceptor. One might expect, therefore, that the carriers near the reducing end will exist in a predominantly reduced form compared with those near the oxidizing end. And this is what is found experimentally (Fig. 9.12). The proportion of the reduced carrier at each step decreases in the predicted order of the components of the respiratory chain. This confirmation of the respiratory chain is especially valuable because it results from experiments on intact, functioning mitochondria.

(d) Redox potentials

Another line of evidence concerning the order of the components of the respiratory chain derives from the measurement of redox potentials. This is a quantitative way of expressing the reducing power of a substance, analogous in some ways to the concept of pK in expressing the strength of an acid (see Chapter 2). In the present context redox potential is useful in two ways. First, it tells us about the capability of one compound to oxidize or reduce another; secondly, it enables us to calculate how much useful energy is available when the two compounds interact.

Oxidation can be considered as the loss of electrons $X_{red} \rightarrow X_{ox} + e$ and reduction as the reverse—the gaining of electrons. If the equation is written to show the removal or acceptance of hydrogen atoms, we may still view it as an electron transfer: $XH_2 \rightarrow X + 2H \rightarrow 2H^+ + 2e$. Of course,

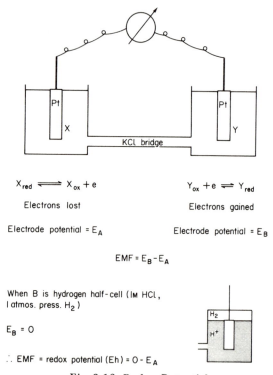

$$X_{red} \rightleftharpoons X_{ox} + e \qquad\qquad Y_{ox} + e \rightleftharpoons Y_{red}$$

Electrons lost Electrons gained

Electrode potential = E_A Electrode potential = E_B

$$EMF = E_B - E_A$$

When B is hydrogen half-cell (1M HCl, 1 atmos. press. H_2)

$$E_B = 0$$

$$\therefore EMF = \text{redox potential } (Eh) = 0 - E_A$$

Fig. 9.13 Redox Potential.

in following the events of the respiratory chain we are obliged to link hydrogen transfer to electron transfer at various points. Given that our substance X can donate electrons it becomes possible to measure this tendency by a suitable electric circuit. At its simplest, two vessels are linked as in Fig. 9.13. Vessel A contains a solution of X, a "reducing" agent, so called because of a strong tendency to lose electrons and B contains a solution of Y, referred to as an "oxidizing" agent because it has a strong tendency to accept electrons. By joining the two vessels, as shown, electrons will flow from A to B and a current will be measured by the galvanometer. In fact, it is the EMF that is measured rather than the current. If one knew the absolute electrode potentials in the two vessels

E_A and E_B then the EMF of the system $\Delta E = E_B - E_A$. In practice one needs to compare the half-cell in A with a reference half-cell, which is always taken as the standard hydrogen electrode. The equilibrium may be written as $\frac{1}{2}H_2 \rightleftharpoons H^+ + e$ and the standard hydrogen half-cell contains hydrogen at 1 atmosphere pressure in contact with 1 M H^+ (i.e., HCl). By definition, its electrode potential is zero. If the other half-cell is more reducing, i.e., losing electrons to the hydrogen half-cell it will have a negative electrode potential and if more oxidizing (i.e., gaining electrons), it will have a positive electrode potential. This potential is referred to as the Redox Potential (Eh). Its magnitude will depend on the concentration of X, more particularly the concentrations of X_{red} and X_{ox}, and also on temperature and pH. When these variables are standardized: 25°C, pH 7, 1 M oxidized and 1 M reduced forms, the potential is referred to as the Standard Redox Potential (E'_0). It is clear from the definition that it is the "midpoint" potential—that produced when the mixture is 50% oxidized. The curve relating Eh to percentage of X reduced is seen in Fig. 9.14. It

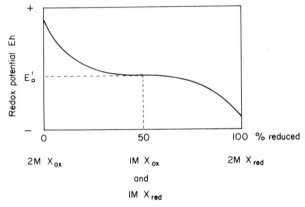

Fig. 9.14 Redox Potential related to percent reduction.

resembles the titration curve for a weak acid, HA, the equation for which (the Henderson-Hasselbalch equation) is pH = pK + log ([A⁻]/[HA]). The redox equation (known as the Nernst equation) has the same form:

$$Eh = E'_0 + \frac{RT}{nF} \ln \frac{[X_{ox}]}{[X_{red}]}$$

where R = the gas constant, 8·31 J K⁻¹ mol⁻¹, T = temp °K, n = number of electrons lost per mole, F = Faraday's Constant, 9·65 × 10⁴ C mol⁻¹. At 25°C the equation becomes

$$Eh = E'_0 + \frac{0\cdot059}{n} \log \frac{[X_{ox}]}{[X_{red}]}$$

The importance of this equation is that it enables us to adjust standard redox potentials to take account of the steady state mixture that obtains within a mitochondrion. Thus for cytochrome c, $E_0' = +0.25$ volts. In the steady state it is 75% oxidized and 25% reduced. Hence $Eh = +0.25 + 0.059 \log 3 = +0.28$. Generally, if the component is more than 50% oxidized, as in this example, the adjustment is a positive increment; if more than 50% reduced, the adjustment is a negative increment.

TABLE 9.5. E_0' values

Reduced form	Oxidized form	E_0' (volts)
pyruvate	acetate + CO_2	-0.70
α-ketoglutarate	succinate + CO_2	-0.67
NADH + H^+	NAD^+	-0.32
dihydrolipoate	lipoate	-0.29
β-hydroxybutyrate	acetoacetate	-0.27
malate	oxaloacetate	-0.17
glutamate	α-ketoglutarate + NH_3	-0.14
$FADH_2$-protein	FAD-protein	-0.05
succinate	fumarate	$+0.03$
ascorbate	dehydroascorbate	$+0.08$
UQH_2	UQ	$+0.10$
Cyt b (Fe^{2+})	Cyt b (Fe^{3+})	$+0.12$
Cyt c_1 (Fe^{2+})	Cyt c_1 (Fe^{3+})	$+0.21$
Cyt c (Fe^{2+})	Cyt c (Fe^{3+})	$+0.25$
Cyt a (Fe^{2+})	Cyt a (Fe^{3+})	$+0.29$
H_2O	$\frac{1}{2}O_2$	$+0.82$

Table 9.5 gives a selection of E_0' values for some substrates and carriers. The higher the position in the list the stronger the reducing power (i.e., tendency to lose electrons). The lower the position the stronger is the tendency to gain electrons. Electrons will normally flow from one pair to another in the order given in this list. For example, NAD^+ is a suitable acceptor for electrons from pyruvate but not from succinate. The order of the respiratory chain components $NADH \rightarrow Fp \rightarrow UQ \rightarrow b \rightarrow c_1 \rightarrow c \rightarrow a \rightarrow a_3$ is consistent with these E_0' values. Some apparent anomalies, e.g., $NADH/NAD^+$ being higher in the list than malate/oxaloacetate implies at first sight that the former pair reduces the latter. But it should be remembered that these values are *standard* redox potentials — for a 50-50 mixture of oxidized and reduced forms and that reversal in the direction of electron flow becomes possible when the proportions are severely altered.

Since we can picture the chain as a cascade of electrons from one carrier to another, a further bonus arises from our consideration of redox potential. This is because it is possible to assess the energy yield between different points in this cascade. The free energy change ΔG from one step to another is related to the difference in *Eh* values. The formula is a simple one

$$\Delta G = - nF\Delta Eh \text{ (n and F being defined in the previous equation)}$$

If we use the *standard* values then $\Delta G^{0\prime} = nF\Delta E_0^\prime$. Applying this to the transfer of electrons from NADH to UQ—the step catalyzed by complex I:

$$\Delta E_0^\prime = +0\cdot10 - (-0\cdot32) \text{ V} = +0\cdot42$$

$$\Delta G^{0\prime} = -2(96,500)\ 0\cdot42 \text{ J}$$

$$= -81,060 \text{ J or } 81 \text{ kJ (19·5 kcal)}$$

We shall need to return to this relationship when we consider the steps at which ATP is generated in the respiratory chain.

9.8. The Physiology of Mitochondria

So far we have considered mitochondria in terms of their ultrastructure and the assembly of the electron carriers that constitute the respiratory chain. In other words, the mitochondrion has been treated as a static object. We must now examine its dynamic role in three related functions: the oxidation of substrates, the generation of ATP and the transport of substances across the inner membrane.

(a) The oxygen electrode

There are several standard techniques for studying mitochondrial respiration. The classical apparatus is the Warburg manometer which permits the oxygen consumption of a tissue preparation to be measured very precisely as a change in pressure. In recent years another technique, the oxygen electrode, has largely displaced manometry as the standard method. The principle of the technique is that of polarography. The apparatus contains two electrodes, one of platinum($-$) and one of silver($+$). The voltage between the two electrodes can be adjusted and the current flowing can be measured. At a certain voltage any oxygen in solution reaching the Pt electrode becomes reduced, $O_2 + 4e \rightarrow 2\ O^{2-}$, and the ions so formed increase the flow of current. The current is extremely small (about 0·1 μA) but, at a potential of 0·6 V, the apparatus is specific for oxygen since other components in the incubation mixture are not reduced. Hydrogen ion, for example, is reduced only if the voltage is raised to 0·8 V. The design of the oxygen electrode is shown in Fig. 9.15. The contents of the vessel are maintained at constant temperature and are continuously stirred. Hence the rate of diffusion of O_2 across the Teflon membrane to the Pt electrode is proportional to its concentration. A flat-bed pen recorder provides a trace of oxygen concentration against time. Each experiment takes only a few minutes and, after washing out, the apparatus is immediately ready for another experiment.

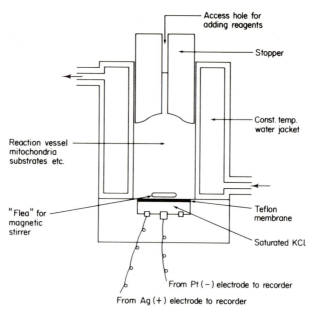

Access hole for adding reagents

Stopper

Const. temp. water jacket

Reaction vessel mitochondria substrates etc.

"Flea" for magnetic stirrer

Teflon membrane

Saturated KCl

From Pt (−) electrode to recorder

From Ag (+) electrode to recorder

Fig. 9.15 The Oxygen Electrode.

(b) Respiratory control

Let us set up a simple experiment using the oxygen electrode (Expt. 1, Fig. 9.16). The chamber contains a solution of isotonic KCl, with inorganic phosphate, and $MgCl_2$, buffered at pH 7·4. At 30°, in equilibrium with air, it contains 0·69 μmole O_2 in 3 ml. If rat liver mitochondria are added, a very small uptake of oxygen is observed—due to endogenous respiration using traces of substrate within the mitochondria. If succinate (in excess) is now added as our exogenous substrate, a measurable rate of oxygen consumption occurs almost immediately. If a little ADP is now added there is a spurt in the rate of O_2 consumption which continues linearly but soon tails off to the previous rate. Why has ADP stimulated the oxidation of succinate? The answer lies in the coupling of phosphorylation (i.e., ATP generation) with oxidation—the phenomenon of oxidative phosphorylation—

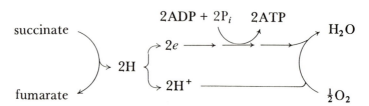

If the mitochondria are completely coupled then no succinate will be oxidized unless ADP and P_i are available. With most preparations of isolated mitochondria the coupling is less than complete, hence a low rate of oxidation proceeded in the absence of ADP but was much increased by its presence. Why was the ADP stimulation short lived? The answer becomes clear by adding a second dose of ADP which immediately restores the rate. This simple experiment demonstrates that [ADP] exerts a control over oxidation. This is obviously a sensible relationship: if ATP

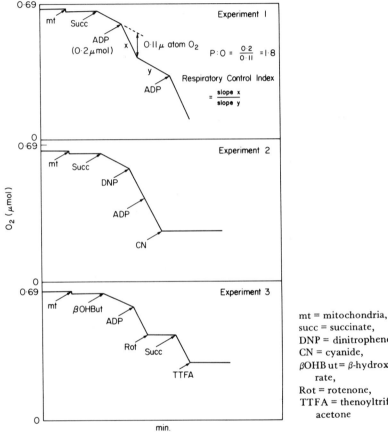

Fig. 9.16 Oxygen Electrode experiments (see text for explanation).

is depleted by processes such as biosynthesis or muscular contraction then the accumulation of ADP can increase the oxidation of substrates so as to regenerate ATP.

The experiment permits us to calculate two quantitative values which describe the degree of coupling. The Respiratory Control Index is the ratio

$$\frac{\text{Rate of } O_2 \text{ consumption with ADP}}{\text{Rate of } O_2 \text{ consumption without ADP}}$$

326

Basic Biochemistry for Medical Students

If completely uncoupled, when ADP exerts no effect, the ratio is 1. For well-coupled mitochondria the ratio is in the range 5-20. In this experiment the ratio of the slopes, shown on trace, was 5. The P : O ratio = $\frac{\text{Moles ATP generated}}{\text{Atoms O}_2 \text{ consumed}}$. If we added an exactly measured quantity of ADP then the amount (not the rate) of O_2 consumed before the ADP was exhausted enables the P : O to be calculated. Here the ratio is

$$\frac{0 \cdot 2 \text{ } \mu\text{moles ADP used}}{0 \cdot 11 \text{ } \mu\text{atoms O}_2 \text{ used}} = 1 \cdot 8,$$

which approximates to a value of 2.

Many substances are known to uncouple oxidative phosphorylation, 2,4-dinitrophenol, for example. If this is added—Experiment 2 in Fig. 9.16—it stimulates oxidation to a maximal rate which is now no longer responsive to additions of ADP.

(c) Phosphorylation sites

The specific inhibitors of each submitochondrial complex (see Table 9.4 and Fig. 9.10) can be used in this type of experiment. Cyanide (Complex IV) or antimycin (Complex III) blocks the oxidation of any substrate. Rotenone (Complex I) blocks NAD-linked oxidations, as shown in Experiment 3, Fig. 9.16, in which β-hydroxybutyrate is the substrate. Succinate is still oxidized but, in turn, it may be blocked by thenoyl trifluoroacetone (Complex II). These reagents permit us to "isolate" different parts of the respiratory chain without disrupting the mitochondria. This is valuable because it becomes possible to locate the steps at which ATP may be generated (Fig. 9.17). Since succinate oxidation yields two molecules and NADH oxidation three molecules of ATP per $2e$ transferred, it is clear that Complex I must be a site of ATP generation. This may be confirmed by blocking the terminal chain with antimycin A and then providing an alternative artificial electron acceptor—phenazine methosulphate ($E_0' = 0 \cdot 08$ V). A P:2e ratio of about unity is then found.

The other sites of phosphorylation can be established by isolating the pathway: succinate → cytochrome c, which involves complexes II and III. Of these, complex II can be eliminated as a phosphorylation site because the overall P:O ratio for succinate oxidation is 2 compared with 3 for NADH oxidation. Cyanide is used to block cytochrome oxidase and the oxidation of succinate can be followed by the reduction of added cytochrome c. The third phosphorylation site can be isolated by providing reduced cytochrome c as the substrate for complex IV. This is achieved indirectly by providing an excess of ascorbate ($E_0' = +0 \cdot 08$ V) together with a trace of a dye, tetramethyl-p-phenylinediamine (TMPD, $E_0' = +0 \cdot 26$ V) which acts as a mediator. The result is a prompt reduction of cytochrome c as soon as it is oxidized by cytochrome oxidase. In each case a P:2e or P:O ratio of about 1 can be determined.

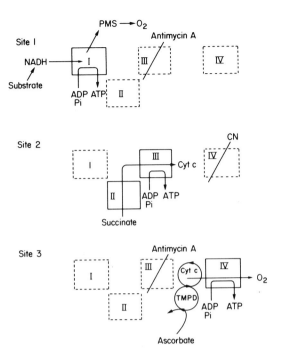

Fig. 9.17 The three phosphorylation sites of the respiratory chain. The three sites can be "isolated" functionally by blocking the unwanted parts of the chain and providing the appropriate electron donor and acceptor system. For site 1, phenazinemethosulphate (PMS) is an acceptor dye which links to O_2. For site 2, succinate is oxidized and the reduction of cytochrome is measured. For site 3, ascorbate and another dye, tetramethyl-p-phenylinediamine (TMPD), act to reduce cytochrome c and oxygen utilization is measured.

(d) The concept of free energy

If the three phosphorylating sites are located as follows:

$$\text{NADH} \xrightarrow[(1)]{} \text{UQ} \xrightarrow[(2)]{} \text{Cyt } c \xrightarrow[(3)]{} O_2,$$

One may go on to ask whether the redox potentials of these components are consistent with the provision of energy required for ATP generation.

The term *free energy* (symbol G) has been employed before (see Chapter 5.8). The concept was introduced by a physical chemist, Willard Gibbs (hence G), to indicate the maximum potential for performing useful work. While it is unnecessary for our purpose to delve deeply into thermodynamics, there are a number of concepts that are relevant to our understanding of bioenergetics.

(i) The absolute energy content for a substance is rarely known. Hence we are concerned only with the *change* in a reaction, denoted by the prefix Δ. If energy is lost from the system the sign is $-$; if energy is gained or consumed the sign is $+$.

(ii) The heat change during a reaction, e.g., when glucose is combusted in a bomb calorimeter, is a measure of the change in enthalpy, ΔH; (strictly speaking ΔH implies an isothermal reaction at constant pressure, as does ΔG). ΔH employs the same convention regarding sign as does ΔG.

(iii) ΔG is a measure of the capacity of the reaction for useful work and it will have a numerical value different from ΔH according to the equation embodying the First and Second Laws of thermodynamics.

$$\Delta G = \Delta H - T\Delta S$$

i.e., ΔG is less than ΔH by a component involving the change in entropy (ΔS) and the absolute temperature T. Entropy is concerned with the ideas of order and chaos. If the products are more disordered than the reactants then entropy is said to increase, and the term $T\Delta S$ is that portion of the heat change, ΔH, which is dissipated in the increasing molecular disorder and which is, therefore, unavailable for useful work.

(iv) If ΔG for a reaction is positive the reaction is referred to as *endergonic* and it consumes free energy. If ΔG is negative the reaction is *exergonic* and free energy is liberated.

Exergonic reactions are spontaneous in the sense that the equilibrium is to the right-hand side of the equation. It does not follow that an exergonic reaction necessarily proceeds rapidly or indeed at any measurable rate, but only that it is feasible without supplying free energy to the system. ΔG is therefore related to the equilibrium constant for a reaction.

Just as redox potentials are defined under standard conditions, ΔG values, standardized at 25°C with the reactants at 1 M concentrations or 1 atmosphere pressure and at pH 7 are signified by the use of the symbol $\Delta G^{0\prime}$. This function is related to the equilibrium constant as follows:

$$\Delta G^{0\prime} = -RT \ln K_{eq} = 2 \cdot 3\, RT \log K_{eq}$$

K_{eq}, for the reaction A + B $\rightleftharpoons$ C + D, $= \dfrac{[C][D]}{[A][B]}$ where the square brackets indicate *activities* (which approximate to concentrations in dilute solutions) of the reactants at equilibrium).

Now, if $K_{eq} = 1$, that is if the reaction is balanced about the midpoint then $\log 1 = 0$ and so $\Delta G^{0\prime} = 0$. If K_{eq} is large, say 1000 indicating that the equilibrium is far to the right then $\Delta G^{0\prime}$ becomes -17 kJ ($-4\cdot1$ kcal), i.e., it is exergonic. If K_{eq} is small, say 0·001 (equilibrium to the left) $\Delta G^{0\prime}$ $= +17$ kJ ($+4\cdot1$ kcal): it is endergonic. This relationship is shown in Fig. 9.18 where a K_{eq} is plotted on a logarithmic scale so that the relationship appears linear.

This relationship concerns *standard* free energy change ($\Delta G^{0\prime}$), defined above. Within a cell the concentrations will differ from those in the definition. So, just as we could trim $E_0^\prime$ values for actual, rather than standard, mixtures of oxidized and reduced forms, we can also adjust the $G^{0\prime}$ value for a particular reaction to take account of the actual concentrations within a cell. The equation is

$$\Delta G = \Delta G^{0\prime} + RT \ln \frac{[C][D]}{[A][B]}$$

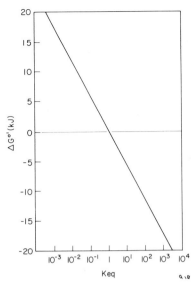

Fig. 9.18 Relationship of $G^{0\prime}$ to the equilibrium constant (K_{eq}).

where the values within square brackets are now the actual, rather than the equilibrium concentrations. If they approximated to those at equilibrium then $\Delta G = 0$, because the two terms on the right become equal but of opposite sign; i.e., systems at equilibrium are incapable of performing any useful work. If we disturb the mixture by raising the concentration of a reactant, say A, then the logarithmic function could become negative and hence the free energy change would become more exergonic. Let us consider the reaction:

$$\text{ATP} + \text{Creatine} \rightarrow \text{ADP} + \text{Creatine Phosphate} \quad \Delta G^{0\prime} = +13 \text{ kJ } (+3 \cdot 1 \text{ kcal})$$

This is, therefore, an endergonic reaction, with the equilibrium to the left-hand side, K_{eq}, from the graph in Fig. 9.18, being 0·005. Now consider a hypothetical situation with each of the reactants and products, except ATP, at a concentration of 10^{-5} M. ATP is then added to a concentration of 10^{-2} M. What happens to the free energy of the reaction? By substituting in the equation we get:

$$\Delta G = +13 + 5 \cdot 7 \log \frac{[10^{-5}][10^{-5}]}{[10^{-2}][10^{-5}]}$$

$$= 13 + 5 \cdot 7 \log 10^{-3}$$
$$= 13 - 17 \cdot 1$$
$$= -4 \cdot 1 \text{ kJ } (-1 \cdot 0 \text{ kcal})$$

So, by manipulating the concentration of one of the reactants we have been able to convert an endergonic reaction to an exergonic one, or to put it another way, we have succeeded in driving the reaction to the right. The same result could be achieved if we devised some means of lowering the concentration of one of the products, e.g., by linking it to another reaction which served to consume it. This is how coupled reactions enable an endergonic reaction to be driven by an exergonic one:

$$A + B \rightarrow C + D \quad (1) \quad \Delta G^{0\prime} = +13 \text{ kJ } (+3\cdot1 \text{ kcal})$$
$$C \rightarrow E + F \quad (2) \quad \Delta G^{0\prime} = -23 \text{ kJ } (-5\cdot5 \text{ kcal})$$

Sum $A + B \rightarrow D + E + F$ (3)

Now, $\Delta G^{0\prime}$ for the combined reaction (3) is the sum of $\Delta G^{0\prime}$ values for (1) and (2).

$$\Delta G^{0\prime} (3) = +13 - 23 = -10 \text{ kJ } (-2\cdot4 \text{ kcal})$$

This illustrates the fundamental additive quality of free energy changes. It derives from the principle that only the starting and finishing points are important in determining ΔG for an overall reaction; it is immaterial by which route the reaction proceeds. Let us take an example. For the reaction,

$$\text{succinate} + \tfrac{1}{2}O_2 \rightarrow \text{fumarate} + H_2O \ \Delta G^{0\prime} = -145 \text{ kJ } (-34 \text{ kcal})$$

This yield of free energy is independent of the mechanism of the oxidation. We could achieve the oxidation by using an artificial hydrogen acceptor—methylene blue (MB).

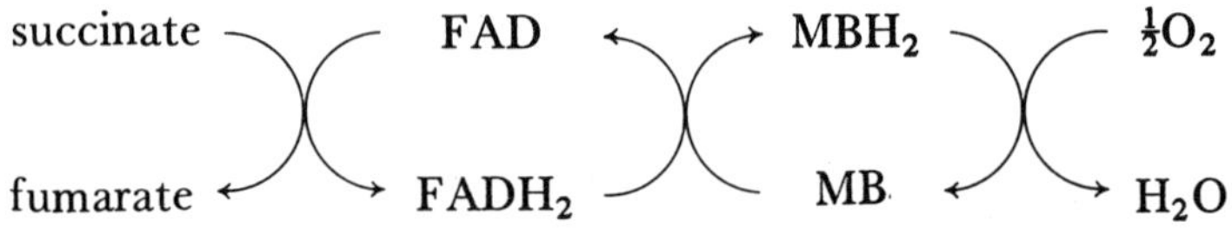

or we might allow it to proceed in intact mitochondria utilizing all the components of the respiratory chain:

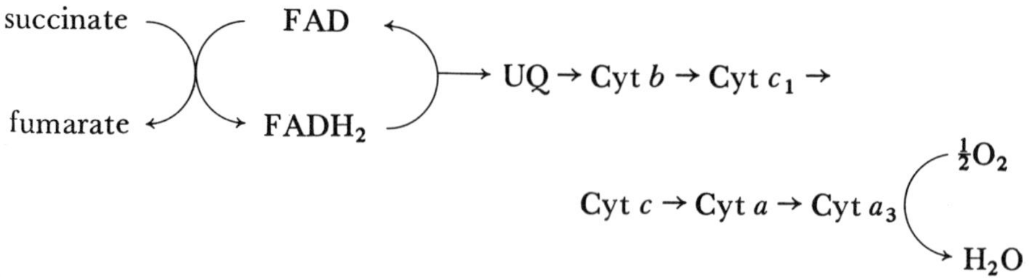

In either case $\Delta G^{0\prime} = -145$ kJ and it represents the sum of the $\Delta G^{0\prime}$ values for each step. The only difference is that in the second case there are more steps than in the first and hence the free energy is released in many smaller packages.

(e) Energy conservation in the respiratory chain

With this thermodynamic background we are now in a position to take a closer look at the bioenergetics of the mitochondrion. The term *energy conservation* refers to the ability of mitochondria to utilize part of the free energy made available by the exergonic reactions of oxidation. The utilization depends on conserving free energy by coupling to an endergonic process. The fundamental endergonic process in the mitochondrion is the generation of ATP by a reaction, reduced to its simplest form:

$$ADP + P_i \rightarrow ATP + H_2O \quad \Delta G^{0'} = +31 \text{ kJ } (+7 \cdot 3 \text{ kcal})$$

In our example in the previous section the free energy released from the oxidation of 2H from succinate amounted to -145 kJ ($-34 \cdot 4$ kcal). If coupling to ATP generation were 100% efficient the P:O ratio would exceed 4. In reality the coupling is less than 50% efficient and the P:O ratio is only 2. The balance of the free energy is not conserved and is dissipated as heat.

We saw earlier that the redox potentials of each of the components of the respiratory chain could be used to predict the free energy changes

$$\Delta G = -nF\Delta Eh$$

If we take values for the standard redox potentials, such as those given in Table 9.4, we can determine the standard free energy changes at each step as we proceed from the substrate to each of the electron carriers in turn and finally to oxygen. The pattern, in Fig. 9.19, reveals how the total energy in the reoxidation of NADH, $\Delta G^{0'} = -220$ kJ ($-52 \cdot 6$ kcal), can be attributed to each interaction along the chain. These "packages" of energy are not of uniform size. There are some sharp, steep, sections of the curve interposed by sections of more gentle gradient. The steep sections correspond to the phosphorylating sites at which sufficient free energy is

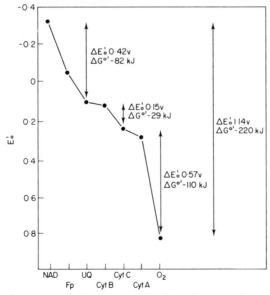

Fig. 9.19 Energy drop during passage of 2*e* along respiratory chain.

released for the synthesis of one molecule of ATP. Using these standard values (E'_0 values) each of the packages of free energy appears to be adequate for the generation of one ATP. Two seem unnecessarily large, particularly the third site (110 kJ (26 kcal)) which in theory should suffice for 3 molecules of ATP. But remember that the E'_0 values are those for a 50-50 mixture of oxidized and reduced forms at M concentrations. We saw that in the steady state the carriers existed in a progressively more oxidized state along the chain. We should not attempt, therefore, to put too precise an interpretation on these calculated values, but to realize that they do provide some essential thermodynamic support for the localization of the sites of phosphorylation.

9.9. Transport of Metabolites across the Mitochondrial Membrane

Now that the specialized functions of the mitochondrion have been explained in some detail we can return to the relationship of the mitochondrion to its intracellular environment. The external mito-chondrial membrane is freely permeable to all but the largest solutes. Were the inner membrane to show the same freedom, all the soluble com-ponents of the matrix would promptly equilibrate with the cytosol with the result that the dilution of substrates would reduce reaction rates by several orders of magnitude. This consideration alone dictates the necessity for an inner membrane which is generally impermeable, but through which selective transport can occur. Selective transport is mediated by carriers which show certain distinctive properties. In the first place, the process exhibits saturation kinetics—exactly comparable to the achievement of a maximum velocity when an enzyme is saturated by its substrate. Secondly, it shows specificity for a substrate or for certain structural features of the substrate. And thirdly, the transport may be susceptible to certain specific inhibitors that do not affect other carriers.

(a) Monocarboxylic acids

Some solutes penetrate the inner membrane without the need for a carrier. These include the basic fuels for mitochondrial oxidation—pyruvate, fatty acids and ketone bodies. These monocarboxylic acids penetrate in their undissociated states:

$$R - COO^- + H^+ \rightleftharpoons R-COOH \quad \longrightarrow\!\!\!|\!\!\longrightarrow \quad R-COOH \rightleftharpoons RCOO^- + H^+$$

In this un-ionized form they do not disturb the ionic balance across the membrane and, since they are wholly metabolized to CO_2 and H_2O, they do not disturb the osmotic balance either.

(b) TCA cycle intermediates

The dicarboxylic and the tricarboxylic acids of the TCA cycle do not

penetrate in the same fashion, probably because the relative concentrations of the undissociated forms are at least 100 times less than in
the case of a monocarboxylic acid. These metabolites must, therefore, be
transported as anions and, in order to preserve electroneutrality across the
inner membrane, an anion must move in the opposite direction. The
scheme is shown in Fig. 9.20. The key to its operation is the specific

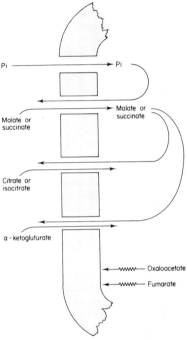

Fig. 9.20 Transport of phosphate and TCA intermediates across the inner mitochondrial membrane.

phosphate carrier which transports phosphate in or out of the matrix. It
crosses in its undissociated state, probably by the acceptance of a proton
from NH_4^+:

$$NH_4^+ \begin{cases} H^+ \\ NH_3 \end{cases} \quad \left.\begin{matrix} H_2PO_4^- \\ \\ H^+ \end{matrix}\right\} H_3PO_4 \quad \longrightarrow \quad H_3PO_4 \left\{\begin{matrix} H_2PO_4^- \\ \\ H^+ \end{matrix}\right. \quad \left.\begin{matrix} \\ NH_3 \end{matrix}\right\} NH_4^+$$

Phosphoric acid and ammonia cross the inner membrane as
uncharged species

This independent phosphate carrier makes phosphate available as a
counter anion to balance the movement of malate or succinate into the
mitochondrion. Fumarate and oxaloacetate are unable to penetrate the

membrane. Citrate, isocitrate and α-ketoglutarate also enter as anions but their carriers use malate rather than phosphate as the counter ion by means of which electroneutrality is maintained.

(c) Shuttle systems between the matrix and the cytoplasm

The inner membrane is impermeable to NAD^+, NADH, Coenzyme A and acyl CoA, since these substances exist in both the matrix and in the cytoplasm, the barrier of the inner membrane effectively separates them into two pools—the intra and extramitochondrial pools. Nevertheless there is a functional need for transfer from one pool to another (Fig. 9.21). Consider NADH arising in glycolysis by the oxidation of cytoplasmic glyceraldehyde 3-phosphate: ultimately this must be oxidized aerobically.

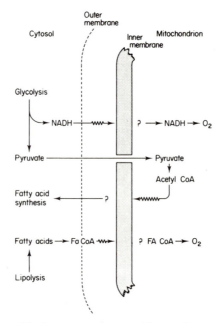

Fig. 9.21 The impermeable inner membrane. Alternative means are necessary if the impermeability of the inner membrane to NADH, acetyl CoA and fatty acyl CoA is to be circumvented.

Consider fatty acid mobilized from adipose tissue and destined to be oxidized in liver or muscle cells. Although the free fatty acids as such can penetrate the inner membrane they will be "activated" by thiokinase in the outer mitochondrial membrane to fatty acyl CoA and must somehow cross the inner membrane to be oxidized in the matrix. Or consider acetyl CoA, produced by pyruvate oxidation within the mitochondrion, some of which will be destined for fatty acid synthesis in the cytoplasm. One's first impression is that the inner membrane is a perverse structure dedicated to frustrating three major pathways—aerobic glycolysis and both the oxidation and synthesis of fatty acids. However, we must remember that the mitochondrion is a well-meaning symbiont and agreeable to an ingenious compromise to overcome each obstacle. Each of these compromises depends on the siting of similar enzymes in each pool—(the intra and extra mitochondrial pools) while permitting the common substrate to shuttle from pool to pool.

So, for NADH, the system uses either glycerol 3-phosphate or malate with the appropriate dehydrogenase present in each pool. The way these two shuttles function is shown in Fig. 9.22. Dihydroxyacetone phosphate

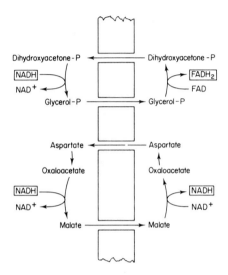

Fig. 9.22 Glycerol phosphate and malate shuttles for aerobic oxidation of cytoplasmic NADH.

reoxidizes NADH, the product, glycerophosphate, enters the mitochondrion where it is reoxidized and returns to the cytosol (playing the part of a counter ion in this carrier system). The glycerophosphate shuttle is unidirectional but the malate-oxaloacetate shuttle is reversible. It has an analogous role, but differs in that oxaloacetate itself cannot cross the membrane but must be transaminated to aspartate.

The carnitine shuttle, which moves long-chain fatty acids into the matrix is shown in Fig. 9.23. Two acyl transferases are involved, one at each face of the inner membrane. The fatty acyl group is transferred from CoA to carnitine

$$(CH_3)_3\underset{+}{N}-CH_2-CH-CH_2-COO^-$$
$$| $$
$$O$$
$$|$$
$$COR \qquad \text{Fatty acylcarnitine}$$

The citrate shuttle on the other hand serves to "move" acetyl CoA out of the mitochondrion. Within the mitochondrion acetyl CoA condenses with

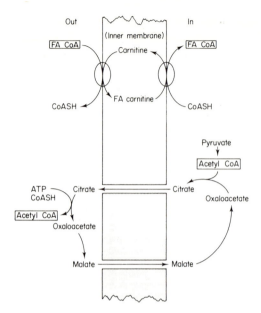

Fig. 9.23 Acyl CoA shuttle systems.

oxaloacetate to form citrate (citrate synthase) which is transported out by a carrier system where it is cleaved by citrate cleavage enzyme to acetyl CoA and oxaloacetate.

(d) ADP and ATP

Since the generation of ATP by harnessing the exergonic reactions of biological oxidation may be regarded as the primary function of the mitochondria, the movement of ADP in and ATP out of the mito-chondrion is no less important. Indeed, since 15 molecules of ATP are generated for each molecule of pyruvate oxidized the transport of the adenine nucleotides is by far the most important carrier function of the inner mitochondrial membrane.

The concept of a counter ion exactly balancing the penetration of an ionized metabolite has been introduced in the previous section. As regards the adenine nucleotides the need is neatly satisfied by the reciprocal movement of ADP in and ATP out of the mitochondrion. The carrier concerned in this process is highly specific for these two nucleotides and the stoicheiometry is precisely a one for one exchange. The result is that the pool of ADP + ATP within the mitochondrion is maintained at a constant level. This particular carrier has been studied extensively, particularly by means of a specific inhibitor called atractyloside. Atractyloside blocks oxidative phosphorylation of external ADP (but not of intramitochondrial ADP) because it prevents the ADP/ATP exchange occurring.

The two nucleotides are exchanged as anions, but it will be apparent that there is a difference in charge because the extra phosphate group in ATP confers an extra negative charge compared to ADP. Since no other counter ion is involved in this carrier system, the imbalance can be compensated only by the generation of a potential gradient across the membrane, for example by the movement of H^+ outwards through the membrane. This is an endergonic process and a portion of the free energy derived from the respiratory chain must be consumed by the ADP/ATP carrier system. It is calculated that $\Delta G^{0'}$ for this process is $+12\cdot5$ kJ ($+3$ kcal).

(e) The transport of cations

The addition of Ca^{2+} ions to a suspension of respiring mitochondria produces rather striking effects. In the first place, Ca^{2+} ions appear to uncouple oxidative phosphorylation and to stimulate respiration, if ADP is the limiting factor. Secondly, the Ca^{2+} ions are found to accumulate within the matrix of the mitochondria, together with phosphate. Indeed, electron micrographs of mitochondria often show electron dense particles of calcium phosphate scattered throughout the matrix. Now this accumulation of Ca^{2+} is an "active", carrier-mediated process and is, therefore, energy dependent. It is inhibited by respiratory chain inhibitors, e.g., CN^-, and by uncoupling agents like dinitrophenol. In the absence of O_2 or oxidizable substrate, it can only continue if ATP is added. The apparent uncoupling effect of Ca^{2+} is more correctly viewed as diverting the energy normally available for ATP generation into the accumulation of Ca^{2+}. The two processes are *alternative* energy acceptor systems in mitochondria.

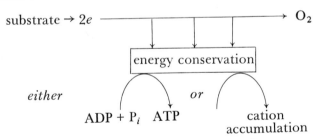

Other cations may also be actively accumulated, for example Na^+ and K^+ ions, but their transport is insignificant compared with Ca^{2+} unless an unusual type of antibiotic is present. Gramicidin and Valinomycin (see Fig. 11.12 for structures) are referred to as ionophorous (ion-carrying) antibiotics. They also behave as uncouplers of oxidative phosphorylation, for the same reason as Ca^{2+} ions—they permit active ion transport to occur and so siphon off energy that would otherwise be available for ATP synthesis. Valinomycin is a cyclic peptide containing D and L valine, lactic and hydroxyisovaleric acids. It binds K^+ (in preference to Na^+) in the centre of the cyclic molecule. The periphery is hydrophobic in properties and is presumed to associate with the lipid environment of membranes perhaps forming an ion-permeable pore or perhaps acting as a mobile carrier.

Since ion transport is linked to respiration it should show a fixed stoicheiometry just like oxidative phosphorylation. For the oxidation of NADH 5 Ca^{2+} and 3 P_i are translocated into the matrix and 6 H^+ are translocated out for each pair of electrons. The ratio of Ca^{2+} to Pi is similar to that in "basic" calcium phosphate or hydroxylapatite.

9.10. The Mechanism of Energy Conservation and Transformation in Mitochondria

We have seen that a significant part of the free energy of oxidation may be conserved in mitochondria and be coupled to the generation of an energy-rich substance (ATP) or to cation transport across the inner membrane. These are alternative endergonic processes which consume the energy of oxidation. There are also other channels in mitochondria which compete with the phosphorylation of ADP, one of which is an energy-driven transhydrogenation of NAD^+ by NADPH. In each instance there must be some mechanism by which the free energy of biological oxidation is transformed into other forms.

The concept of "energy rich" compounds introduced by Lipmann and symbolized by the squiggle $\sim$ is deeply entrenched in biochemical thought and writing. The place of ATP, or rather AMP$\sim$P$\sim$P, as a "middling" member of the class of energy-rich compounds—below "top" members like phosphoenolpyruvate, diphosphoglycerate and creatine phosphate but higher than "bottom" members like cyclic 3,5'-AMP, glucose 1-phosphate and glucose 6-phosphate—has been explained in Section 6.1.

The mechanism by which ATP is generated by the oxidative step in glycolysis was worked out over 30 years ago. This example of "substrate-level" oxidative phosphorylation (see Fig. 9.24) shows the essential role of a thiol group in the enzyme. Oxidation generates a thio-acyl bond RCO$\sim$ S—E, later cleaved by inorganic phosphate yielding another energy rich intermediate R—CO $\sim$ⓟ (1,3-diphosphoglycerate) and finally transferring the 1-phosphate group to ADP. Overall, the reaction is exergonic.

The simplicity of this substrate-level mechanism has led to one hypothesis designed to explain respiratory chain phosphorylation—the "chemical-coupling" or the "chemical-intermediate" hypothesis—first put forward by Slater in 1953. The essential concept is that oxidation in the respiratory chain leads to one or more energy-rich intermediates, designated A$\sim$I, X$\sim$I, and that energy is then conserved as a phosphorylated intermediate X$\sim$P capable of reacting with ADP to form ATP.

Before we look at this scheme in more detail, let us consider the main experimental observations which this and any rival hypothesis must be

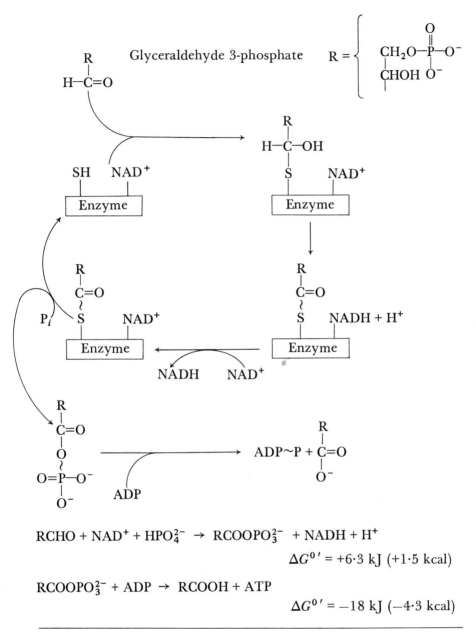

$$RCHO + NAD^+ + HPO_4^{2-} \rightarrow RCOOPO_3^{2-} + NADH + H^+$$

$$\Delta G^{0\prime} = +6 \cdot 3 \text{ kJ } (+1 \cdot 5 \text{ kcal})$$

$$RCOOPO_3^{2-} + ADP \rightarrow RCOOH + ATP$$

$$\Delta G^{0\prime} = -18 \text{ kJ } (-4 \cdot 3 \text{ kcal})$$

Sum: $RCHO + NAD^+ + HPO_4^{2-} + ADP \rightarrow RCOOH + ATP + NADH + H^+$

$$\Delta G^{0\prime} = -11 \cdot 7 \text{ kJ } (-3 \text{ kcal})$$

Fig. 9.24 Glyceraldehyde 3-phosphate dehydrogenase—an example of substrate-level oxidative phosphorylation.

able to explain. Firstly there is the effect of uncouplers, like dinitrophenol, which prevent *any* energy conservation taking place. Secondly, there is the existence of mitochondrial ATPase—an enzyme catalyzing the hydrolysis of ATP: $ATP + H_2O \rightarrow ADP + P_i$. This is a reaction which is the reverse of ATP synthesis and not, therefore, a normal function of coupled, respiring mitochondria. Nevertheless the enzyme is of great interest because it is activated when dinitrophenol uncouples oxidative phosphorylation and it is generally agreed that the same protein is responsible for catalyzing one of the steps normally involved in ATP generation. Thirdly, there are inhibitors which block phosphorylation but not electron transport—for example Oligomycin. In coupled mitochondria such an inhibitor will reduce respiration as well as phosphorylation—the same effect as when ADP is depleted (Fig. 9.25). If dinitrophenol is now

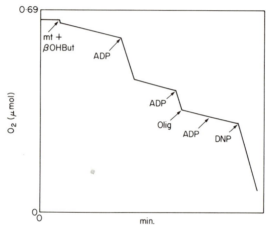

Fig. 9.25 Oligomycin inhibits and dinitrophenol uncouples oxidative phosphorylation. This is an oxygen electrode experiment. The start is similar to those in Fig. 9.16. The mitochondria are coupled as shown by the stimulation of oxidation when ADP is added. When oligomycin (Olig) is added coupled oxidation is inhibited and ADP no longer stimulates. Dinitrophenol (DNP) releases this inhibition by uncoupling oxidation from the (inhibited) phosphorylation.

added the secondary block of oxidation is overcome, without affecting that of phosphorylation—indeed the ATPase activity revealed by dinitrophenol is inhibited by Oligomycin. Fourthly, there are the other energy-accepting systems which are alternatives to phosphorylation, in particular cation translocation—Ca^{2+}, Na^+, or K^+ (see Section 9.9). These effects are inhibited by dinitrophenol but not by Oligomycin. In other words cation translocation taps off energy at a stage before Oligomycin-sensitive phosphorylation is involved.

The basis of the chemical coupling hypothesis is shown in Fig. 9.26. It provides explanations for the experimental observations noted above. A common energy-rich intermediate is able to supply energy for phosphorylation as well as other energy-consuming functions such as cation translocation. When dinitrophenol uncouples, it promotes the breakdown of this or an earlier energy-rich intermediate. Oligomycin blocks at a "lower" point concerned only with ATP synthesis. ATPase activity is stimulated by dinitrophenol because it disturbs the steady state conditions and enables the phosphorylating steps to move in the reverse direction, which is thermodynamically favourable.

This scheme postulates the existence, albeit a transient one, of several distinct energy-rich intermediates, some of which are site-specific ($A_1 \sim I$ etc.) and others common to all three sites ($X \sim I$ and $X \sim P$). Unfortunately no such intermediates have ever been identified, let alone isolated. It is not possible to say which, if any, of the respiratory chain carriers correspond to A_1, A_2 and A_3, nor whether the reduced or oxidized forms take part in the energy-rich compound. And there is no clue as to the identity of I or X. Of course, it may be that these energy-rich inter-

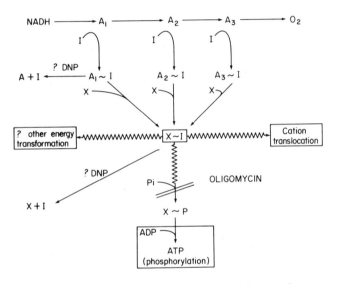

Fig. 9.26 The chemical coupling hypothesis. A_1, A_2 and A_3 are electron carriers in the respiratory chain which form energy-rich intermediates with I (? a protein) $A_1 \sim I$ etc. Another reaction with X (? another protein) yields the *common intermediate* for all three energy conserving sites, $X \sim I$. This is a source of energy for several transforming functions including phosphorylation. On this hypothesis dinitrophenol leads to the hydrolysis of either $A \sim I$ or $X \sim I$ with the destruction of the energy-rich bond. Oligomycin inhibits the formation of the first phosphorylated intermediate $X \sim P$.

mediates are extremely labile and cannot survive experimental procedures designed to identify them. But it is clear that this lack of corroboration is a serious weakness in the chemical coupling hypothesis.

An essential aspect of energy conservation in mitochondria is the necessity for an intact inner membrane. With few exceptions submito-chondrial particles are capable only of oxidation and/or electron transport and have lost the ability to generate ATP. This is true for the complexes I-IV, each containing different portions of the respiratory chain, which are obtained by lipid solvent and detergent treatment of the inner membrane. The only fragments which possess phosphorylating ability are those

produced by mild procedures—sonic vibration or phospholipase C treatment. Racker has shown that, by these treatments, vesicles are formed
from the inner membrane such that the inner, or matrix, surface is on the
exterior (Fig. 9.27). This inner surface has a surface covering of spherical
particles about 8·5 nm diameter. Preparations of such vesicles are capable of
oxidation and energy conservation. They behave like intact mitochondria
when treated with dinitrophenol or oligomycin and they exhibit ATPase
activity. It is possible to treat them by mechanical shaking with small glass

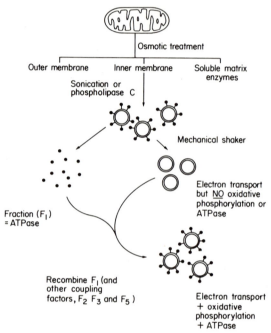

Fig. 9.27 Inner membrane preparations—separation and recombination of F_1 spheres.

beads so that the spherical particles are shaken off. After such treatment
the vesicles are capable only of electron transport but not phosphorylation. ATPase activity has disappeared too, and is now identified in
the fraction containing the spheres—usually known as coupling factor F_1.
Recombination of the spheres with the membrane fraction restores
ATPase activity to the spheres, but full restoration of phosphorylation
requires some other soluble proteins—further coupling factors F_2, F_3 and
F_5. One or more of these factors is involved in conferring Oligomycin
sensitivity which is not a feature of the isolated F_1.

Now this type of experiment tells us, firstly that the inner membrane
must be more or less intact for oxidative phosphorylation and secondly,
that the final part of the phosphorylating mechanism, identifiable as an
ATPase when working in reverse, is located in the spheres on the matrix
surface of the inner membrane. Whether these spheres (F_1 particles) or
any of the coupling factors can be assigned precise roles in the chemical

coupling hypothesis is still very uncertain. Nor is it clear why an intact inner membrane should be so essential, since this is not a fundamental requirement of the chemical coupling theory.

The dominant role of the inner membrane in energy conservation was one of the factors which led Mitchell in 1961 to put forward an alternative hypothesis. This is known as the chemiosmotic hypothesis. It seeks to provide an explanation in terms of an ion-impermeable membrane across which electron transport generates a proton (H^+) gradient which, in turn, "drives" ATPase in the reverse direction thus generating ATP. The proton gradient is central to Mitchell's theory. Protons are involved in the respiratory chain at each point where a hydrogen carrier links with an electron carrier. They are liberated in one direction:

$$\text{e.g., } FADH_2 + 2Fe^{3+} \rightarrow FAD + 2Fe^{2+} + 2H^+$$

and consumed in the opposite direction

$$\text{e.g., } 2Fe^{2+} + \tfrac{1}{2}O_2 + 2H^+ \rightarrow 2Fe^{3+} + H_2O$$

If the respiratory chain loops from one surface of the inner membrane to the other, one can postulate the consumption of protons on one side and their secretion on the other. The scheme in Fig. 9.28 shows how NADH oxidation requires a three-loop chain, with alternating hydrogen

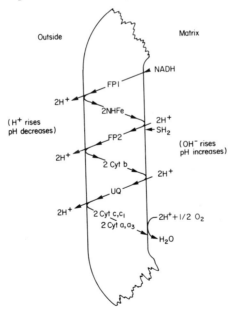

Fig. 9.28 Chemiosmotic hypothesis—proton translocation across the inner membrane. The order of the components is slightly unconventional—nonhaem ion (NHFe) is separated by two flavoproteins (FP1, FP2) and ubiquinone (UQ) is situated after rather than before cytochrome b. This order is not inconsistent with redox values. The result of the oxidation of 1 NADH is the transfer of 6 protons from the inner (matrix) side to the outside. ←SH$_2$ indicates that succinate oxidation would be achieved by a two-loop system with the movement of 4 protons.

and electron carriers arranged so that 3 pairs of protons are consumed from the matrix side and expelled on the outer side. Succinate or fatty acyl CoA oxidation could be accommodated by a "two-loop" chain. The result of this proton movement is to cause a rise in pH (an excess of OH^-) on the matrix side and a fall of pH (an excess of H^+) on the outside of the membrane. Mitchell has used a sensitive and rapidly responding pH recording apparatus and has shown that when a "pulse" of oxygen is admitted to an anaerobic suspension of mitochondria there is a rapid fall in the pH of the medium. The stoicheiometry is exactly that predicted in the scheme ($6H^+$ per $2e$) and the fall in pH may be blocked by the usual respiratory chain inhibitors—cyanide, Antimycin A etc.

The result of this proton translocation is to set up a proton gradient and an electron gradient across the membrane:

outside			matrix
high [H^+]	inner membrane		high [OH^-]
electropositive			electronegative

Each gradient exerts a force which tends to drive protons back into the matrix (protonmotive and electromotive forces).

The ATPase reaction, like other hydrolases, is strongly exergonic. To reverse it, so as to generate ATP, involves a dehydration, which is strongly endergonic. Mitchell proposes that the proton gradient provides the energy to drive the reversed ATPase reaction. We know that the ATPase is located on the matrix side of the membrane (the F_1 spheres, see above) and because of its asymmetric location it is susceptible to the forces arising from the proton and electron gradients across the membrane. A scheme is shown in Fig. 9.29. The dehydration occurs on the matrix side,

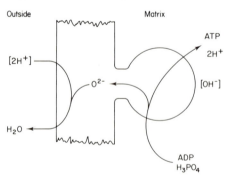

Fig. 9.29 ATP generation by reversed ATPase driven by the proton gradient across the inner membrane. The ATPase, located in the inner spheres, acts, in reverse, to dehydrate ADP and P_i. The elements of water thereby extracted diverge: $2H^+$ drawn to the matrix by the excess [OH^-] while O^{2-} crosses the membrane to react with the excess H^+ on the outside of the membrane. The reason for postulating O^{2-} rather than OH^- is that $2H^+$ are involved for each ATP generated. It is probable that O^{2-} is transported across the membrane by means of carriers, X and I, which ionize X^- and IO^- and which might be identified with coupling factors.

two protons emerging on this side to be neutralized by the excess hydroxyl ion, while the oxide ion (O^{2-}) is transported (probably by a pair of carriers, omitted from the scheme) to the outside where it reacts with two protons, present in excess on this side of the membrane.

The chemiosmotic hypothesis is also competent to explain the main experimental observations by which the chemical coupling hypothesis was tested. Dinitrophenol uncouples because it is able to penetrate the membrane freely and, as a weak acid, it acts as a proton carrier which destroys the proton gradient. In other words it spoils the proton-impermeable membrane which is essential for energy conservation. ATPase activity becomes manifested only when the proton gradient is released by uncoupling with dinitrophenol. Oligomycin inhibits a carrier or coupling factor (X or I) concerned in translocating the O^{2-} ion. And, finally, cation transport is readily accounted for by the proton and electromotive forces set up across the membrane. In addition to supplying explanations to these points the Mitchell hypothesis has an important advantage over the chemical coupling hypothesis. Unlike the latter, it explains why energy conservation demands an intact inner membrane. At the present time mitochondrial experts are still fairly equally divided between those who support each of the two main theories and those who maintain an open mind. The debate and the experimentation continues apace: so far each apparently mortal blow has been parried by an ingenious experiment or interpretation by the other side.

10

Extramitochondrial Oxidation

The activities of mitochondria described in Chapter 9 account for nearly all of cellular respiration and, with the exception of two energy-conserving reactions in glycolysis, for all the ATP generated in the cell. The contribution of oxidative enzymes located elsewhere in the cell is quantitatively small in terms of oxygen consumption but not insignificant in the metabolism of specific substrates. In this chapter we shall be concerned with examples of two types of oxidoreductase—containing either copper or flavine nucleotide and with the oxidative roles of two intracellular structures—peroxisomes and the endoplasmic reticulum. We shall not be primarily concerned with various NAD- and NADP-linked dehydrogeneses present in the cytosol except in so far as they relate to these topics.

10.1. Copper-containing Oxidoreductases

Copper can serve as an electron carrier by reversible oxidation and reduction $Cu^+ \underset{+e}{\overset{-e}{\rightleftarrows}} Cu^{2+}$.. Enzymes utilizing copper in this way can be directly reoxidized by oxygen. The general reaction is:

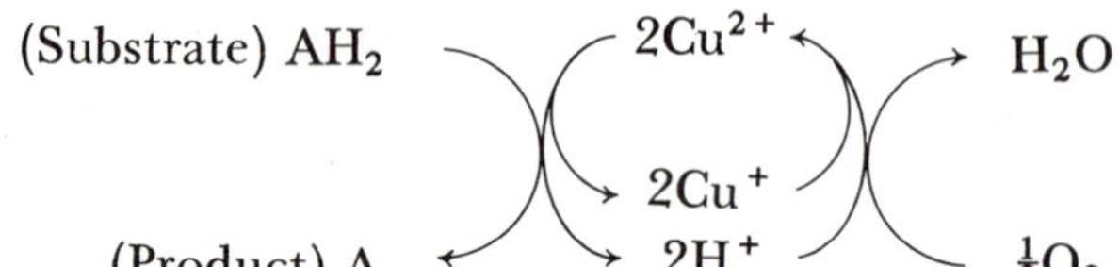

Several enzymes in this class are listed in Table 10.1. In plants the common ones are polyphenol oxidases that oxidize catechol and other polyphenolic compounds to quinones, which may in turn polymerize to form dark pigments. The browning of the cut surface of a potato or an apple is due to this type of enzyme. In mammals the most important example is tyrosinase:

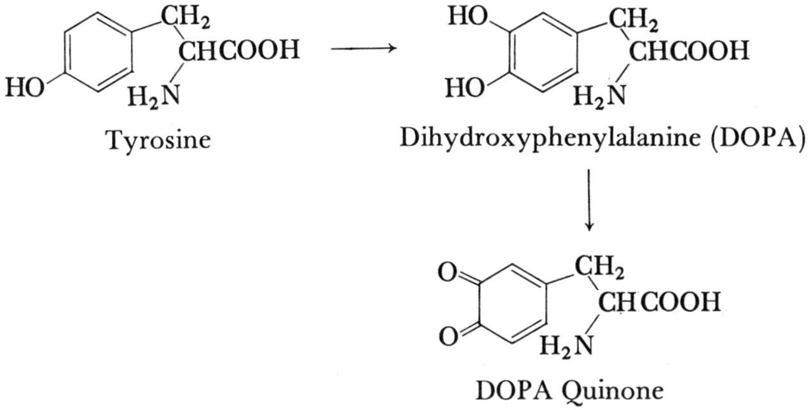

Tyrosine   Dihydroxyphenylalanine (DOPA)

DOPA Quinone

It catalyzes two oxidative steps, first to DOPA and then to DOPA quinone. DOPA is the precursor, in neurones of the adrenal medulla and elsewhere in the nervous system, of catecholamines. DOPA quinone is the starting point for a complex series of reactions by which a black polymer, melanin, is elaborated by melanocytes in the skin and other pigmented tissues. The condition of albinism is due to a genetic deficiency of tyrosinase.

TABLE 10.1. Extramitochondrial oxidases

Copper containing oxido-reductases	Flavoprotein oxido-reductases
Reduced carrier (Cu^+) reoxidized by $\frac{1}{2}O_2$ to form H_2O	Reduced carrier ($FADH_2$) reoxidized by O_2 to form H_2O_2
tyrosinase	xanthine oxidase
ascorbate oxidase	D-amino acid oxidase
urate oxidase	glycine oxidase
caeruloplasmin (a blue plasma protein)	monoamine oxidase
phenol oxidases (plants)	glucose oxidase (fungi)
(cf. also cytochrome a_3)	

10.2. Flavoproteins

A number of important mitochondrial flavoproteins were described in Chapter 9. The extramitochondrial flavoproteins differ in a fundamental way—the reduced flavine group is directly reoxidized by oxygen to form hydrogen peroxide.

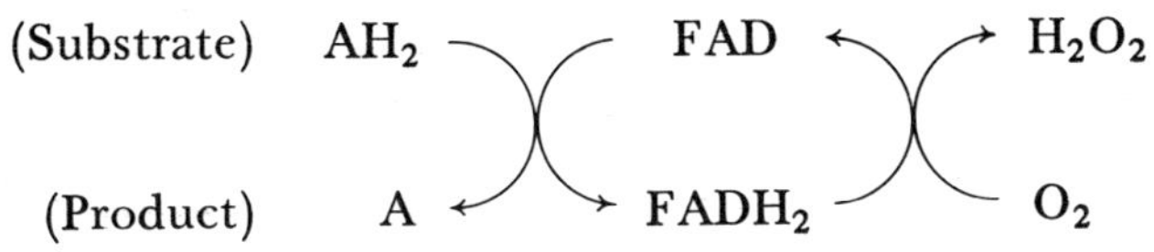

A few enzymes of this type are also listed in Table 10.1. Some of them contain metals—xanthine oxidase, for example, an enzyme important in purine metabolism which oxidizes hypoxanthine to xanthine and xanthine to uric acid, contains Fe and Mo. Hydrogen peroxide is potentially toxic and its formation as the product of this type of reaction demands an explanation concerning its removal. Two types of enzyme are capable of utilizing H_2O_2—peroxidase and catalase. Both are haem proteins and both catalyze essentially similar reactions.

Peroxidase   Catalase

$$XH_2 \quad\rightharpoondown\quad H_2O_2 \qquad\qquad H_2O_2 \quad\rightharpoondown\quad H_2O_2$$

$$X \quad\leftharpoondown\quad 2H_2O \qquad\qquad O_2 \quad\leftharpoondown\quad 2H_2O$$

In the peroxidase reaction another substrate XH_2 is oxidized as H_2O_2 is reduced. In the catalase reaction two molecules of H_2O_2 are involved in the same relationship. Catalase has the distinction of being among the most active enzymes known—it has a turnover number of 5×10^6 (i.e., moles H_2O_2/min/mole enzyme).

10.3. Peroxisomes

These organelles were originally called microbodies by electron microscopists. In liver cells they appear to be about a quarter of the volume of a mitochondrion and about one third as frequent. They are roughly spherical and characteristically show a crystal-like lattice at their centre (Fig. 10.1). Their function was obscure until 1965 when a purified fraction free of mitochondria and lysosomes was obtained. Because their function is concerned with the production and destruction of hydrogen peroxide the name peroxisome has come into general use for these organelles.

Peroxisomes contain an odd collection of copper-, FAD- and haem-containing enzymes, some of which are listed in Table 10.2. Urate oxidase is located at the centre of the peroxisome and accounts for the crystalline core seen in electron micrographs.

The role of the peroxisome is still a matter of some speculation. One theory is that it represents a primitive, premitochondrial system for biological oxidation. The FAD lactate oxidase has an NAD counterpart (lactate DH) in the cytosol and it has been suggested that a shuttle system operates whereby NADH can be reoxidized independently of the mitochondria. A hypothetical scheme is shown in Fig. 10.2. Pyruvate, lactate,

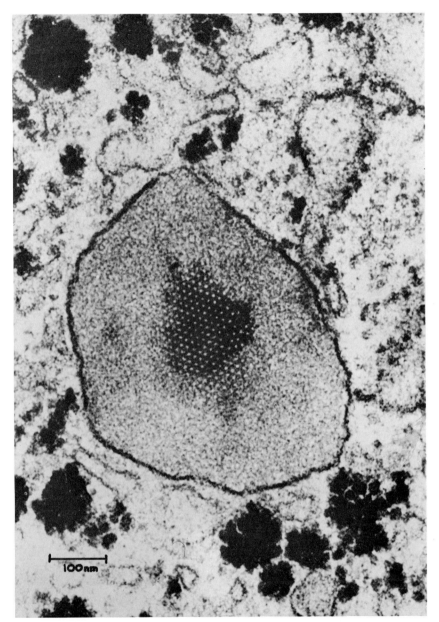

Fig. 10.1 Electron micrograph of a peroxisome (microbody). Guinea pig liver. The crystalline lattice in the centre of the organelle is clearly visible. (From Roodyn.)

ethanol and acetaldehyde shuttle across the peroxisome with the overall reaction:

$$2NADH + 2H^+ + O_2 \rightarrow 2NAD^+ + 2H_2O$$

Two molecules of NADH are thus oxidized but no energy is conserved. Although feasible, the usefulness of such a system is difficult to understand.

TABLE 10.2. Some enzymes identified in Peroxisomes

Enzyme	Prosthetic group	Function
Urate oxidase	(Cu)	Urate → allantoin
Lactate oxidase (an α-hydroxyacid oxidase)	(FAD)	lactate → pyruvate
D-amino acid oxidase	(FAD)	$RCHNH_2COOH \rightarrow RCOCOOH$
Catalase	(haem)	$2H_2O_2 \rightarrow 2H_2O + O_2$

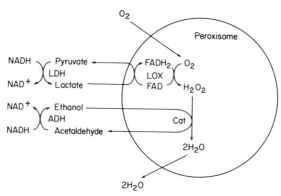

Fig. 10.2 Possible role of peroxisomes in reoxidizing cytoplasmic NADH. LDH, ADH = Lactate DH and Alcohol DH in cytoplasm; LOX = (FAD) Lactate oxidase in peroxisome; Cat = Catalase acting as a peroxidase;

overall: $2NADH + 2H^+ + O_2 \rightarrow 2NAD^+ + 2H_2O$ but no energy conserved.

10.4. Biological Oxidations in the Endoplasmic Reticulum (Microsomes)

When a tissue, such as liver, is homogenized and fractionated by differential centrifugation the high-speed pellet, known as the microsomal fraction, is pink in colour and contains many small membranous vesicles derived mainly from rough and smooth endoplasmic reticulum (ER). The pink colour is due to two haem proteins—cytochrome b_5 and cytochrome P450. They are important components in a microsomal electron transport chain, the function of which is still only partially clear. These two cytochromes are present in roughly equal amounts in liver and each may exceed the liver content of cytochrome c by a factor of three.

Cytochrome b_5 has an absorption spectrum and redox potential (E_0' −0·02) characteristic of the b group. It is firmly bound to the ER membrane, to which some flavoproteins are also attached. These flavoproteins can transfer electrons from NADH or NADPH to cytochrome b_5:

NADH ⟶ Fp ↘
                    cyt b_5 ⟶ cyt c
NADPH ⟶ Fp ↗

Both electron transport chains are quite active, particularly that from NADH and therefore provide another means of reoxidizing these reduced coenzymes. It is also well-established that *in vitro* cytochrome c can act as an acceptor of electrons from reduced cytochrome b_5. If this pathway operates in the intact cell it represents a link between microsomal and mitochondrial oxidation. However, there is no good evidence that this occurs *in vivo* and it may be that the free energy of this chain is utilized for ion transport or some other function in the ER.

Cytochrome P450 is so named because of its strong absorption band at 450 nm which appears when microsomes are exposed to carbon monoxide. This ability to bind CO is also a feature of haemoglobin and cytochrome a_3—two haem proteins that have the ability to react with oxygen. It is clear that cytochrome P450 also shares this property. The presence of this oxygen-combining protein in the ER is now known to be related to various reactions in the ER which utilize oxygen. Such reactions are catalyzed by enzymes known as *oxygenases*, the characteristic feature of which is that molecular oxygen enters into the reaction. Either both atoms are incorporated into the substrate:

$$A + O_2 \rightarrow AO_2$$

or only one atom:

$$AH + O_2 \rightarrow AOH + [O]$$

In the second case another acceptor for [O] is involved, usually NADPH. This latter reaction is an example of a hydroxylase, sometimes called a mixed function oxygenase because it requires two different acceptors for each oxygen atom. Hydroxylations of all kinds of lipid-soluble substrates are carried out in the ER. These include steroids, e.g., cholesterol, oestradiol, cortisol and bile acids, the biosynthesis of which involves several specific hydroxylations—as well as many foreign compounds including a wide range of drugs. Although there are many specific hydroxylases for particular substrates, each hydroxylase shares in a common system by which oxygen enters the reaction. This is explained in Fig. 10.3. There are three components—the hydroxylase (shown here as a steroid hydroxylation but equally applicable to a barbiturate) the oxygen carrier, cytochrome P450, which only functions in the reduced (Fe^{2+}) form and must be reduced by an electron transport chain comprising non-haem iron, flavoprotein and NADPH. This cytochrome P450 system also participates in several other oxidative reactions by which drugs are metabolized. Although in liver this system is located in the ER, in adrenal cells cytochrome P450 is located in the mitochondria in which the respiratory chain provides electrons to regenerate the reduced form.

The importance of this system in drug metabolism is apparent from two general observations. Firstly, the administration of a barbiturate, say phenobarbitone, over a few days leads to a striking increase in the amount

Overall the hydroxylation can be written as:

Steroid $\xrightarrow[\text{NADPH} + \text{H}^+]{\text{O}_2}$ Hydroxy Steroid $\xrightarrow[\text{NADP}^+]{\text{H}_2\text{O}}$

RH   ROH

In detail it is a complicated mechanism in which $\frac{1}{2}$(NADPH + H$^+$) is required at two points: first to provide H to form H$_2$O from one atom of O$_2$, secondly to provide an electron (e) to reduce cytochrome P450 to the Fe^{2+} state. The latter step needs a chain of electron carriers including flavoprotein (Fp) and nonhaem Fe.

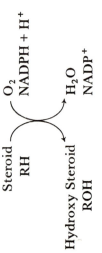
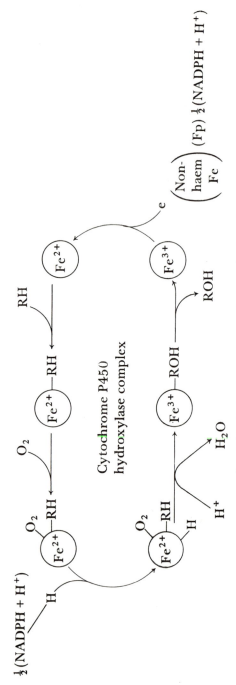

Fig. 10.3 Role of cytochrome P450 in hydroxylations.

of smooth ER in liver cells accompanied by a four-fold rise in cytochrome P450, in a flavoprotein which reoxidizes NADPH and in the hydroxylation rates of many substrates, including steroids. This is an example of a drug-induced enzyme synthesis (see also Chapter 5.11(b)). Increased tolerance to drugs results from this effect. Secondly, an experimental drug, usually known as SKF 525A, is an inhibitor of cytochrome P450. When it is administered to animals, it inhibits drug metabolism generally, i.e., the rate of inactivation of many drugs is much reduced. The result is that the duration of drug action can be prolonged—an effect which may be useful but may also be potentially dangerous.

11

Membranes and Transport of Materials

11.1. Introduction

The very concept of discrete cells implies the presence of limiting boundaries surrounding these entities, but the nature of this boundary was for a long time very ill-defined. At first, it was assumed that the cell wall, as most obviously seen in plants, constituted the limiting membrane. But when experiments revealed firstly that dyes could penetrate this wall but not actually pass into the body of the cell, and secondly, that suspension of plant cells in strong sucrose solutions caused the inner matrix to shrink away from the rigid wall, some re-evaluation of the original ideas became necessary. We now know that the cell wall and the cell membrane are two distinct organelles, the first a somewhat passive structural element and the second a much more dynamic entity. They also serve quite different functions, which are intimately related to the environment of the organism. This is best illustrated in the case of bacteria, which often exist in media of widely differing osmolarity. In instances where the extracellular environment is significantly more dilute than the intracellular fluid, were it not for the restraining influence of a rigid cell wall, the resulting inrush of water would cause the cells to burst. Since most animal systems possess homeostatic mechanisms to control their cellular environment, the cell wall is unnecessary and hence usually absent. The cell membrane is not designed to provide a mechanical buttress but constitutes instead an important organizational element in the cell. Firstly, it acts as a permeability barrier, controlling the transfer of both matter and information between external and internal compartments. Secondly, it serves as a structural support to facilitate the correct interaction of multicomponent enzyme systems.

Whilst this discussion of membrane form and function will be devoted almost entirely to the plasma membrane, it is important to appreciate that most eukaryotic cells also possess intracellular membranes. These are essential to the structure of such elements as the endoplasmic reticulum, the Golgi apparatus, the nucleus, mitochondria and various forms of vesicle. All membranes, nevertheless, seem to have a uniformity in organization upon which is superimposed the diversity in composition and function associated with different cell types and organelles.

11.2. Role of the Lipid Fraction

Since the lipid components of cells (other than adipocytes) are largely confined to the membranous elements, it is fitting that we should begin by examining the role of lipids in this structural feature. As was explained in Chapter 8, phospholipids contain two distinct portions which together profoundly influence the behaviour of these molecules in the aqueous environment of the cell. One portion, usually considered non-polar, is composed of the two long hydrocarbon chains of the esterified fatty acids whilst the second, polar portion, has charged groups associated with phosphate and the attached bases. If one adds to the surface of an aqueous medium a drop of a solution of phospholipid in some organic solvent and allows the solvent to evaporate away, the phospholipid fraction becomes organized into a monomolecular film in which the individual molecules are oriented in a very specific manner. In such a surface film the polar or hydrophilic portions of the phospholipid molecules face the water phase and the non-polar or hydrophobic portions project up into the air.

Considering for a moment the energetic characteristics of hydrophilic and hydrophobic groupings, it is easily seen that the structural organization of this film maximizes the contact of the polar head group with water (a polar solvent) and at the same time minimizes the contact between water and the fatty acid chains. The overall result is to achieve a system of minimum free energy.

In 1925, Gorter and Grendel utilized this phenomenon in an attempt to obtain some estimate of the thickness of the plasma membrane of the red blood cell. Their experiment consisted of extracting all the phospholipid fraction from a known number of erythrocytes (which contain no cytoplasmic membranes of any kind), and measuring the total surface area occupied by a monomolecular film derived from that quantity of phospholipid. The area so obtained was nearly twice that of the total surface area of the original erythrocyte population. They, therefore, concluded that the membrane was comprised of a *Bimolecular leaflet* of phospholipid molecules arranged (as shown in Fig. 11.1) in compliance with the fundamental energetic principles developed earlier.

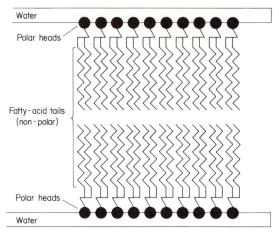

Fig. 11.1 Bimolecular Lipid Leaflet.

Nearly half a century later this basic structure is still thought to be correct although, ironically, repetition of their work has revealed that the extraction of phospholipid was far from complete but the error was compensated for by similar errors in the estimation of the surface area of the intact erythrocyte.

So far we have considered only the result of the application of phospholipid to the surface of an aqueous medium. If instead such material is introduced into the bulk phase, several other structural forms are obtained—in particular, extensive sheets or lamellae occur (smectic mesophase). These may be treated with potassium permanganate, osmium tetroxide, uranyl acetate, or phosphotungstic acid and examined by electron microscopy. The picture that always emerges shows a repeating pattern of dark bands each separated by a lighter, electron transparent area. This is the typical trilaminar appearance of cell membranes which is taken to indicate the presence of a lipid bilayer. The dark lines correspond to the hydrophilic head groups with which the electron dense fixatives become associated and the light regions the interior or hydrophobic compartment of the membrane. Although this pattern is ubiquitous from the simple lipid system discussed above to the highly developed membranes of the mitochondrion, it is pertinent to point out that the dimensions of these areas may vary widely, depending upon the composition and degree of hydration of the membrane. A reasonable "rule of thumb", ascribes to the lighter area a width twice that of a single dark line, the overall thickness of the 3-layered structure generally being of the order of 10 nm.

It is also possible to orient the synthetic lipid layers so that they may be amenable to X-ray diffraction, a technique which, it will be recalled, reveals the electron density profile of regularly repeating units. A typical example of such a profile is illustrated in Fig. 11.2 and this again confirms the presence of the bimolecular leaflet form. This pattern is important in that it serves as a basic reference source for diffraction studies on natural membranes.

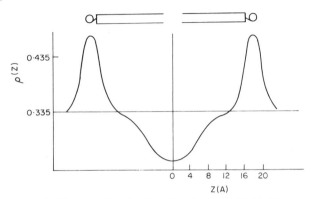

Fig. 11.2 Electron Density Profile of a pure Lipid Bilayer.

This X-ray technique has also revealed a further aspect of lipid organization which is assuming increasing importance in our understanding of membrane function. Investigations of the diffraction patterns obtained from lipid bilayers at different temperatures shows that the molecules can undergo phase transitions. Below about 15°C to 20°C the fatty acid chains of the phospholipids are closely packed in a hexagonal crystalline array. Above this temperature (called the transitional or melting temperature), these hydrocarbon units have much greater freedom of movement and are considered, therefore, to be in a liquid or fluid-like state. As one would expect, this phase transition results in lateral expansion in the plane of the membrane, a decrease in the thickness of the bilayer and a net increase in volume per lipid molecule. The exact temperature at which the phase change occurs depends on the length and degree of saturation of the esterified fatty acids. In general, the larger and more saturated the hydrocarbon chain the higher the temperature of the transition.

This property of fluidity has been further investigated using electron spin resonance (ESR) in conjunction with spin labels. The technique is based on the ability of certain chemical compounds which possess unpaired electrons (i.e., are free radicals) to absorb microwave energy at a characteristic magnetic field strength. A typical spin label (usually a nitroxide) has the structure $O \leftarrow N$ ⬡ . This material will give rise to a characteristic ESR signal—which will depend, amongst other things—on the position of the molecule relative to the applied magnetic field. If such a spin label is attached to the fatty acid portion of a phospholipid and is incorporated into the bilayer, one can deduce from the ESR spectra the degree of freedom of movement within the phospholipid molecule. Consistent with the X-ray data, it is found that the fatty acids are either static, as in a crystal, or mobile, as in a liquid, depending upon the temperature at which the spectral measurements are carried out. The technique is also powerful enough to allow further important parameters of bilayer organization to be defined. For example, if the nitroxide group is placed at various distances from the esterified carboxyl group on the fatty acid, it is possible to deduce that the hydrocarbon chain is extended in a direction perpendicular to the membrane surface. Moreover, the further the spin label is into the bilayer, the more motion it is able to undergo. This suggests that the fatty acid side chains of the phospholipids are relatively fixed at the membrane surface but are able to move in a "whipping" form of motion through an arc or cone in the hydrophobic portion of the bilayer.

So far we have discussed only the relative mobility of the hydrocarbon chain. What of the movements of the whole phospholipid molecule? McConnell and his colleagues have found that spin labelled derivatives of phosphatidyl choline (PC), at temperatures above that of the transition point, are able to diffuse rapidly in the plane of the membrane at a rate of about 1 μm/second. In contrast, the same molecules do not "jump" very quickly from one side of the bilayer to the other. Indeed the rate of exchange (or "flip-flop") between the two faces of the membrane corresponds to a half-time of approx. 6 h at 30°C, a rate about 10^9 times slower than lateral diffusion. This is hardly surprising in view of the unfavourable situation—energetically speaking—created by having charged species in a hydrophobic environment such as would arise with rapid "flip-flop". These general principles of lipid mobility can be applied to both artificially produced and natural membranes subject only to slight quantitative differences in the relative rates. For example, in the fusion of two membranes such as might occur when intracellular vesicles combine with the plasma membrane, one would expect that intermixing of the lipids from the two sources would occur and this, in fact, is what is usually observed.

One of the consequences of this slow "flip-flop" is to inhibit the randomization of lipid between the internal and external faces of the membrane. Therefore, any asymmetric distribution of phospholipid species, created in the course of membrane biosynthesis should tend to persist. Is this postulated asymmetry a reality in natural membranes? The answer seems to be yes. There is good evidence from experiments with specific reagents that can label phospholipids and from using phospholipases that degrade phospholipids that PC is on the external surface of the membrane while phosphatidyl serine (PS) and phosphatidyl ethanolamine (PE) are located on the cytoplasmic side of the erythrocyte membrane. PE and PS are only available to the reagent or the enzyme if the membrane is first fragmented. This observation is interpreted as reflecting an inherent asymmetry in phospholipid distribution; PS and PE being located in the cytoplasmic surface of the membrane with PC and sphingomyelin assigned to the external surface. Consistent with this hypothesis, one finds that the PC and sphingomyelin content is roughly equal to the amount of the amino containing residues present. Although these studies on the asymmetry of lipid distribution have been confined to the erythrocyte, it seems likely that a similar situation may exist in many other cell types.

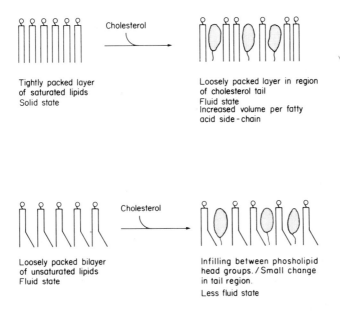

Fig. 11.3 The Influence of Cholesterol on Lipid Fluidity.

This account of membrane structure has so far dealt only with the charged lipids. However, a neutral fraction also exists, principally represented by cholesterol, which may comprise anything from zero % (in bacterial membranes) to nearly 50% of the total membrane lipid (in some animal cells). The influence of cholesterol particularly on the fluidity of the membrane has been examined by X-ray, NMR (nuclear-magnetic resonance), and spin label techniques. The data from these experiments are consistent and can best be illustrated as shown in Fig. 11.3; the "kink" in the hydrocarbon portion represents a change in the position of the fatty acid chain often associated with the presence of a double bond.

In other words, cholesterol has the property of modulating the state of the bilayer such that the fluidity of saturated components is increased and that of unsaturated, reduced. A good example of its effects is seen in nerve myelin where the presence of cholesterol in a highly saturated bilayer reduces the rigidity of the resulting membrane.

Since most natural membranes are composed of a spectrum of fatty acids which differ in their degree of unsaturation, the bilayer will, at any one temperature probably contain both liquid and crystalline phases (just as, at room temperature, saturated animal fat is usually solid and unsaturated vegetable oil, liquid). The equilibrium between these two states can be altered by the pH and ionic composition of the medium, by the presence of cholesterol as indicated above, or by external influences such as the dissolution of certain anaesthetics in the membrane. This co-existence of solid and liquid phases confers on the lipid fractions the ability to be almost infinitely compressible, a property of profound importance to the regulation of membrane-associated processes. We shall return to this aspect later.

In summary, therefore, we see that the lipid portion of membranes is organized into an asymmetric lipid bilayer. Within this leaflet the molecules are free to undergo intramolecular and lateral motion, which is governed by their chemical structure, and the temperature of the medium.

11.3. Role of Proteins

On the basis that membranes are comprised solely of lipid, the surface tension of a cell should be roughly comparable to that of an oil droplet. In fact, the tension at the cell surface of a sea urchin egg was found to be considerably lower than that of an oil droplet. This observation precipitated a series of experiments which culminated in the Davson-Danielli model for membrane structure. Their model incorporated the concept of the bimolecular lipid leaflet and attributed the reduced surface tension observed to the presence of protein molecules coating the ionic surfaces of the bilayer. In a later version of this model, Stein and Danielli proposed that the proteins existed largely in the β-pleated sheet conformation and that there were protein-lined channels through the membrane. The essential features of these proposals were also to be found in Robertson's concept of the generalized "unit membrane". Robertson further postulated that the protein distribution on the inner and outer surfaces of the bilayer was not, in fact, symmetrical—a point to which we will return (see Fig. 11.4).

Let us examine this model in the context of the human erythrocyte membrane. If the membrane is solubilized by a detergent (sodium dodecyl sulphate, SDS) and the components separated by gel electrophoresis (Fig. 11.5), its protein components exhibit as wide a variation in molecular size

Exterior

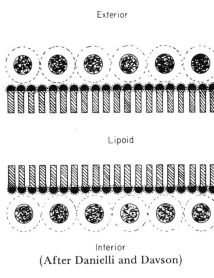

Lipoid

Interior
(After Danielli and Davson)

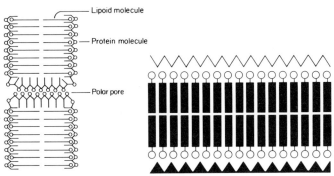

After Stein and Danielli   After Robertson (1964).

Fig. 11.4 Early Membrane Models.

as cytoplasmic proteins. Treatment of the isolated membrane with the divalent ion-chelator, EDTA, results in the release of components I, II and V and they disappear from the gel pattern. The addition of 0·1 M NaCl removes component VI. Both these effects are consistent with ionic interactions of some form, perhaps with the phospholipid headgroups as suggested by Danielli.

On the other hand, several proteins, particularly component III, can only be solubilized from the membrane with detergents which possess the basic amphipathic structure exhibited by phospholipids. This result suggests that some membrane proteins have a hydrophobic surface to which detergent molecules must bind before the protein can be released into the aqueous phase. Components IV (a) and (b) require for their dissolution, strong denaturing agents. These are thought to unfold the polypeptide chain, thereby exposing buried hydrophilic side-chains, which render these proteins water-soluble.

In the light of these observations, some refinement of the above models

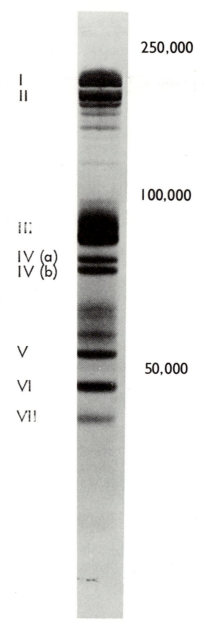

Fig. 11.5 SDS-Polyacrylamide Gel Pattern of Protein Components from Human Erythrocytes.

is called for, particularly as they relate to protein-lipid interactions. However, the basic concepts in these models remain essentially correct.

We know that proteins contain both polar and non-polar side-chains and, therefore, in principle should be capable of interaction with both the hydrophobic and hydrophilic portions of the bilayer. The first of these possibilities was not really anticipated in the earlier models. It is also easily appreciated that the degree of integration of the protein into the lipid matrix will depend on the relative amounts of charged and uncharged residues on the surface of the molecule. Several different forms of this association may be envisaged as seen in Fig. 11.6.

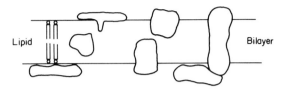

Fig. 11.6 Integration of proteins into the lipid bilayer.

Thus, by determining how much detergent a protein, in its native conformation, is able to bind, we can arrive at a rough estimate of how much of its surface is submerged in the hydrophobic portion of the bilayer.

It is also important to ascertain which proteins are on the outside of the membrane, which on the inside and which, if any, traverse the entire bilayer. The experimental design in this case is similar to that described earlier in connection with the asymmetric distribution of phospholipid. It consists of treating intact cells with a group-reactive reagent (e.g., for an amino group) which because of its membrane impermeability will only label those proteins accessible on the outer surface. Repetition of the procedure with membrane fragments, will now permit polypeptides on both faces to react with the reagent. Where a protein spans the membrane those areas of the polypeptide labelled from the outside will be different from those labelled from the inside and can be distinguished by conventional peptide mapping techniques. Instead of using chemical labels of this type one may also enzymically modify the polypeptide chains. A good example of this technique is the use of lactoperoxidase (which does not enter the cell) to catalyze the iodination of exposed tyrosyl residues.

$$H_2O_2 \quad 2\ I^- \qquad \qquad + \qquad \underset{\underset{\underset{protein}{|}}{CH_2}}{\overset{OH}{\bigcirc}} \quad \longrightarrow \quad \underset{\underset{\underset{protein}{|}}{CH_2}}{\overset{OH}{\bigcirc}}{}^{I} \quad + \quad HI$$
$$2\ OH^- \quad I_2$$
$$\text{Lacto-peroxidase}$$

Still a third method employs specific antibodies prepared against purified membrane proteins. These will combine with the membrane at faces where the antigen is located. Finally, one can prepare closed vesicles from the original erythrocyte membrane which are all either inside out or right side out. By determining which enzymes, associated with these vesicles, express their activity to externally added substrate, one can ascertain on which side of the original membrane they are normally found.

Experiments of this kind have allowed the assignment of most of the major protein species of the erythrocyte membrane to its internal face. The enzyme acetylcholinesterase contrastingly is externally located and at least one principal component—Band III in Fig. 11.4—seems to span the bilayer.

Evidence to suggest that protein molecules do integrate deeply into the bilayer has been obtained also from two other important sources. X-ray

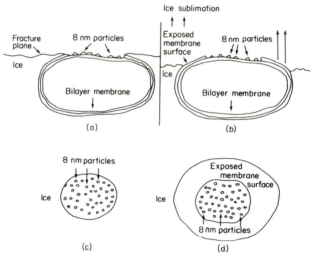

Fig. 11.7 Freeze-fracture and Freeze-etch procedure (After Tourtellotte). Diagrams b and d show the technique of fracture followed by deep-etching (resulting from the sublimation of ice) which allows a surface view of the surrounding membrane face.

diffraction studies indicate that the electron density in the hydrophobic portion of the bimolecular leaflet is both higher, and exhibits a different profile, than can be accounted for on the basis of fatty acid side-chains alone. Further support comes from "Freeze cleaving" in conjunction with electron microscopy. In this important new technique, a cell membrane preparation is deeply frozen and the resulting solid block of material struck sharply with a knife to produce a crack or fracture. The fracture plane runs along the line of least resistance which, in the case of frozen membrane specimens, occurs between the two lipid layers as shown in Fig. 11.7. This process, therefore, allows us to look inside the bilayer itself. The preparation may now be shadowed with a thin film of heavy metal, e.g., platinum, and examined face-on in the electron microscope (Fig. 11.8). Particles can be seen distributed over the smooth background of the lipid matrix. These particles represent proteins or protein aggregates and correspond to those components which either traverse the membrane or are deeply embedded in the hydrocarbon layer. They may vary in size from 5 to 15 nm and are often more prolific on the cytoplasmic face.

Fig. 11.8 Intramembranous Particles seen as a result of the Freeze-fracture technique.

The picture of the membrane which now emerges suggests that proteins may be associated with the lipid fraction through hydrophobic and/or hydrophilic interactions of variable strength. Proteins which may be easily released by manipulation of the ionic composition of the medium are often referred to as *extrinsic* components. Those thoroughly integrated into the lipid matrix and requiring detergents for solubilization are regarded as *intrinsic*. Of course, there is likely to be every form of intermediate between these two extremes. Recent results with spin labelled phospholipids also indicate that those species which are located in the bilayer may be surrounded by a tightly bound layer of lipid, distinct from the bulk phase and exchanging only slowly with it. It is probable that similar kinds of protein-protein interactions also occur. Finally, as we have seen, Robertson's proposal of an asymmetry in the distribution of protein species has been completely validated as shown in the model (Fig. 11.9).

In most instances, only a small proportion (25% to 30%) of the total external surface area of the membrane will be occupied by proteins, appearing rather like icebergs in a lipid sea. We know that these lipids can undergo rapid lateral diffusion, but what of the proteins, particularly those on the outer surface of the membrane?

When the type of freeze-fracture experiment described earlier was performed on erythrocytes that had been incubated either at pH 5·0 or pH 7·0 respectively (prior to freezing), an unexpected phenomenon was observed. The pH 7·0 treated cells exhibited a random distribution of the intramembranous particles whereas in the pH 5·0 treated cells these nodules were associated in large aggregates. The aggregation could be reversed if the pH 5·0 cells were returned to pH 7·0 before the freeze-cleaving experiment. These observations, therefore, suggested that the proteinaceous particles had the ability to move in the plane of the membrane. More dramatic evidence for this phenomenon came from experiments by Frye and Edidin who studied the distribution of characteristic proteins on the surfaces of two different cell types. The positions of these proteins could be seen by treating the cells with specific antibodies to which fluorescent compounds had been attached.

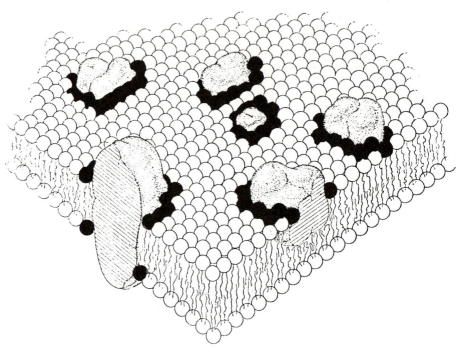

Fig. 11.9 The Fluid Mosaic Model of Membrane Structure after Singer and Nicolson. (The dark spheres represent phospholipid tightly bound to protein.)

The experiment consisted of inducing the fusion of human with mouse cells to form a heterokaryon and, at various time intervals, examining the distribution of their specific surface antigens. They found that immediately after fusion, the hybrids had mouse antigen distributed over one half of their surface, and human antigens over the other. With time, gradual mixing of the two types occurred until, within 40-60 minutes, complete randomization of the cell-specific proteins had taken place over the entire surface (area) of the heterokaryon. Further experiments showed that the proteins were able to diffuse freely in the plane of the membrane, independent of the presence of ATP.

Not only can proteins diffuse laterally, some may also rotate (in the plane of the membrane), as experiments with the visual pigment protein rhodopsin have indicated. However, it seems unlikely that any major "tumbling" through the bilayer will occur since the energy required to drag large hydrophilic areas across the hydrophobic lipid environment (of the hydrocarbon region) may be prohibitive. This does not rule out slight movements perpendicular to the plane of the membrane much in the way a cork might bob in water. The evidence to date, therefore, strongly implies that proteins or protein aggregates embedded in the bilayer can undergo several forms of motion. Next one must ask whether this diffusion always occurs freely. Several observations indicate that the answer is no—at least in some instances.

In the case of the acetylcholine receptor of nerve cells, for example, extensive investigation has revealed that this protein is located only in the region of the synaptic cleft. The receptor is apparently restrained from moving to other areas of the membrane. Berlin and his colleagues have studied the effect of phagocytosis on the ability of macrophages to transport various nutrients. Since phagocytosis involves the inclusion of large quantities of plasma membrane it was argued that randomly distributed proteins such as those involved in transport phenomena would also be internalized during the process. Thus one would expect that the initial transport capacity of the cell would decline due to a decrease in the number of mediators. This was found not to occur, suggesting that these proteins were either separated from or could be removed from that part of the membrane undergoing inclusion. Both alternatives imply some constraint on the free lateral movement of protein molecules, perhaps by anchoring them to some form of structural unit located on the inner surface of the membrane. Strong candidates for this role exist in a polypeptide complex similar to that responsible for the contractility of muscle cells and/or the microtubule proteins which (amongst other functions) comprise the spindle apparatus in mitotically dividing cells.

To summarize, therefore, we have built up a picture of protein molecules associated with the lipid fraction of the membrane and/or with one another. The intermolecular forces responsible for the association may be either hydrophobic, hydrophilic or both. The extent of this interaction may determine not only the degree to which they are embedded in the bilayer but also their freedom of movement in the membrane.

11.4. The Involvement of Carbohydrate

As the technique of electron microscopy became more refined, a so-called "Fuzzy coat" surrounding the external surfaces of most, if not all, mammalian cells became increasingly obvious. Evidence to date tends to support the presumption that carbohydrate moieties are largely responsible for this morphological feature. Whilst such carbohydrate could represent either material in the process of excretion from the cell or be of extracellular origin but loosely adhering to the cell surface, it now appears more likely that the majority of these sugar residues are intrinsic constituents of the membrane. Moreover, since they are always attached to the protein and lipid components, we ought now to define the composition of biological membranes in terms of *lipid, glycolipid, protein* and *glycoprotein*.

Reaction between the reducing groups of carbohydrates and the amino or hydroxyl groups in amino acid residues of proteins leads to N-, or O-glycosidic linkages and examples of these have been observed in the glycoproteins associated with membranes. Although a limited number of sugars (six or seven) account for the majority of the attached carbohydrate residues, several others are also found on rare occasions, making it difficult to establish hard and fast rules on the carbohydrate composition. Fucose and sialic acid, when present, are generally found at the non-reducing ends of the chains. These carbohydrate chains may vary in length from simple disaccharides to units containing 20-30 residues. Moreover, they may be linear or highly branched. The sugar content of a glycoprotein also exhibits a great deal of variability. For example, 65% of the mass of the major glycoprotein in human erythrocytes is carbohydrate, whilst the minor glycoprotein (Band III) only contains 5-8% carbohydrate.

The glycolipid portion is always based on sphingosine, the monosaccharide unit being glycosidically linked to the terminal hydroxyl of the ceramide. The frequency of different sugars closely follows that found in glycoprotein, the only possible exception being the relative scarcity of mannose.

Starting with the early observations of Robertson, one of the characteristics of membrane structure has been the asymmetric distribution of both the lipid and protein components. When the "Fuzzy coat" was first reported, this asymmetry was again obvious, suggesting that the carbohydrate moieties were confined to the outer face of the bimolecular leaflet. This proposal has since been confirmed in a variety of ways. The first of them made use of the enzyme neuraminidase which cleaves off the negatively charged sialic acid residues. When Eylar performed this experiment with human erythrocytes the electrophoretic mobility of the cells decreased and he was able to ascertain that all the sialic acid present on the cell surface was accessible to the enzyme. A second line of evidence came from work with lectins, proteins extracted from plants which have the ability to react with specific sugar residues. These proteins were seen to bind only to that side of membrane fragments which corresponded to the outer surface of the cell. Very similar experiments using blood group specific antibodies gave the same conclusion and will be discussed at greater length below.

The fact that cells are covered with carbohydrate-containing molecules and that these structures can vary enormously in both complexity and composition, raises the question as to what function they serve. A few examples are given below.

The previous development of techniques in blood transfusion and the present development of transplantation, highlight the crucial antigenic role for sugar residues at cell surfaces. The competence of the immune system relies on its ability both to recognize as alien (and hence potentially lethal) a foreign substance introduced into the system, and then to neutralize its influence. The most obvious example of this process is the common ABO blood group classification. Individuals possessing A-type erythrocytes will possess anti-B antibodies in their sera. Hence transfusion with blood containing type B erythrocytes will result in massive cell agglutination. The basis for this differentiation lies in a single sugar residue (Fig. 11.10) for whereas the characterizing oligosaccharide chain on type A red blood cells terminates in an N-acetyl galactosamine residue, type B erythrocytes possess carbohydrate units ending with galactose. In this case the structure of the monosaccharide is the sole governing factor, since A and B type antigenicity is associated with both the glycolipid and the glycoprotein fractions of the membrane. Many other blood group systems are also present, particularly the MN and Rhesus systems, the second of which can give rise to the serious problem of maternal-foetal incompatibility.

The ability of the immune system to differentiate "self from non-self" on—in this case—the basis of carbohydrate structure, demonstrates simply but effectively the problems of rejection inherent in transplantation. For each cell will possess on the outer surface of its membrane an array of genetically defined histocompatibility antigens characteristic of that individual. Thus the importance of accurate tissue typing can be appreciated, particularly when one considers that there may exist upwards of 100 different specificities, capable of eliciting an antigenic response if recognized as non-self by the recipient.

In 1907, Wilson described a system which may serve as an important prototype in defining the basis for cell-cell recognition. The experiment consisted of separately dissociating a yellow and a red species of sponge into their respective cell types and then mixing the two populations. He subsequently found that the cells reassociated into colour specific aggregates. It now appears that the sponge cells possess macromolecular assemblies associated with their cell surface which are capable of mediating specific intercellular adhesion. Since these aggregation factors have a high carbohydrate content, it prompted a search for a simple sugar which might inhibit the process. In the case of one species of cell such an inhibitor was found in glucuronic acid. Interestingly, this monosaccharide had no effect on the aggregation of a different species of sponge. Thus it seems that the specificity in cell-cell aggregation lay in the presence of

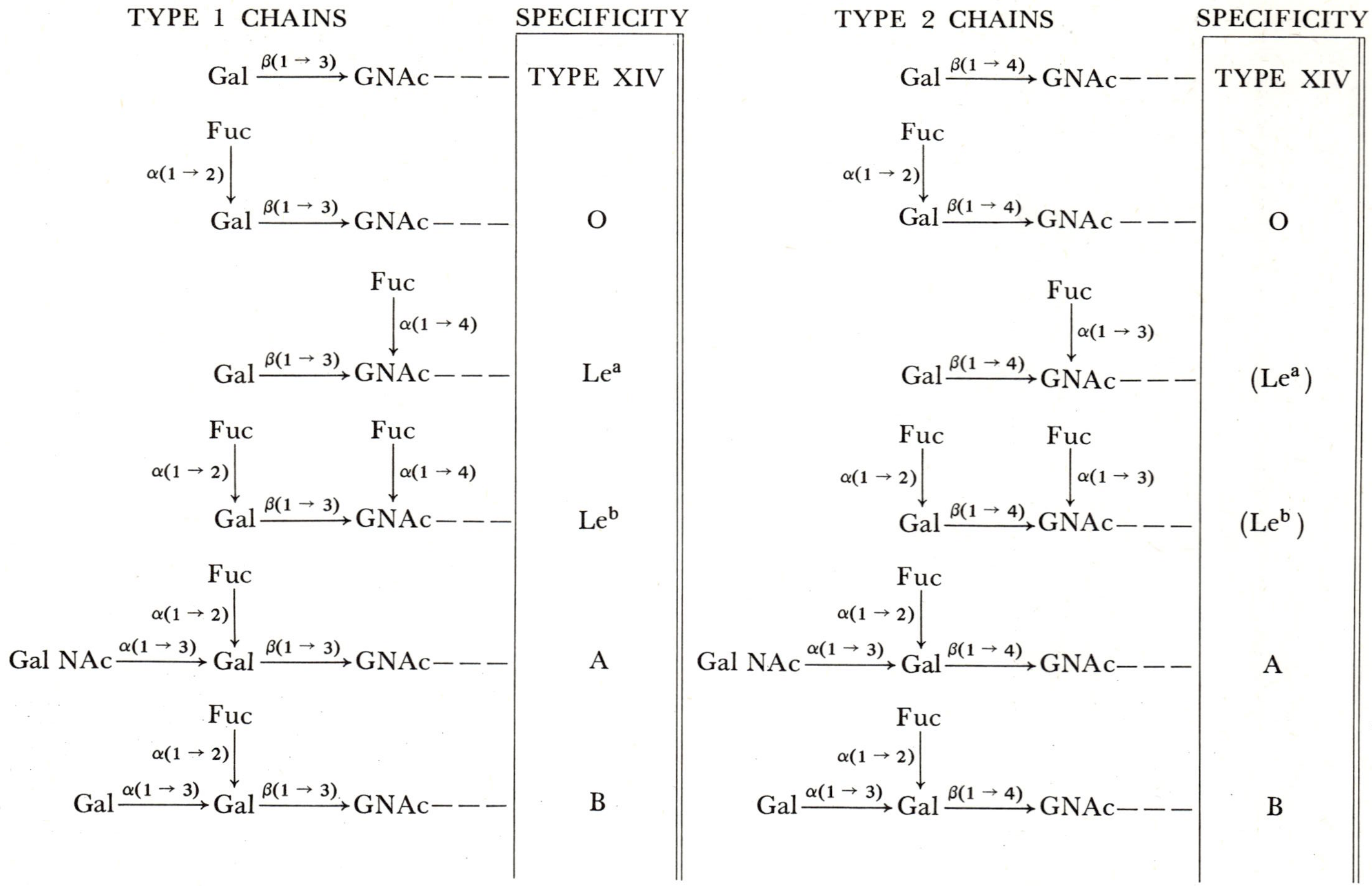

TYPE 1 CHAINS

SPECIFICITY	Structure
TYPE XIV	Gal $\xrightarrow{\beta(1\to3)}$ GNAc---
O	Fuc $\xrightarrow{\alpha(1\to2)}$ Gal $\xrightarrow{\beta(1\to3)}$ GNAc---
Leᵃ	Gal $\xrightarrow{\beta(1\to3)}$ GNAc--- ; Fuc $\xrightarrow{\alpha(1\to4)}$
Leᵇ	Fuc $\xrightarrow{\alpha(1\to2)}$ Gal $\xrightarrow{\beta(1\to3)}$ GNAc--- ; Fuc $\xrightarrow{\alpha(1\to4)}$
A	Gal NAc $\xrightarrow{\alpha(1\to3)}$ Gal $\xrightarrow{\beta(1\to3)}$ GNAc--- ; Fuc $\xrightarrow{\alpha(1\to2)}$
B	Gal $\xrightarrow{\alpha(1\to3)}$ Gal $\xrightarrow{\beta(1\to3)}$ GNAc--- ; Fuc $\xrightarrow{\alpha(1\to2)}$

TYPE 2 CHAINS

SPECIFICITY	Structure
TYPE XIV	Gal $\xrightarrow{\beta(1\to4)}$ GNAc---
O	Fuc $\xrightarrow{\alpha(1\to2)}$ Gal $\xrightarrow{\beta(1\to4)}$ GNAc---
(Leᵃ)	Gal $\xrightarrow{\beta(1\to4)}$ GNAc--- ; Fuc $\xrightarrow{\alpha(1\to3)}$
(Leᵇ)	Fuc $\xrightarrow{\alpha(1\to2)}$ Gal $\xrightarrow{\beta(1\to4)}$ GNAc--- ; Fuc $\xrightarrow{\alpha(1\to3)}$
A	Gal NAc $\xrightarrow{\alpha(1\to3)}$ Gal $\xrightarrow{\beta(1\to4)}$ GNAc--- ; Fuc $\xrightarrow{\alpha(1\to2)}$
B	Gal $\xrightarrow{\alpha(1\to3)}$ Gal $\xrightarrow{\beta(1\to4)}$ GNAc--- ; Fuc $\xrightarrow{\alpha(1\to2)}$

Fig. 11.10 Structures of the Non-reducing Termini of the ABO and Lewis Blood Group Antigens.

certain sugar residues on the surface of one cell, capable of interacting with some form of receptor on another.

This kind of cell-sorting phenomenon has its direct analogy in more complex vertebrate systems. The fundamental contribution in this field was made by Moscona and his colleagues who discovered that single cell preparations from dissociated chicken embryonic tissue would, given a suitable recovery time, form organ-specific aggregates. More recently this reassociation phenomenon has been investigated with cultured tumour cells from a mouse teratoma or embryoma. The important conclusion from these experiments was that the presence of certain, as yet un-identified, sugar residues on the cell surface were necessary for specific reaggregation. With chicken embryonic neural retina cells, a β-galactosyl residue is apparently implicated in the cellular adhesion process.

This brief illustration of possible roles for the carbohydrate fraction of the cell (plasma) membrane serves to introduce the important concept of an information content inherent in these sugar residues. How this information is "read" and interpreted by surrounding cells remains to be determined. So too does the relevance—if any—of the changed pattern of the carbohydrate-containing components seen when a cell becomes cancerous.

11.5. Transport

Although the cell membrane constitutes a barrier between compartments, nutrients and metabolites have still to be able to pass from one side of the membrane to the other.

Let us therefore examine the transport characteristics of synthetic lipid bilayers with a view to establishing a few general rules for cell-permeability. Such artificial membranes may be prepared from pure lipid in two principal ways. If phospholipid introduced into an aqueous medium is sonicated, small closed vesicles (or liposomes) surrounded by the typical bimolecular lipid leaflet quickly form. Alternatively, it is possible to construct a bilayer between two aqueous compartments by painting a solution of phospholipid dissolved in organic solvent over a small orifice in a plastic sheet separating the two aqueous solutions. The film of material over the hole rapidly thins out as the solvent goes into the bulk phase, leaving behind a partition sealed with a lipid bilayer comparable to that seen in natural membranes. We are now in a position to examine the permeability of the bilayer to various substances.

This is simply done by placing the material to be studied, e.g., glucose, Na^+, Cl^- etc. on one side of the lipid barrier and monitoring its appearance on the other using either radiotracers or electrical conductivity measurements. The results indicate that the phospholipid bilayer is relatively impermeable to all substances of polar character, e.g., glucose or ions. On the other hand, lipid soluble material will easily cross this barrier. Thus the ability of a substance to pass across the bilayer, defined as its *permeability coefficient, P,* will be some function of its relative solubility in the non-polar lipid medium compared with the water phase (called its *partition coefficient, β*).

More explicitly $P = (D/d)β$, where D = diffusion coefficient and d, the thickness of the bilayer.

(a) Facilitated diffusion (passive transport)

If one now performs the same kind of experiment only measuring in this instance the uptake of $[^{14}C]$glucose by the erythrocyte, a different result is obtained. Glucose is seen to pass very rapidly into the cell down its concentration gradient. This movement is directly concentration dependent but only up to a certain value at which point the system is apparently saturated (see Fig. 11.11.)

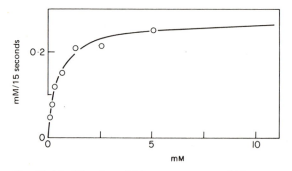

Fig. 11.11 Kinetics of Glucose transport (After Stein).

Moreover, by introducing other hexoses, e.g., galactose into the medium, the rate of glucose uptake is competitively inhibited. In contrast, structurally dissimilar substances, e.g., disaccharides and amino acids, have no effect, indicating a high degree of specificity in the transport mechanism. Very similar results (differing only in the values for K_m and V_{max}) are obtained when the cells are pre-loaded with $[^{14}C]$ glucose and the efflux of radioactivity measured, thus movement can take place in both directions. Where the concentration of glucose is equivalent on both sides of the membrane, exchange diffusion occurs. This form of biological transport is entirely independent of ATP or other chemical energy source (i.e., is Passive) and since it represents, in essence, an acceleration of the normal diffusion process, it has been termed *Facilitated Diffusion.* Notice, however, that this mechanism is not capable *per se* of mediating the accumulation of substances against their concentration gradients.

Since we know that the glucose molecule cannot easily pass through the hydrocarbon layer, the explanation for the above results must lie in the intervention of a transport mediator able to circumvent the energy barrier.

The typical Michaelis-Menten kinetic behaviour of glucose flux, which allows us to compute values for K_m and V_{max}, for the system, suggests that some form of enzyme (a permease) is involved in such processes. This conclusion has recently been supported by the isolation of a protein from the human erythrocyte membrane which facilitates the movement of several anions in an exactly analogous fashion. In both instances, reagents which are known to chemically modify amino acid side chains will inhibit transport. Further evidence for the implication of proteins in transport phenomena has also been obtained from the study of bacterial mutants defective in certain permease systems.

In searching for a model which might describe the mechanism by which facilitated diffusion was occurring, much attention has been paid to a series of lipid-soluble antibiotics of fungal origin. When added to the synthetic membranes mentioned earlier, these substances have the ability to specifically increase the permeability of the lipid bilayers to certain ions. Valinomycin, which is highly specific for K^+, has the cyclic structure shown in Fig. 11.12. It assumes a "doughnut" type configuration whereby the non-polar side chains are exposed to the hydrocarbon portion of the bilayer and the charged residues point inwards to form the binding site for a potassium ion. The system is thus energetically very stable. Since each molecule of Valinomycin acts independently, and the complex is not large enough to span the bilayer, it is presumed that the antibiotic acts as a *carrier* shuttling the potassium ions across the bilayer one at a time, from regions of high concentration to those of low $[K^+]$. Consistent with this proposal the width of the lipid barrier does not appear to affect significantly the efficacy of transport. Moreover, since the mechanism is dependent upon free diffusion of the complex in the membrane, it is a necessary requirement that the hydrocarbon portion of the bilayer be fluid. Cooling below the melting temperature of the lipid will effectively abolish any transport.

The second or *channel*-type mechanism is exhibited by Gramicidin A, (Fig. 11.12), a 15 residue peptide composed of alternating D and L amino acids. The molecule forms a helix of approximately three and a half turns, stabilized by hydrogen bonds and of length equivalent to one half the width of the bilayer. The carboxyl oxygens point towards the interior of the helix and the hydrophobic side-chains towards the lipid phase. Cation transport then occurs when two Gramicidin molecules form a head to head dimer, thereby creating a pore down the middle of the two helices and extending across the entire width of the bilayer. In contrast to the situation with Valinomycin, the effectiveness of Gramicidin depends on membrane width but, as one would expect for a pore, is relatively unaffected by freezing the lipid. Other pore forming antibiotics such as Amphotericin B rely on side by side aggregation of individual molecules and are specific for anions. Thus we see that, notwithstanding the

HCO-L-Val-Gly-L-Ala-D-Leu-L-Ala-D-Val-D-Val-L-Trp-D-Leu-L-Trp-D-Leu-L-Trp-NHCH$_2$CH$_2$OH

1   2   3   4   5   6   7   8   9   10   11   12   13   14   15

Gramicidin A

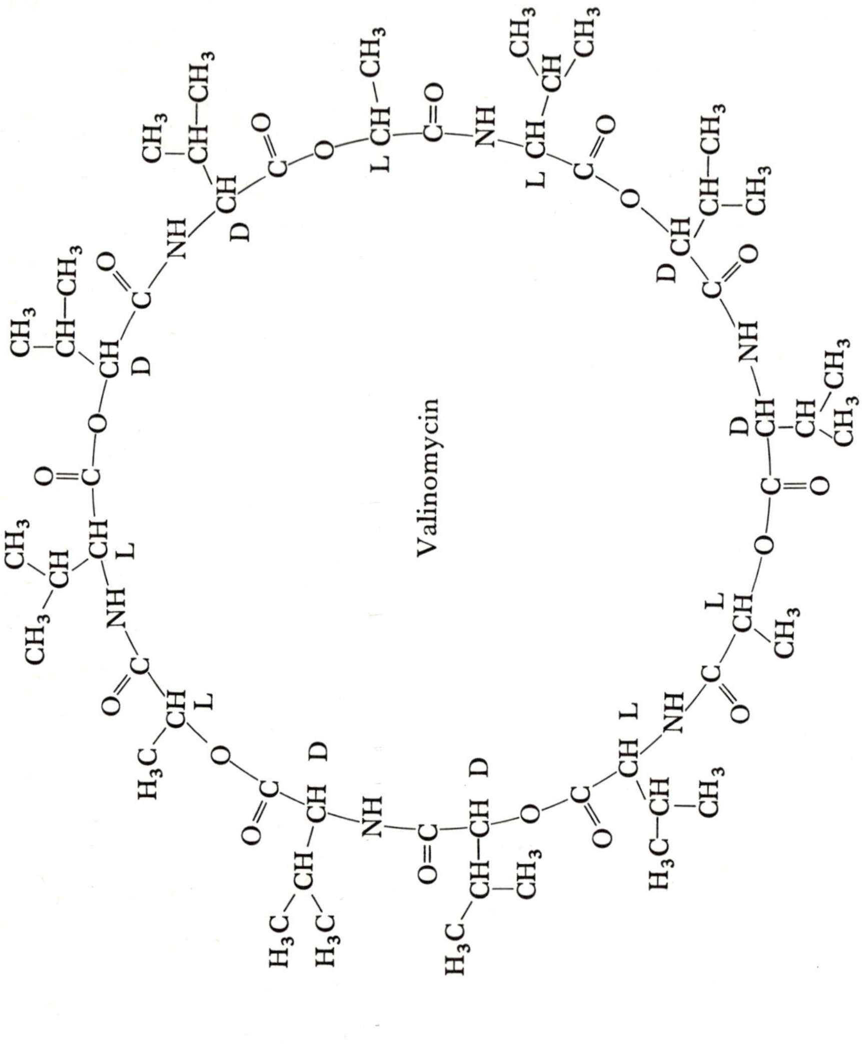

Valinomycin

Fig. 11.12 Structure of Valinomycin and Gramicidin A.

difference in mechanism, the antibiotics exhibit the most important criteria for transport mediators—namely the ability to increase the permeability of the lipid bilayer to specifically defined substances.

We can now ask which of these two mechanisms best describes the situation found in natural membranes. The answer to this question is at this time highly controversial. Any analysis of the kinetic properties of the two systems is subject to ambiguous interpretation and the obvious temperature change experiments are complicated by two factors. Firstly, there is the lack of any sharp or well-defined phase transition in the lipid fraction of mammalian membranes. Secondly, as one lowers the temperature of incubation of any enzyme system below its optimum, the activity of the enzyme will decline. So too with transport proteins whose activity similarly depends on subtle conformational changes. Thus it can easily be appreciated that variations in transport rate as a function of temperature cannot as yet be unambiguously related to the mechanism of the transport process. Notwithstanding these problems, certain considerations lend a measure of support to some form of the channel hypothesis for facilitated diffusion—at least in some instances.

(1) The anion transport protein of the human erythrocyte spans the entire lipid bilayer, exposing different parts of the polypeptide chain at the external and internal surfaces of the membrane. Moreover, attaching a large lipid-insoluble lectin to one side of the transport protein apparently does not inhibit anion translocation.

(2) Somewhat better evidence is obtained by a study of the transmission of the electrical impulse along the nerve axon. This process involves dramatically increased permeabilities of the axonal membrane to sodium and potassium ions. The movement of these ions occurs passively, mediated by so-called "sodium" and "potassium" gates, opened in response to small localized changes in the distribution of electrical charge. Theoretical calculations based on the number of "sodium" gates present in the membrane, suggest that the diffusion rate of a carrier through the lipid matrix would not be fast enough to account for the number of sodium ions entering the axon in a given time interval. No such objections can be raised against the pore mechanism. It must be stressed, however, that these pores, since they have to be capable of both specificity and control, will be complex structures not just simple channels.

(b) Active transport

The processes we have discussed up to this point have dealt with situations in which the concentration of substrate was high and where simple equilibration across the cell membrane took place. Bacteria, however, often exist in conditions of very low nutrient concentration and for their survival have to be able to scavenge every available molecule of substrate. Similarly in more complex organisms, substances have to move against the prevailing concentration gradients. Since this movement requires the expenditure of metabolic energy, it is known as *Active Transport*. Of course, in the simple example mentioned above, the ultimate benefit to the system of active nutrient uptake is vastly greater than the energy loss sustained in the transport process.

(i) ION TRANSPORT

Let us return to the case of impulse conduction along a nerve axon. The rising phase of the action potential is caused by the opening of pores through which sodium ions rush into the cell. The depolarization stage results from the closing of these sodium gates and the opening of channels through which potassium ions exit from the cell. In both instances, the movement is passive, relying upon the existence of two concentration gradients, sodium being 140 mM outside and 20 mM inside and potassium 120 mM inside and 5 mM outside. Thus the transmission of the impulse is absolutely dependent upon the ability of the cell to maintain the asymmetric distribution of these cations. This is achieved by pumping those sodium ions, which have entered the cell during the passage of the action potential, out again into the extracellular medium. At the same time, potassium is brought in. Such movements are against the prevailing concentration gradients and require energy. Therefore, treatments which disrupt energy production will be reflected by an increase in intracellular sodium and a decrease in potassium. Most vertebrate cells exhibit a similar asymmetry in the distribution of these cationic species. The ion gradients so created can be used for a variety of purposes, ranging from the control of osmolarity to coupled transport.

The maintenance of these concentration differences is the province of the membrane-bound enzyme $Na^+ - K^+$ dependent adenosine-triphosphatase $[Na^+ - K^+$ ATPase]. The protein present in nerve and red blood cell membranes exhibits a strange stoicheiometry, since for every three sodium ions translocated out of the cell two potassium ions are brought in. During this process one molecule of ATP is cleaved to ADP and inorganic phosphate. The overall reaction—which is reversible—can be visualized as:

$$3Na_{in}^+ + 2K_{out}^+ + ATP \rightleftharpoons 3Na_{out}^+ + 2K_{in}^+ + ADP + P_i$$

If any one of these three constituents is absent, no transport will occur. In addition the enzyme is inhibited by the cardiac glycosides ouabain and strophanthidin.

In our earlier discussion on membrane structure attention was drawn to the "sidedness" of proteins which span the bilayer. This property is beautifully demonstrated in the case of $Na^+ - K^+$ ATPase. ATP is only hydrolyzed when exposed to the cytoplasmic face of the membrane and the enzyme for activity requires both intracellular sodium and extracellular potassium. The external sodium and internal potassium ion concentrations have little influence on the transport process. Ouabain inhibits only from the outside and its effects can be partially reversed by high external concentrations of potassium. Since one can also bind specific antibodies to the interior face of the molecule without affecting its transport capacity, the protein does not rely on any form of tumbling or shuttling through the bilayer for its activity. All these characteristics are nicely consistent with the postulated physiological role for the enzyme.

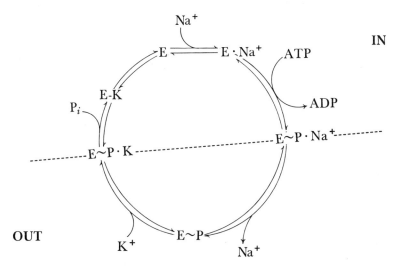

Fig. 11.13 Coupling of ATP hydrolysis with translocation of sodium and potassium ions.

The exact mechanism by which ATP hydrolysis can be coupled to the translocation of ions is still unclear but certain details of the reaction have been elucidated. In the presence of sodium and $[\gamma^{32}P]\,MgATP$ (i.e., Ad-Rib-(P)-(P)-^{32}(P) as Mg^{2+} salt) and in the absence of potassium, the labelled phosphate group of ATP is found covalently bound to the enzyme. If potassium is included, this labelling is effectively abolished. Similarly, addition of potassium to the phosphorylated enzyme, causes the release of the phosphate grouping. We can, therefore, picture the events as shown in Fig. 11.13. A purified preparation of the enzyme gives two bands corresponding to molecular weights of 130,000 and 45,000 respectively on SDS-polyacrylamide gel electrophoresis. The larger polypeptide contains both the binding site for ouabain and is phosphorylated at an aspartyl residue, as indicated above. The smaller molecular weight component is glycosylated but as yet no specific role has been ascribed to it. Indeed the two polypeptides are often found in such widely varying molar ratios that one wonders whether the smaller component is actually part of the transport system. Detergent binding studies reveal that only 10% to 15% of the surface area of the enzyme participates in hydrophobic interactions. We can, therefore, propose that most of the enzyme lies outside the lipid bilayer. However, in the absence of further information, it is worthless to speculate concerning the conformational changes involved in the translocation process.

Although up to this point we have been discussing $Na^+ - K^+$ counter-transport in the nerve, muscle and red blood cells, some evidence suggests that the enzyme or a close relative may be catalyzing forms of co-transport in other tissues (i.e., transport of both substances in the same direction). In the kidney, for example, coupled transport of sodium and chloride may occur by means of a so-called Electrogenic Sodium Pump. Since a major function of the kidney tubule is the reabsorption of ions, this possibility is not unreasonable.

The active transport of ions is not confined solely to sodium and potassium. There also exists a powerful but little-understood chloride pump in the membrane of oxyntic cells which is responsible for the hydrochloric acid content of gastric juice. The action of cholera toxin is believed to be mediated by the activation of a similar chloride pump in the intestine. The thyroid gland can actively accumulate iodide ions. In the category of divalent cations, both Mg^{2+} ATPases and Ca^{2+} ATPases have been reported. The calcium pump is generally responsible for maintaining the low intracellular concentrations of the ion, except in the case of certain tissues, such as the intestine, where it apparently mediates calcium accumulation from the lumen of the gut. Perhaps the best characterized Ca^{2+} ATPase, however, is that present in an intracellular membrane system—the sarcoplasmic reticulum—of muscle cells.

Muscular contraction is currently thought to be initiated by the release, in response to electrical stimuli, of calcium ions stored at high concentration in the sarcoplasmic reticulum. Conversely, muscle relaxation is reliant upon the subsequent recapture of these ions by the same organelle. The uptake occurs against a concentration gradient and is carried out by an enzyme of molecular weight 100,000. Like other transport proteins that have been adequately characterized, the Ca^{2+} ATPase spans the bilayer and can be correlated on freeze fracture with 9-10 nm intramembranous particles. The enzyme is lipid-requiring and in common with the $Na^+ - K^+$ ATPase is phosphorylated in the course of the reaction

$$2Ca^{2+}_{out} + ATP + E \rightleftharpoons Ca^{2+}_2 \cdot E \sim P + ADP$$

$$Ca^{2+}_2 \cdot E \sim P \rightleftharpoons E + P_i + 2Ca^{2+}_{in}$$

(ii) SUGAR TRANSPORT

We looked earlier at the way in which glucose entered the erythrocyte by means of passive diffusion. This mechanism is also operative in many other tissues exposed to the circulatory system where the concentration of glucose is high, e.g., liver, muscle and brain. On the other hand, both the small intestine and the proximal kidney tubule have to be able to accumulate the sugar against a concentration gradient. As a result, a special energy-dependent transport mechanism is found in the mucosal membrane of the epithelial cells lining these organs.

Most of the characterization of this system has been achieved using small closed sacs made of everted portions of intestine, or with small strips of mucosa, effectively only one cell layer thick. When glucose uptake into the everted sac was studied, the following observations were made:

(1) Although there was no absolute requirement for sodium ions in the extracellular medium, the rate of sugar transport into the sac was markedly stimulated as a function of the sodium ion concentration.

(2) Potassium ions inhibited glucose uptake.

(3) The transport system functioned poorly under conditions of low energy production.

In another experiment a small piece of intestine was used to separate two aqueous compartments containing equivalent amounts of sodium and glucose. The electric current passing across the tissue during the transport of glucose was then measured by voltage clamp techniques (Fig. 11.14). This

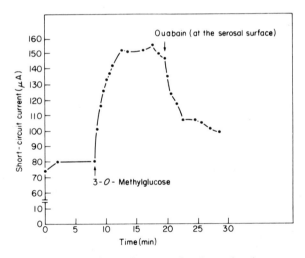

Fig. 11.14 The Time-course of the increase in short-circuit current following the addition of sugar and the subsequent effect of added inhibitor. (After Schultz and Zalusky.) 3-O-methylglucose is actively absorbed like glucose.

current corresponded exactly to the chemically measured flux of sodium ions, implying that not only did these ions stimulate glucose uptake but that they were themselves also transported. Furthermore, adding phloretin (an inhibitor of passive glucose transport) introduced at the mucosal surface has a similar effect. All these pieces of data have been combined by Crane into a theory of Na^+-glucose co-transport (see Fig. 11.15).

It is postulated that the sodium ion first equilibrates with the transport protein at the external surface of the membrane (the site at which phloretin blocks) where the sodium concentration is high. This binding increases the affinity of the carrier for glucose (i.e., reduces its K_m) and the now fully loaded enzyme changes conformation to expose both substances to the inside of the cell. The sodium ion will reequilibrate with the intracellular medium which, on account of its low sodium content, means in effect that the cation will come off the protein. The resulting change in substrate affinity then reverses the original binding of glucose. Since potassium is a competitive inhibitor of the sodium effects, the high intracellular concentration of the former ion may further stimulate both the release of sugar inside the cell, and the return of the carrier to its original conformation.

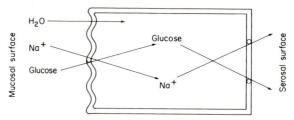

Fig. 11.15 The movement of glucose and Na^+ across the epithelial cell layer of the intestine.

As it is visualized, the transport system is carrying out simple equilibration, the accumulation of sugar depending primarily on the downhill flow of sodium ions and perhaps aided by the inhibitory influences of high intracellular potassium. It, therefore, follows that where the concentration of sodium is the same on both sides of the membrane, equilibration rather than accumulation of glucose will occur. Where one is able experimentally to create conditions such that the usual ionic distribution is reversed, glucose flows out of the cell even against its own concentration gradient.

According to this hypothesis, therefore, the active uptake of glucose is, under normal conditions, absolutely dependent upon the asymmetric distribution of sodium and potassium ions across the plasma membrane. We know that this is achieved by the ATPase. Hence the energy required for glucose accumulation by the epithelial cells of the intestinal mucosa and the proximal tubule is used not in the transport of glucose itself but for the maintenance of the two ion gradients. Therefore, ouabain which inactivates the sodium pump will ultimately also inhibit glucose transport. Once within the cell, sugar then passes into the blood by facilitated diffusion down the newly created gradient.

Although we have dealt exclusively with glucose uptake, most other monosaccharides appear to be transported by a similar mechanism. But the carrier is not common to all sugars, for whilst glucose and galactose share the same transporter, fructose apparently does not. In the case of "glucose-galactose maladsorption"—an inborn error of hexose transport—glucose and galactose uptake is seriously impaired but the transport of fructose is unaffected.

Disaccharides are not transported *per se* but are first cleaved into their respective monomers by enzymes located on the outer surface of the brush border membrane. The lactose intolerance syndrome, in which lactose present in milk cannot be utilized by the individual, is associated with the inability to cleave the disaccharide rather than with any defect in the absorption of glucose and galactose.

(iii) AMINO ACID AND PEPTIDE TRANSPORT

Unlike glucose, the concentration of amino acids in the blood is not thought to be great enough to permit an adequate supply to tissues such as the brain, muscle, liver, etc., using only the mechanism of facilitated diffusion. It is a good example of conservative biological efficiency that the mechanism developed for amino acids appears to be identical to that for glucose in the intestine—namely co-transport with sodium. Since the properties of this system have been detailed earlier, let us look instead at the specificity of the transport proteins for their respective substrates.

Most of this evidence has been derived from *in vivo* and *in vitro* studies on the kinetic parameters of intestinal accumulation and renal tubular reabsorption of amino acids. These investigations have generally been concerned with the ability of some amino acids to inhibit competitively the uptake of certain others—the unproven inference being that where inhibition exists and is competitive, both substances are using the same carrier molecule. Thus the presence of lysine will depress arginine absorption but leave the uptake of alanine, leucine and glutamic acid relatively unaffected. The pattern which emerges from this type of study indicates that amino acids with similar chemical structures will generally share a common transporter. In some instances, e.g., lysine, cystine, glycine, proline, it appears that more than one transport system may exist for a particular amino acid.

The second major source of information on transport specificity has come from the characterization of several inherited defects in absorption. The classical example of this type of phenomenon is cystinuria, a condition in which the concentration of cystine, arginine, lysine and ornithine in the urine is markedly elevated due to a failure in the renal reabsorption mechanism. Since cystine alone among this group is relatively insoluble in water, it often precipitates in the ureter or bladder, leading to the formation of cystine stones, which in turn give rise to urinary tract obstruction or infection.

Similar inabilities to transport the neutral amino acids (Hartnup disease), proline, hydroxyproline and glycine (iminoglycinuria), and methionine have all been reported.

In some instances the uptake of amino acids, which are thought to share carriers, are not identically affected. This has been taken to suggest that a multiplicity of transport proteins exist. Great care must be exercised in such interpretations, however, since genetically directed amino acid replacements which prohibit the binding of one substrate need not necessarily also exclude a second of somewhat different structure. Many of these defects although initially observed by derangements in renal tubular reabsorption are also observed as abnormalities of intestinal uptake. [Some of the anomalies may also be explained by the peptide transport systems discussed below.]

Research into these inborn errors of transport has confirmed that small peptides as well as free amino acids can be absorbed from the lumen of the small intestine in quantitatively significant amounts. In Hartnup disease, for example, the uptake of phenylalanine is almost nil. When the dipeptide phenylalanylphenylalanine is administered, however, the level of phenylalanine in the portal blood rises. A similar occurrence is seen in cystinuria where, although the transport system for the dibasic amino acids is defective, these residues do appear in the blood when suitable peptides are administered to the patient. Other evidence supporting this phenomenon was obtained from both *in vivo* and *in vitro* studies on the rate of amino acid absorption. The uptake of glycine, for example, was much slower using a given dosage of the free amino acid than when introduced as glycylglycine or better still glycylglycylglycine. Again, glutamic acid was taken up much more rapidly when in the form of glutamyl methionine than as the free acid.

The transport mechanism is obscure. Hydrolysis of the peptide to its constituent residues may be an integral part of the process, occurring either at the outer or inner face of the membrane. The evidence to date is weak but tends to favour the latter site. In any event, the absorbed material appears in the portal blood only as the free amino acids. These are then transported into other cells by the processes detailed earlier.

Larger polypeptides generally are unable to cross the lipid bilayer. In certain rare instances, however, such transport must occur. The best characterized example is seen in Diphtheria toxin, a 63,000 molecular weight protein secreted by Corynebacterium diphtheriae. The pathogenicity of this organism is due to the progressive enzymically catalyzed inhibition of host cell protein synthesis by the toxin.

The single polypeptide chain of the toxin can be cleaved by small amounts of trypsin into two functional portions. One of these (the A segment) is responsible for the inhibition reaction. The B fragment, of much more hydrophobic character, binds to a specific receptor at the cell surface. These receptors are thought to be galactose-containing glycolipids whose presence confers on the host cell its susceptibility to the diphtheria toxin. The A fragment by itself cannot bind to the membrane and thereby enter the cell—and the B portion is enzymatically inactive. The mechanism by which the A polypeptide, or perhaps even the whole toxin, passes through the lipid bilayer is unique and unexplained. It appears likely, however, that some toxic plant proteins such as abrin and ricin; cholera, tetanus and botulinus toxins may all possess two similar functional entities, one which binds to the membrane of susceptible cells and the second responsible for the actual intracellular reaction. It may also be that

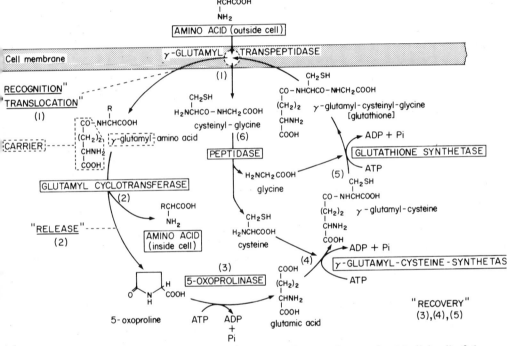

Fig. 11.16 The transport of amino acids across the membrane of epithelial cell of the proximal kidney tubule as envisaged by Meister.

the mechanism of entry into the cells may be shared by certain naturally occurring polypeptide hormones and by the immunoglobulins of colostrum which pass, apparently intact, from gut to blood.

Recently a quite new system for amino acid transport in the renal tubule and perhaps also in other tissues such as the brain has been proposed by Meister. The essential features of the scheme are given in Fig. 11.16. It is suggested that the membrane-bound glycoprotein γ-glutamyl transpeptidase catalyzes the transfer of the γ-glutamyl group from glutathione to the amino acid being absorbed. The γ-glutamyl-amino acid complex is then released into the interior of the cell. The recognition site for the amino acid being transported may lie with the transpeptidase itself, which would imply the presence of isoenzymes. Alternatively there may be binding proteins exhibiting the specificities we have noted above, which operate by transferring their substrates to the transpeptidase. Since the regeneration of glutathione requires ATP, the active nature of the transport process is easily appreciated. However, the role of Na^+ is not at all obvious. Being new, there are many unanswered questions about this system, which only further research will resolve.

(c) Other transport processes

In addition to ions, amino acids and sugars, mammalian diets also contain fat, water and the essential vitamins. These are all taken up by the epithelial cells of the intestinal tract and conveyed to the various target tissues.

The process of lipid absorption has already been detailed (Chapter 8) and it seems likely that the fat soluble vitamins (A, D, E and K) follow much the same path. Bile salts on the other hand may be actively taken up by a variant of the Na^+ co-transport system. The water soluble vitamins (ascorbic acid and the B group other than B_{12}) are thought to undergo facilitated diffusion, as is also the case with the more complex anions, sulphate, nitrate, phosphate, etc. Vitamin B_{12} on the other hand, is absorbed by a special mechanism which requires the intervention of "intrinsic factor", a protein secreted from the gastric mucosa. The complex formed between this component and B_{12} is a necessary pre-requisite for the subsequent active transport of the vitamin. The mechanism of this absorption is, however, obscure.

Perhaps one of the most interesting transport processes is that involving the movement of water across cell membranes. Governed only by osmolarity, water flow is never in itself considered energy dependent. Two examples of this are the ease with which erythrocytes lyse due to the influx of water when placed in hypotonic solution and the general dehydration resulting from cholera. Cholera toxin induces massive secretion of chloride ions from the mucosal cells into the gut and water then flows down the osmotic gradient. Due to the polar character of the water molecule it is not to be expected, despite some views to the contrary, that this substance will diffuse at any significant rate through the lipid matrix of the membrane. Indeed the impermeability of the distal tubule and the collecting duct of the nephron, attests to this conclusion. Solomon and his colleagues working with the erythrocyte have good evidence that water transport occurs through 0·4-0·5 nm channels in the membrane. It is likely that this situation will also prevail for other tissues. One would then presume that the impermeability of the kidney membranes mentioned above is due to the closure of such pores in the absence of the antidiuretic hormone, vasopressin. The function of the hormone is then to open these channels and so allow the body to retain water.

Emphasis has been placed at various points in this discussion that transport processes generally utilize small molecules, sugars, amino acids, etc., rather than polymeric precursors. Where the uptake or release of much larger substances is involved, different mechanisms come into play. The processes of endocytosis and exocytosis are represented diagrammatically in Fig. 11.17. In both instances, whole areas of membrane are involved rather than individual protein components.

With the so-called "gap junctions" formed between cells, large protein-lined pores seem to exist through the bilayers of both cells. Substances of molecular weight as large as 9000-10,000 may often pass from one cell to another via these specialized channels. This situation is not to be confused with the pores in the nuclear envelope through which proteins and nucleic acids migrate to and from the nucleus. Figure 11.18 illustrates that these pores are formed by a double membrane and do not represent any hole through the bilayer.

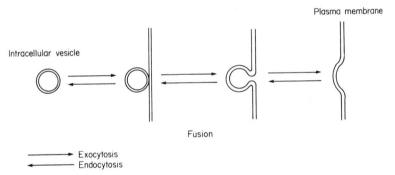

Fig. 11.17 The process of Exo- and Endocytosis. Exocytosis (also called reverse pinocytosis) involves the fusion of intracellular vesicles with the plasma membrane and the extrusion of the vesicular contents. Endocytosis involves the invagination and "pinching off" of the plasma membrane, thereby enclosing extracellular material.

(d) Control of transport processes

The model for transport which we have developed so far involves membrane proteins thoroughly integrated into a fluid phospholipid bilayer. It is suggested that in many if not all instances, these proteins will span the bilayer and by subtle structural changes effect the movement of specific substrates from one side of the membrane to the other. Since these conformational changes will inevitably take place in a lipid *milieu*, let us return to our starting point and examine the influence of lipid organization on transport processes.

You will recall that phospholipids undergo temperature dependent phase transitions between crystalline and liquid-crystalline states. For homogeneous phospholipid species this transition will be sharp. Using *E. coli* mutants (auxotrophic) with a requirement for unsaturated fatty acids, it is possible, by including selected fatty acids in their growth medium,

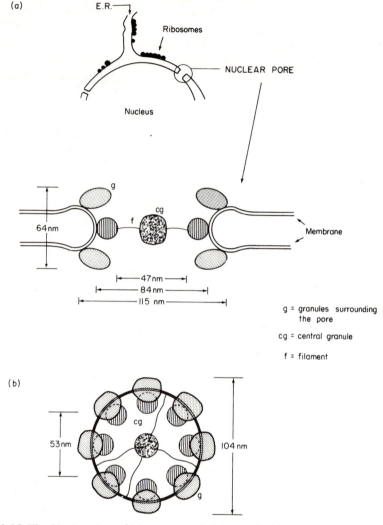

Fig. 11.18 The Nuclear Pore (After Kay and Johnston). (a) side view of two magnifi-
cations, (b) surface view.

(e.g., *trans* 18 : 1 and *cis* 18 : 1), to obtain bacteria with defined
membrane lipid compositions. When the rate of sugar transport was
monitored as a function of temperature, the plots shown in Fig. 11.19
were obtained. The important observation was that these breaks in the
curves were in good agreement with the melting temperatures observed for
vesicles made from their isolated phospholipids. In other words, the sharp
phase change from the fluid to the solid state of the lipid, corresponded
with an abrupt fall in the activity of the transport system. These results
were interpreted as suggesting that the transport proteins found it more
difficult to undergo the conformational changes necessary for their
activity when their lipid milieu was solid rather than fluid.

Let us examine the system more carefully, using *E. coli* supplemented with elaidic acid. This time two breaks in the profile (see Fig. 11.20) were observed. In the phase diagram for a system composed of phospholipids containing this fatty acid, these breaks corresponded to the temperatures at which (a) the solid state first appears and (b) all the lipid was in the solid

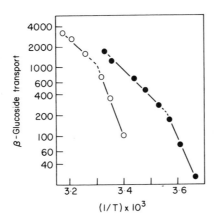

Fig. 11.19 Arrhenius plots for the effect of temperature on the rate of β-galactoside transport by an essential fatty acid auxotroph grown at $37°$ in linoleic acid (black circles) and elaidic acid (white circles). [After Wilson, G. and Fox, C. F.]

phase. Between these two temperatures both fluid and solid phases exist, making for an increase in the *isothermal lateral compressibility* of the membrane lipids. That is, the application of any form of pressure to the membrane will result in the conversion of some fluid lipid to solid lipid. Since the volume occupied by a phospholipid molecule when in the fluid phase is larger than when in the solid state, this conversion will cause a corres-

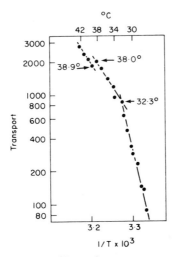

Fig. 11.20 Arrhenius plot for the effect of temperature on β-galactoside transport by cells grown at $37°$ in a medium supplemented with elaidic acid. [After Linden *et al.*]

ponding decrease in volume per fluid lipid molecule, thereby effectively relieving the pressure. In the light of these observations it is possible to interpret the sudden jump in transport rate which occurred at 38°C, and the second break at 32°C. The higher temperature corresponded to the on-set of phase separation (i.e., (a) above). If one assumes that transport activity is dependent upon conformational changes in the protein (equivalent to volume changes), then these volume fluctuations can be more easily accommodated, as we have seen, in a system containing both solid and fluid states. The compressibility of membrane lipids is much reduced when they are all solid below 32°C or all fluid above 38°C, hindering conformational changes in protein molecules and thus decreasing the transport rate. If one grows wild type bacteria at several different temperatures and then examines the transition points of their membrane lipids an interesting property is observed. The melting temperature increases—within limits—with the growth temperature. In other words, the bacteria are able to maintain the overall fluidity characteristics of their membranes by changing the fatty acid composition. At low growth temperatures a higher proportion of unsaturated and short chain species are present. At higher temperatures, the longer and more saturated fatty acids predominate. The co-existence of solid and liquid phases is, there-fore, an important property of membrane function, and may represent the basis by which substances such as anaesthetics and hormones modulate membrane behaviour.

The role played by calcium in the structure of the membrane at present is obscure. It appears to form an ionic bridge between individual phosphatidyl serine molecules such that a solid phase of phosphatidyl serine aggregates is generated. With the neutral lipid, phosphatidyl choline, no such effect is observed. In some instances, calcium may also shift the Rate of Transport against I/T profiles considered above, to the right. The cation certainly is an important and tightly bound constituent of the membrane, mediating perhaps the various ionic interactions between the lipid head groups and membrane associated proteins. For this reason, its effects are likely to be very diverse.

12

Biosynthesis of Nucleic Acids and Proteins

12.1. Introduction

It is usual for the general principles of the way in which nucleic acids and proteins are synthesized, and the location of these activities within the cells, to be considered under cell biology. Here, therefore, we are concerned with the more biochemical aspects of the subject which lies at the core of what has in recent times been named "molecular biology". There are three major reasons why it is appropriate for a student of medicine to know something of this subject.

(a) It is an extremely active growing point in biochemistry and has attracted the interest not only of those who have set out to specialize in biochemistry but also microbiologists, geneticists, cell biologists, chemists and physicists. Thus an extraordinary galaxy of talent has been centred on the subject at the international level and very considerable advances in our understanding have been made. The subject has important implications for the possible creation of life in "the test-tube" and the origin of life and a good deal has been written about the "biological time-bomb". It is right that a medical student should have a basic understanding of the subject so that he can more soberly place in perspective subsequent discoveries.

(b) Many aspects of our understanding of the control of metabolic reactions and the mechanism of action of hormones at the molecular level depend on a knowledge of the mechanism of the synthesis of proteins and nucleic acids.

(c) Many effective drugs have in the past been discovered as the result of massive screening programmes and their precise mode of action has not been elucidated until many years after their discovery. A major omission in our armoury of drugs concerns the control of viral infection. There is a belief, therefore, that we need to understand more about the replication of viruses before drugs can be designed to control the process. All this is wrapped up in the development of drugs to control the growth of tumours since there is evidence that viruses may play an important role in the aetiology of cancer. In short an understanding of the subject helps us to explain the mechanism of action of existing drugs and may help in the design of new drugs to prevent those diseases that are so far difficult to control.

Medical students are sometimes reluctant to hear about biological systems other than the human body. This is a particularly mistaken attitude in the case of our present subject for many of the advances have come from a study of bacteria, particularly the simple bacterium *E. coli*. One of the major conclusions at the present time, from all the work done on protein and nucleic acid biosynthesis with a broad variety of living organisms, is that the fundamental mechanisms involved are very similar. Hence one can extrapolate with some justification from the results obtained with bacteria to animal cells. As we shall see, however, there are some differences and since these differences are important in explaining the differential action of antibiotics we shall from time to time go into some aspects in greater detail.

12.2 Physiological Aspects of Protein Synthesis

Until about 1940 it was the general view that the body only indulged in protein synthesis when new protein was required. This new protein might be for (a) growth, including the growth of the foetus (b) turnover of cells, such as occurs with the red blood cells and those that are damaged in the mucosal wall of the gut (c) for the production of excreted protein by the animal as occurs in lactation in mammals or eggs in birds.

Around 1935 there had been suggestions by Borsook from nitrogen balance studies that this was not the whole story and that even in the adult animal there was a considerable amount of protein being synthesized. This was proved by Schoenheimer by feeding animals with amino acids labelled with the heavy isotope of nitrogen, ^{15}N. He showed that even in the adult the bodily proteins became labelled. These experiments led to the concept of the "Dynamic state of the body constituents" and in particular to a realization that many, if not all, of the proteins of the body were being constantly broken down and re-synthesized. Thus, if the concentration of serum albumin is constant this is because the rate of synthesis and breakdown is balanced. When in infections the serum albumin concentration decreases, this may be due either to the rate of synthesis being reduced or the rate of degradation being increased.

We now realize that the rate at which different proteins "turnover" varies markedly. The rate of turnover is usually expressed as the half-life, i.e., the time for half the molecules of a particular protein to be broken down and replaced by new molecules. For the plasma proteins the half-life is on average about 10 days but this can vary under different conditions. In contrast the adult human, except in the case of pregnancy or wound healing, synthesizes virtually no collagen and it is not normally degraded. This is important with respect to the administration of radioactive isotopes to children, for such isotopes will be incorporated into the collagen and will not be replaced as they will be in the case of plasma proteins. Some of the brain proteins also seem to be biologically very stable.

Just as the whole animal may synthesize protein that is retained or excreted, so this can occur at the level of the cell. For example the reticulocyte synthesizes haemoglobin that is retained by the cell, but other cells such as the acinar cells of the pancreas synthesize zymogens, which are exported. The parenchymal liver cell synthesizes all the plasma proteins except for γ-globulin and probably as much of the synthesized protein is exported from the liver cell as is retained. The γ-globulin is synthesized by plasma cells in the lymph glands. This introduces another point, namely that the synthesis of the various proteins required by the body is shared as between the cells of the different tissues.

Chemical pathologists employ a group of tests known as liver-function tests. In essence these measure the albumin/globulin ratio of the serum. On the basis that the liver is the sole site of synthesis of serum albumin but does not synthesize on a quantitative basis most of the globulin, the albumin/globulin ratio is a measure of the ability of the liver to synthesize serum albumin. In a pathological state, where the liver function is defective, the ratio may be expected to change from normal. This is also indicated by demonstrating more directly a drop in the concentration of albumin in the serum.

12.3. Prosthetic Groups

You will be aware that the major component of the proteins is the polypeptide chain which consists of amino acids linked together through peptide bonds. It is clear, therefore, that a major concern is the way in which these peptide bonds come to be formed. However, proteins also contain prosthetic groups and these are very important for their biological activity. Some of the prosthetic groups are associated with the polypeptide chains by weak bonds, as for example the haem in haemoglobin. In such cases the polypeptide chain and prosthetic group can be easily dissociated and reassociated and no doubt this also occurs in the cell.

In other cases the prosthetic groups are covalently linked to the polypeptide chains. This applies in glycoproteins and phosphoproteins. Here specific enzymes are required and this may well be an important potential control point in the formation of a protein but we will not be concerned with it here.

12.4. Nature of Amino Acids Incorporated into Proteins

Among the hydrolysis products of different proteins one can detect 24 different amino acids. Are all these incorporated into polypeptide chains or are some modified after the peptide bond has been formed? We now know that only 20 different amino acids are incorporated as listed below.

A list of those amino acids that participate in the formation of poly-peptide chains

Glycine	Tryptophan	Lysine
Valine	Tyrosine	Arginine
Alanine	Phenylalanine	Histidine
Leucine	Glutamic acid	Methionine
Isoleucine	Glutamine	Cysteine
Threonine	Aspartic acid	Proline
Serine	Asparagine	

Inspection of the list leads to the following points.

(a) Most of the common amino acids are included.

(b) There are some unexpected inclusions. Thus the amides asparagine and glutamine which are derived from aspartic acid and glutamic acid appear in the list.

(c) There are some unexpected omissions. Thus phosphoserine and phosphothreonine which occur in phosphoproteins are formed from selected residues of serine and threonine in the polypeptide chain. In fact phosphoprotein kinases are responsible which utilize ATP as the substrate. Similarly hydroxyproline and hydroxylysine are not included in the list. Again certain residues of proline and lysine in the chain are hydroxylated.

12.5. Protein Conformation is Determined by the Primary Structure

It will be recalled that the structure of a single polypeptide chain can be envisaged as primary (sequence of amino acid residues in the chain), secondary (coiling within the chain) and tertiary (interaction of coiled chain). Since it is now clear that the primary structure determines the secondary and tertiary structures the process of protein biosynthesis is only concerned with this aspect of protein structure. The cell does not have to possess special mechanisms to ensure that the proteins take up the correct secondary and tertiary structure.

12.6. The Biosynthesis of Multi-chain Proteins

Many proteins contain more than one polypeptide chain. Sometimes the chains are associated by weak forces as with haemoglobin and in such cases it is presumed that the chains are synthesized in the cell in close juxtaposition and easily associate. In other cases the chains are covalently linked. An example of such a structure is immunoglobulin, the structure of which has already been considered in Chapter 4 (see Fig. 4.7).

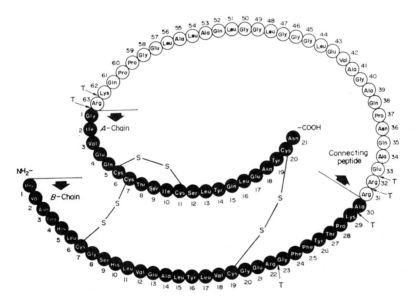

Fig. 12.1 The primary structure of pig proinsulin. The amino acid residues in the A and B chains of insulin are shown by dark circles. ←T indicates peptide bonds that fulfil the specificity requirements of trypsin. The enzyme(s) that are responsible for the conversion of proinsulin to insulin in the cell have not yet been characterized. (From Grant, P. T. and Coombs, T. L. *Essays in Biochemistry*, vol. 6.)

Although the Heavy and Light chains are linked by S-S bridges they may be easily dissociated and reassociated. The two kinds of chain are separately synthesized in the cell and then associated together before the immunoglobulin is secreted.

A much more fascinating problem exists in the case of insulin. It will be recalled that this consists of two chains linked by two S-S bridges. Attempts to associate the two chains together in free solution never yielded significant amounts of biologically active insulin so that it was difficult to envisage the circumstances whereby they interacted in the correct manner in the cell. The problem was solved by two groups in the U.S.A. who showed that insulin was synthesized as one long chain, called proinsulin, the structure of which is shown in Fig. 12.1. The long peptide containing residues 31-63 inclusive in the figure is excised to give insulin consisting of two chains. The connecting peptide ensures that the chains in insulin are linked in the manner which gives the molecule its biological activity. Attempts are now being made to locate the precise position in the cell at which the conversion of proinsulin to insulin takes place and the nature of the specific enzyme involved

12.7. Possible Interconversion of one Protein into Another

Before much was known about the biosynthesis of proteins there was speculation that various proteins in the body were converted into one another by minor modifications to the polypeptide chains. Two possibilities were to the fore, the conversion of serum albumin to α_1-globulin because in infections the serum concentration of the one is lowered and the other raised, and the conversion of serum proteins to tumour tissue proteins. It is now clear that these ideas were quite wrong and that for amino acids to exchange between different proteins the proteins must first be degraded to free amino acids and the new protein synthesized from the newly available amino acids.

12.8. The Concept of a Template for the Assembly of Amino Acids

If polypeptide chains were to be synthesized in the cell from amino acids by traditional enzymic methods, we might expect a whole battery of specific enzymes. Thus

$$aa_1 + aa_2 \xrightarrow{\text{enzyme } x} aa_1 aa_2$$

$$aa_1 aa_2 + aa_3 \xrightarrow{\text{enzyme } y} aa_1 aa_2 aa_3$$

with enzymes x and y being quite specific with respect to amino acid and peptide. It was soon apparent that a phenomenal number of different enzymes would be required and since enzymes are proteins it became clear that the space in the cell was not big enough to accommodate them.

From such arguments it was deduced that the amino acids were more likely to be assembled on a template. The amino acids might be assembled temporally either in a manner which bore no resemblance to their order in the polypeptide chain or in an order that reflected precisely their order in the chain. In fact without any knowledge of the nature of the template it was determined experimentally that the amino acids were assembled temporally in the order in which they occurred in the chain. Moreover, the polypeptide chain can be shown to be assembled from the N to the C terminus as normally written by convention.

12.9. The Integration of Nucleic Acid and Protein Synthesis

It is useful to sketch the place of DNA and RNA biosynthesis in the biosynthesis of protein.

replication transcription RNA

DNA

translation

polypeptide chain

post translational
modifications

finished protein

We have already mentioned some of the post translational modifications, e.g., the hydroxylation of proline or the addition of phosphate groups to existing polypeptide chains. The other important steps are replication, transcription and translation and each will be considered in turn.

12.10. Enzymes Involved in the Biosynthesis or Replication of DNA

It will be recalled from Chapter 3, that the essential features of DNA are that it consists of a polymer of nucleotides, that the nucleotides are characterized by the presence of one of 4 bases (2 purines and 2 pyrimidines), that DNA normally contains two polynucleotide chains and that these are anti-parallel (i.e., one is $5' \rightarrow 3'$ and the other $3' \rightarrow 5'$).

Since the polynucleotide chains are polymers it follows that energy will be required to make the links just as has already been shown in the case of glycogen. That the energy would be utilized in the form of an "activated" precursor was predictable but it was Kornberg who in 1955 showed how the process was enzymically possible. He isolated an enzyme called DNA polymerase and showed that the precursors were the triphosphates of the deoxy nucleosides. The essential reaction is:

d TPPP
d GPPP
d APPP $\xrightarrow[\text{polymerase}]{\text{DNA}}$ $[dTp - dGp - dAp - dCp]_n + PP_i$
d CPPP

The utilization of triphosphates with the elimination of PP_i is a rather common reaction in the synthesis and degradation of polymers. Not only does this occur in the synthesis of glycogen but also in fatty acid metabolism and, as we shall see, in the activation of amino acids.

The reaction causes the synthesis of the new polynucleotide chain in the direction $5' \rightarrow 3'$. Thus

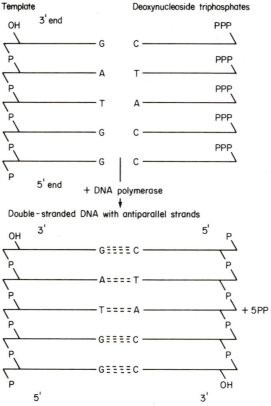

The reaction requires the presence of some DNA or at least short lengths of polynucleotide. The DNA is here serving in the main as a template and is not acting merely as a primer as glycogen does in glycogen synthesis.

The sequence of bases in the newly synthesized polynucleotide reflects the base sequence of the template DNA and as expected the product is the complement of the template. This is shown in Fig. 12.2.

Fig. 12.2 The biosynthesis of DNA by DNA polymerase using a single strand of DNA as template.

The template and new strand are of course anti-parallel with respect to each other. Kornberg's polymerase enzyme is more effective when the template DNA is denatured, that is the two chains have been separated. If such denatured DNA is used as template then the newly synthesized polynucleotide not only has the correct base sequence but also it has many of the expected physical properties of DNA. If, however, a native double stranded DNA is used as template the properties of the new polynucleotide differ quite markedly from DNA and in particular it has not been possible to demonstrate biological activity of DNA synthesized *in vitro* using such a native template.

The original method for demonstrating the biological activity of DNA stemmed from the classical work of Griffith who in 1931 demonstrated the "transformation" of *Pneumococcus*. He showed that if a mutant strain of *Pneumococcus* which had lost its ability to cause septicaemia in mice was mixed with a heat-killed preparation of a virulent strain and then injected into mice they succumbed to septicaemia. Later in 1944, Avery, MacLeod and McCarty showed that the active substance in the heat-killed extract was DNA. More recently a simpler method has been adopted to demonstrate the biological activity of DNA. This is based on the fact that bacteria can be infected by viruses which are called "bacteriophage" or "phage" for short. The genome of the phage may either be DNA, (DNA-phage) or RNA, (RNA-phage). The DNA of a phage will cause a bacterium to be infected and this will lead to lysis.

Since it proved impossible to demonstrate the biological activity of DNA synthesized in the presence of a native DNA template, Kornberg, in collaboration with Sinsheimer, set out on another approach. Sinsheimer had discovered a phage (ϕX174) that contained a DNA which was circular and single-stranded. When this DNA was used as template with the Kornberg enzyme a biologically active phage DNA was produced. A major step in this work was the discovery of another enzyme called DNA ligase. This enzyme was discovered by the work of radiation biologists. One cause of the cytotoxic effects of X-rays is the production of single-strand breaks in DNA as shown below.

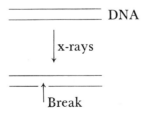

The breaks could be repaired by the action of ligase as shown in Fig. 12.3. Through the action of ATP the 5'P is converted to a high energy compound which reacts with the 3'OH.

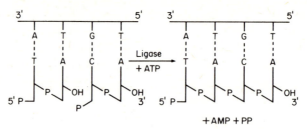

Fig. 12.3 The mechanism of action of DNA ligase in repairing the break in a DNA strand which is hydrogen bonded to an unbroken strand.

While the ϕX174 DNA is single stranded Sinsheimer showed that it went through a double stranded form during its replication. We may label the original strand of ϕX174 plus (+) and the new strand minus (−). The steps in the synthesis of new + strands under *in vitro* conditions are shown in Fig. 12.4. Thus it is possible to use the DNA polymerase of Kornberg in conjunction with the ligase to cause the formation of a biologically active DNA.

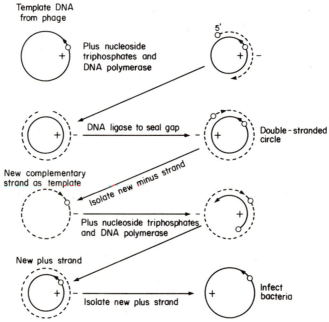

Fig. 12.4 The role of DNA polymerase and DNA ligase in the biosynthesis of the bacteriophage ϕX174 DNA *in vitro*.

12.11. Biosynthesis of DNA in the Cell

The original experiments of Meselson and Stahl showed that DNA was synthesized in the cell by a semi-conservative mechanism as shown in Fig. 12.5. It is called semi-conservative because in the first generation one of

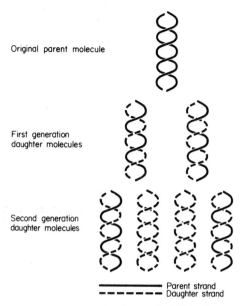

Original parent molecule

First generation
daughter molecules

Second generation
daughter molecules

―――――― Parent strand
― ― ― ― ― Daughter strand

Fig. 12.5 The semi-conservative replication of DNA *in vivo* as proved by the experiments of Meselson and Stahl. One of the strands of the parent DNA is retained in each of the first generation daughter molecules.

the parental strands of DNA is present in each of the daughter molecules. In *E. coli* DNA is circular and Cairns has shown that a native double stranded circular DNA has to replicate as shown in Fig. 12.6.

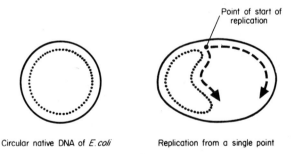

Point of start of
replication

Circular native DNA of *E. coli*   Replication from a single point

Fig. 12.6 The replication of the double stranded DNA of *E. coli in vivo* as suggested by the experiments of Cairns. The replication of both strands is initiated at the same point.

Replication starts at a single point on both strands and the direction of growth of each new strand is in the same direction. This evidence came from careful studies in which the circular DNA was visualized by autoradiography of [^{3}H] thymidine labelled DNA. Now the Kornberg polymerase enzyme effects synthesis in only one direction $5' \rightarrow 3'$ so how can the results of Cairns be accommodated? A possible explanation is shown in Fig. 12.7.

According to this hypothesis one of the strands is synthesized continuously and the other in short lengths as it becomes exposed. The ligase then causes the link-up of the short lengths. Certainly it can be shown using mutants, which lack the ligase, that more short lengths of newly synthesized DNA can be isolated than in normal cells.

More recently Cairns has isolated a mutant of *E. coli* which lacks the Kornberg DNA polymerase and yet appears to replicate its DNA normally.

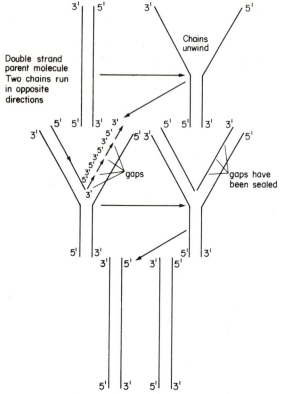

Fig. 12.7 A possible method of replication of double stranded DNA *in vivo* involving the Kornberg DNA polymerase and DNA ligase. One chain is replicated continuously and the other in a series of fragments which are then linked together.

This and other results now prove that the Kornberg DNA polymerase is not the enzyme responsible for DNA replication in the cell. Rather it is concerned with the repair mechanism when bases of DNA become changed e.g., through u.v. irradiation. The mutant cells lacking the enzyme are more sensitive to u.v. radiation. Much current research is aimed to characterize the enzyme which effects chromosome replication. It seems likely that it is membrane bound but in any event the mechanism of action is as described for the Kornberg polymerase and it is very probable that both strands are replicated as shown in Fig. 12.7.

12.12. Transmission of Antibiotic Resistance

An important current medical problem involving a knowledge of DNA replication is that of conjugation in bacteria and the effect of this process on the transmission of resistance of bacteria to antibiotics.

A bacterium such as *E. coli* which is normally harmless, when present in the gut, may develop resistance to certain antibiotics. Most probably this involves the synthesis of an enzyme which degrades the antibiotic, hence new information in the form of a gene has been incorporated into the DNA. This resistant *E. coli* is able to transfer this resistance to a pathogenic bacterium such as the *Salmonella* responsible for typhoid fever. Moreover, the transfer allows the passage of a number of distinct genes simultaneously and so the *Salmonella* may become resistant to a range of unrelated antibiotics. It is clear that what happens is that a class of extra-chromosomal genetic elements called "transmissible R" or resistance factors are responsible. The factors contain DNA. Thus we have a donor cell, or male cell, and a recipient, or female cell. The mode of transmission of resistance is shown in Fig. 12.8. The *Salmonella* having received a piece of DNA from the *E. coli* through the sex pilus has conferred on it resistance to a range of antibiotics and, moreover, has been provided with a sex pilus through which it can transfer its resistance to other *Salmonella*. The *E. coli* loses its resistance and sex pilus.

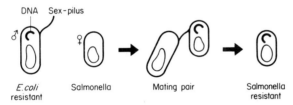

Fig. 12.8 The basic mechanism of bacterial conjugation between a strain of *E. coli* which is resistant to a group of antibiotics and a strain of *Salmonella* that acquires resistance through the transfer of DNA from the *E. coli*.

12.13. The Enzymes Involved in the Biosynthesis of RNA and Transcription of DNA

It will be recalled from Chapter 3, that in general the structure of RNA bears a fairly close resemblance to that of DNA. Again there are usually 4 bases but this time uracil replaces thymine. The sugar is ribose instead of deoxyribose. RNA does not normally exist as double complementary strands like DNA but the single strand of a given RNA often loops in such a way that hydrogen-bonded regions are formed. RNA is in general much smaller than DNA and a broad range of molecular sizes are found. Some species of RNA are of a size that allows the sequence of bases in the chain to be determined. The methods used for such sequence studies are in principle rather similar to those already explained for the determination of the primary sequence of proteins in Chapter 4.

RNA is synthesized by polymerases which again utilize the nucleoside triphosphates. Thus we have

$$
\begin{matrix}
\text{ATP} \\
\text{UTP} \\
\text{CTP} \\
\text{GTP}
\end{matrix}
\quad \xrightarrow[\text{+DNA}]{\text{polymerase}} \quad
[\text{Ap} - \text{Gp} - \text{Cp} - \text{Up}]_n + \text{PP}_i
$$

The DNA serves as a template and the new strand is synthesized in the direction $5' \rightarrow 3'$ as was the case for DNA synthesized by the Kornberg polymerase. Thus the enzyme is named DNA dependent RNA polymerase. The idea of RNA being synthesized on a template of DNA will seem reasonable when one remembers the general scheme already described which showed that an important intermediate step in the transfer of information from DNA to the primary sequence of a polypeptide involved the transcription of DNA to RNA.

The RNA polymerase differs in many respects from the DNA polymerases. Thus it prefers the DNA template to be native and double stranded. In this case only one strand of DNA is used as a template for the formation of a particular strand of RNA and so the synthesis is *asymmetric*. It is also *conservative* because both the parent strands are conserved. It appears that either of the strands of DNA can be used as a template and so we have the possibility of the following process.

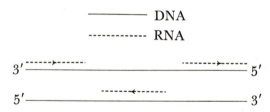

It should be noted that two adjacent strands cannot be transcribed. It seems that the two strands of DNA must divide a little at the point where the RNA is being synthesized. The polymerase is a very large enzyme consisting of four subunits and it is possible, by electron microscopy, in some cases to observe its presence on the chromosomes.

Both the DNA of the normal cell and the DNA of a virus are transcribed by this DNA dependent RNA polymerase. However, we have already mentioned that some viruses contain no DNA, the viral genome consisting of RNA. In such cases the viral RNA is replicated by an RNA dependent RNA polymerase, sometimes called an RNA replicase, which again utilizes nucleoside triphosphates. The incoming viral RNA will be single stranded and in order for it to replicate it will have to pass through a replicative form. Thus the enzymic reaction can be envisaged as occurring in two steps.

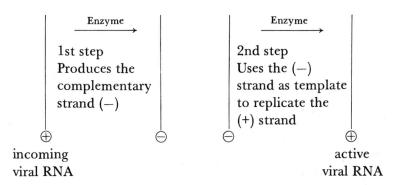

1st step
Produces the
complementary
strand (−)

incoming
viral RNA

2nd step
Uses the (−)
strand as template
to replicate the
(+) strand

active
viral RNA

The same enzyme catalyzes each step but different protein factors are involved. In the second step the synthesis will again be asymmetric to allow the production of the + strand. Since this RNA replicase is only concerned with viruses it is an obvious point at which to try to develop anti-viral drugs.

12.14 RNA directed DNA Polymerase

This is a recently discovered enzyme. Since it is of potential clinical importance a brief explanation of its discovery and possible role in the cell is justified.

Just as the genome in phage can be either DNA or RNA so it is with the viruses that infect man. Some are DNA viruses such as those causing Herpes or Trachoma while others are RNA e.g., poliovirus. For a long time it has been known that in experimental animals certain viruses can cause a normal cell to become malignant. Such viruses are known as oncogenic and may be either DNA or RNA viruses. We cannot be certain at this time that viruses are actually oncogenic in humans but it is a reasonable assumption that some tumours in man are of viral origin.

A particularly interesting phenomenon in the conversion of a normal cell to a malignant cell by an RNA virus is that the virus appears in each new generation of cells; the virus is transmitted in a predictable and stable fashion just like any normal genetic characteristic. It seems that the RNA virus is first copied into DNA and the DNA is then integrated with that of the cells own chromosomes.

Working with an RNA virus that causes tumours in chickens, the Rous Sarcoma virus, Temin, an American scientist, detected in the cells a DNA polymerase that used the viral RNA as a template. The reactions appear to occur in three steps.

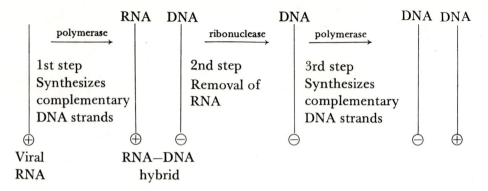

Similar RNA directed DNA polymerases have been found in cells infected by many different RNA viruses and in particular in leukaemia cells and also in the viruses themselves. More recently a similar polymerase has been found in proliferating tissue such as foetal tissue.

At one time it seemed that the new polymerase was confined to oncogenic RNA viruses and there was hope of finding ways of inhibiting the polymerase and thereby preventing cancer. Although this now looks less likely the discovery of this polymerase is of real importance in cancer research. It is fascinating to realize that transcription can be reversed and that the copying is not always from DNA to RNA but can occur in the other direction too:

$$
\text{DNA} \underset{\substack{\longleftarrow\\ \text{RNA directed DNA polymerase}}}{\overset{\substack{\text{DNA dependent RNA polymerase}\\ \longrightarrow}}{\text{TRANSCRIPTION}}} \text{RNA}
$$

12.15. Translation: General Characteristics of Cell Free Protein Synthesis

We are now in a position to consider the process of translation in which the information contained by the base sequence of RNA can be utilized to order the amino acids in the correct sequence in a polypeptide chain. The major advances in our understanding of this process have come from studies utilizing subcellular fractions and their ability to effect protein synthesis *in vitro*. If a cell, such as a liver cell, is disrupted and the various particulate constituents are separated by differential centrifugation one can obtain a fractionation as shown in Fig. 12.9. Of the four fractions that result three contain particles. If one wants to discover which fraction is most active with respect to protein synthesis, then one can inject an animal with a ^{14}C-labelled amino acid and then determine the location of the radioactive protein in the different fractions prepared from liver at various times after the injection. The results obtained by one of the pioneers in this work, Paul Zamecnik, a Professor of Medicine at

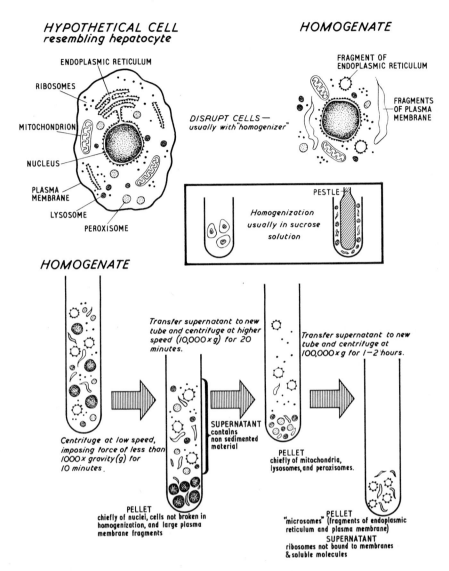

Fig. 12.9 The fractionation of subcellular particles from rat liver. The liver cell is disrupted by grinding and the particulate material is separated by differential centrifugation to give a nuclear, mitochondrial, microsomal and supernatant fraction. (Modified from *Cells and Organelles*, Novikoff, A. B. and Holtzman, E.)

Massachusetts General Hospital, Boston, are shown in Fig. 12.10. It will be seen that the fraction which contains the most radioactive protein is the microsome fraction. In preparations from liver this fraction is rich in pieces of the rough-surfaced endoplasmic reticulum, that is membranes to which ribosomes are attached.

Having established the site of the radioactive protein the next step was to incubate the microsome fraction isolated from the liver of an untreated rat with radioactive amino acid and find out whether it was possible to demonstrate the incorporation of ^{14}C-amino acid into radioactive protein. This indeed did occur provided the following conditions pertained.

(a) In addition to the microsome fraction, containing ribosomes, some of the soluble components of the cytoplasm were also needed.

(b) A source of energy was required in the form of ATP (or a high-energy compound capable of generating ATP).

(c) In addition to ATP some GTP was also essential.

Bearing in mind the above conditions we now discuss what is happening in the *in vitro* system.

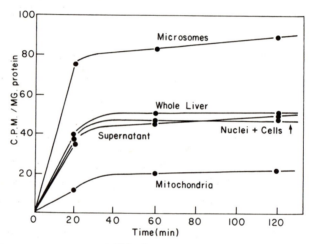

Fig. 12.10 The incorporation of [^{14}C] leucine into the protein of the subcellular fractions of rat liver. The animal was injected with the amino acid and pieces of liver were removed at the times shown, the liver disrupted and the particulate fractionated by differential centrifugation. (From Keller, Zamecnik and Loftfield, 1954, *Histochem. and Cytochem.* 2, 378.)

12.16. Formation of "Activated" Amino Acid

The formation of a peptide bond will obviously require energy and it always seemed likely that this would involve an "activated" amino acid. This activation might involve either the NH_2 or the COOH group of the amino acid but Hoagland established that it was the COOH group that was activated. You will see in Fig. 12.11 that the ATP reacts with the amino acid to form an amino acid adenylate, again with the elimination of pyrophosphate. There is a specific enzyme for each of the 20 different amino acids. This reaction would explain the need for ATP in the cell-free system for protein synthesis.

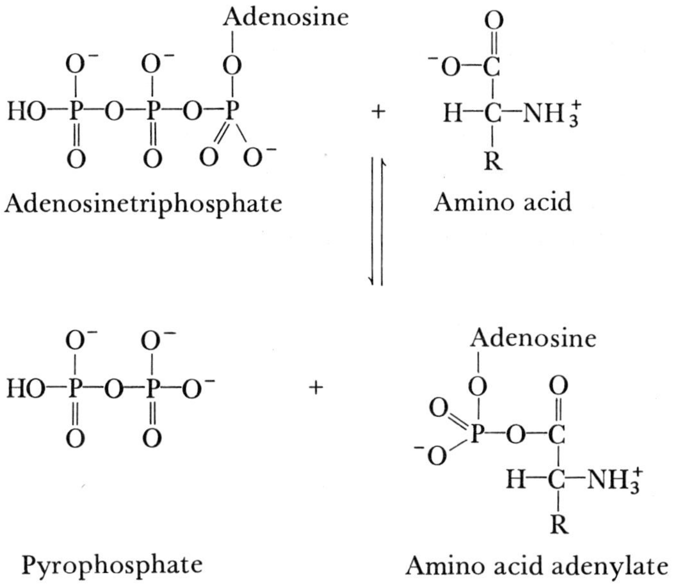

Adenosine

$$HO-P-O-P-O-P \quad + \quad H-C-NH_3^+$$

Adenosinetriphosphate   Amino acid

Pyrophosphate

$$HO-P-O-P-O^-$$

Adenosine

Amino acid adenylate

Fig. 12.11 The "activation" of an amino acid by reaction with ATP to form an amino acid adenylate and pyrophosphate.

12.17 Transfer-RNA and the Attachment of Amino Acid

The discovery of the next step in the formation of a peptide bond lay in the finding of transfer-RNA. If a ^{14}C-labelled amino acid is incubated with soluble cytoplasm in the presence of ATP and then trichloroacetic acid is added, a precipitate is obtained. An analysis of this precipitate shows that the ^{14}C-labelled amino acid has not been incorporated into protein but has been attached to RNA. Hoagland showed that the RNA to which the amino acid was attached was of low molecular weight and was located in the soluble cytoplasm. It was for this reason that it was first named soluble or S-RNA but we will use the name transfer or t-RNA. The t-RNA represents most of the RNA in the soluble cytoplasm and is 10-15% of the total RNA of the cell. There is at least one specific t-RNA for each of the 20 different amino acids. Each has a molecular weight of about 29,000 and contains 75-85 nucleotides. The amino acids are attached to the 3'OH group at the terminus of the polynucleotide by an ester bond. This is a bond of about the same stability as a peptide bond. The transfer of the activated acid to the t-RNA is shown in Fig. 12.12. Although the different t-RNA are specific for the particular amino acids it was soon apparent that all had the same triplet of bases at the 3' terminal namely, cytosine-cytosine-adenine and that the amino acid was always attached to adenosine.

It is worth pointing out at this stage that various of the cell organelles cooperate in the synthesis of protein. This is summarized in Fig. 12.13.

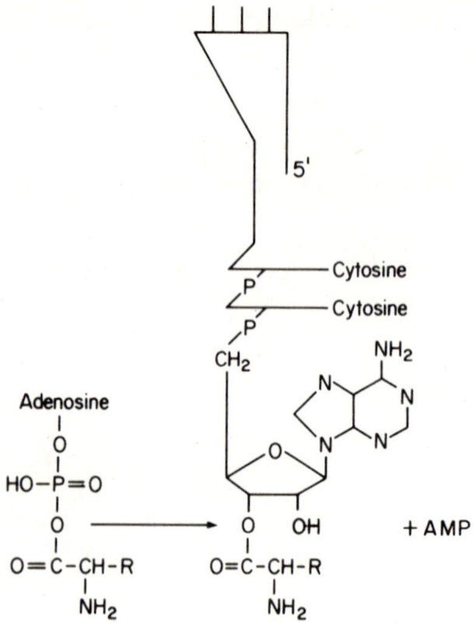

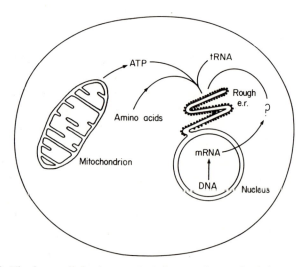

Fig. 12.12 The transfer of amino acid from the amino acid adenylate to the 3′ terminal end of t-RNA with the release of AMP.

Fig. 12.13 The intracellular interactions in protein synthesis in an animal cell.

12.18. Ribosomes

We have seen why ATP and the soluble cytoplasm are necessary for protein synthesis in a cell-free system. We can now ask what part the microsome fraction plays. The answer concerns the ribosomes attached to the membrane of the endoplasmic reticulum.

If we examine the structure of the whole range of living cells we find that they can be divided into two types, the prokaryotes and the eukaryotes. The prokaryotes lack a membrane-bounded nucleus and contain no mitochondria or chloroplasts. They are in the main represented by the bacteria. The eukaryotic cells contain a nucleus with membrane as well as mitochondria and in some cases chloroplasts. They are represented by animal and plant cells including fungi.

Both types of cell contain ribosomes, but those isolated from the eukaryotes (animal cells) are twice as large as those from the prokaryotes (bacteria). It is usual to characterize ribosomes by the rate at which they sediment in a centrifugal field and quantitatively this is represented by the sedimentation coefficient or S value. The value of S depends not only on the size of the particle but also on its shape and density; and so is not directly proportional to size. The S value of the ribosomes from animal cells is about 80 and so they are known as 80S ribosomes and those from bacteria as 70S ribosomes. (The one is in fact about twice the size of the other.)

Ribosomes appear to contain only RNA and protein, 80S ribosomes containing approximately equal amounts of each and 70S ribosomes RNA and protein in the ratio 2 : 1. In order to differentiate the RNA in the ribosomes from messenger-RNA and transfer-RNA we call it ribosomal-RNA.

Both 70S and 80S ribosomes consist of two subunits. This can be shown either by electron microscopy or by treating the ribosomes with solutions containing only very low concentrations of Mg^{2+} ions. Under these conditions the ribosomes are dissociated into their subunits and these can be separated by ultracentrifugation.

One of the subunits from the ribosomes is about twice the size of the other so that 70S ribosomes give subunits with S values of 50S and 30S and 80S ribosomes give subunits with S values of 60S and 40S.

We can summarize the position as follows.

RIBOSOMES

Prokaryotes	Eukaryotes
$70S \begin{cases} 30S \\ 50S \end{cases}$	$80S \begin{cases} 40S \\ 60S \end{cases}$
RNA : Protein	RNA : Protein
2 : 1	1 : 1

12.19. Messenger-RNA

The idea that a template could ensure that the amino acids for the synthesis of a polypeptide become linked in the right sequence has been mentioned before. Following the characterization of the ribosomes, and other evidence that suggested that the template would be composed of

RNA, it was first thought likely that ribosomal-RNA would itself serve as the template. This idea became increasingly untenable as more information concerning ribosomal-RNA became available. The experiments that first upset the idea were those involving the infection of *E. coli* with a DNA phage. Immediately after infection the normal DNA synthesis of the cell ceases and virtually all the DNA synthesized is phage DNA. The whole protein-synthesizing apparatus of the cell is directed to the synthesis of phage protein. Since the latter is quite different in composition from the normal bacterial protein one would expect the template RNA to change its base sequence after viral infection but this did not in fact occur.

Because of the need for the template RNA to change its base composition when the type of protein being synthesized is changed it was always predicted that template RNA would have a high rate of turnover. In other words it would be constantly being synthesized and degraded so that its composition could be rapidly changed. In fact ribosomal-RNA turned out to be metabolically very stable.

For the above reasons, stable structure and stable metabolism, it became apparent that ribosomal-RNA could not act as the template. From kinetic experiments it was predicted by Brenner that template-RNA would be an RNA, present in small amounts, short-lived and associated with the ribosomes. He termed this messenger-RNA, since its role would be to carry information from the DNA in the nucleus (where it arises by transcription by the DNA dependent RNA polymerase) to the cytoplasm (where it would associate with the ribosomes). The ribosomes were likened to a tape-recorder being charged with a messenger-RNA tape. This brilliant hypothesis was to be fully vindicated soon after it was put forward by a fascinating discovery from another scientist.

It was Nirenberg, working at that time in the Cancer Institute of the National Institute of Health in the U.S.A., who was studying the effect of various RNA fractions on the ability of ribosomes to synthesize polypeptides. You will recall the experiments of Zamecnik that demonstrated the ability of the microsome fraction from rat liver to effect the synthesis of protein. He later showed that bacterial ribosomes could replace the microsomal fraction in these experiments. Nirenberg showed that certain RNA fractions stimulated *E. coli* ribosomes to incorporate ^{14}C-amino acids into protein but he could not characterize the protein that was synthesized. He, therefore, added a synthetic polynucleotide containing the single base uracil and called polyuridylic acid (poly-U). This gave an enormous boost in the incorporation of only one of the 20 amino acids, namely phenylalanine. It was soon shown that poly-U had effected the synthesis of polyphenylalanine by the cell-free system. The significance of this finding was clear: a specific polynucleotide added to a preparation of ribosomes had effected the synthesis of a specific polypeptide.

12.20. Coding

A long time before the concept of messenger-RNA was established it had seemed very likely that the template for protein synthesis would consist of RNA. Crick argued that the base sequence of template RNA would "code" for the order of amino acids in a polypeptide. Since there are only 4 different bases in RNA and there are 20 different amino acids in a protein it was obvious that there could not be a 1-base-1 amino acid code. A 3-base-1 amino acid code would give a possibility of 64 different amino acids that was more than enough. Crick therefore predicted that the code would consist of triplets of bases coding for specific amino acids and that there would be more than one triplet for one amino acid, i.e., the code would be "degenerate". He also showed that if one had a long row of bases in an RNA, which was to be used to code for a series of amino acids, the message encoded in the RNA would have to be read from a fixed point. Thus the bases in RNA might be in the following order:

$$5' \qquad\qquad\qquad 3'$$
$$\text{U G A C C U A G U C C}$$

The resulting polypeptide would obviously be of a different composition if the reading started at U or G or A at the 5' end. Crick also predicted that the code would be comma-less, i.e., there would not be untranslated bases between the triplets.

From the results of Nirenberg it was immediately clear, because of this ground-work by Crick, that the triplet of bases for phenylalanine was UUU. Such a triplet is known as a "codon". By incubating the ribosomes with other polynucleotides of known base sequence and a variety of other experimental methods the "meaning" of all 64 possible codons was determined.

The complete code is shown in Fig. 12.14. You will see that out of the 64 codons 61 code for an amino acid. The other three "nonsense" codons will be discussed later. As an example of the polypeptide synthesized by a polynucleotide of known sequence we can consider.

$$\text{U A U C U A U C U A U U U A U}$$
$$\text{Tyr} \quad \text{Leu} \quad \text{Ser} \quad \text{Ileu} \quad \text{Tyr}$$

12.21. The Relevance of the Code to Events in the Cell

It could be argued that the story of the code thus far is ingenious but is not necessarily relevant to the way in which proteins are synthesized in intact cells. This now seems unlikely for the following reasons. Firstly, by working with subcellular extracts from a very wide range of cells and using various synthetic messengers it became obvious that the code was

RNA-Amino acid code

First base of codon:	Second base of codon: U	C	A	G	Third base of codon:
U	UUU ⎤ Phe UUC ⎦ UUA ⎤ Leu UUG ⎦	UCU ⎤ UCC ⎥ Ser UCA ⎥ UCG ⎦	UAU ⎤ Tyr UAC ⎦ UAA ⎤ Term. UAG ⎦	UGU ⎤ Cys UGC ⎦ UGA Term. UGG Try	U C A G
C	CUU ⎤ CUC ⎥ Leu CUA ⎥ CUG ⎦	CCU ⎤ CCC ⎥ Pro CCA ⎥ CCG ⎦	CAU ⎤ His CAC ⎦ CAA ⎤ GluN CAG ⎦	CGU ⎤ CGC ⎥ Arg CGA ⎥ CGG ⎦	U C A G
A	AUU ⎤ Ileu AUC ⎥ AUA ⎦ AUG Met + Init.	ACU ⎤ ACC ⎥ Thr ACA ⎥ ACG ⎦	AAU ⎤ AspN AAC ⎦ AAA ⎤ Lys AAG ⎦	AGU ⎤ Ser AGC ⎦ AGA ⎤ Arg AGG ⎦	U C A G
G	GUU ⎤ GUC ⎥ Val GUA ⎥ GUG ⎦ + Init.	GCU ⎤ GCC ⎥ Ala GCA ⎥ GCG ⎦	GAU ⎤ Asp GAC ⎦ GAA ⎤ Glu GAG ⎦	GGU ⎤ GGC ⎥ Gly GGA ⎥ GGG ⎦	U C A G

Fig. 12.14 The genetic code.

universal. Secondly, one can examine a number of variants of a single protein such as haemoglobin. The specific sequence of amino acids in the polypeptide chains of haemoglobins isolated from patients all over the world has now been determined and small changes, the results of genetic mutation, have been found which account for the changes in the physical properties of the haemoglobins that have allowed them to be detected. The first and best known example of a mutant haemoglobin was that from patients suffering from sickle cell anaemia. This was discussed under the structure of proteins in Chapter 4. You will recall that the difference between HbA and HbS is the substitution of a glutamic acid for a valine. According to the genetic code this could arise as follows.

HbA	HbS
Glu	Val
GAA	GUA
or	or
GAG	GUG

In each case one adenine has been replaced by a uracil. In other words the change can be explained on the basis of a single-point mutation in the DNA which is transcribed into RNA. The same explanation holds for all the other haemoglobin variants so far examined. This is very satisfactory for it would have been difficult to accept the possibility of a multi-point mutation affecting adjacent bases. The evidence from haemoglobin fully supports the relevance of the genetic code to the synthesis of haemoglobin in man.

More recently Sanger has developed direct methods for determining the sequence of messenger-RNA, in particular the sequence of RNA of a phage. He was able to correlate the base sequence of messenger-RNA with the sequence of a protein for which it codes, in this case the coat protein of the phage. Again the code is fully vindicated and, moreover, it is seen that more than one codon is used for a particular amino acid even within the messenger-RNA for one protein.

12.22. Direction of Reading of the Code

We have seen that proteins are synthesized from the amino to the carboxy terminus. It would be possible for the code in the messenger-RNA to be translated from either $5' \rightarrow 3'$ or $3' \rightarrow 5'$. Fortunately it could be shown that the messenger-RNA is always read in the $5' \rightarrow 3'$ direction, i.e., the way in which the base sequence is written by convention. Thus we have:

RNA : 5′ pA pC pG pU 3′

peptide: NH$_2$⋯⋯⋯⋯⋯⋯⋯⋯⋯⋯COOH

12.23. The Role of Transfer-RNA

It is now appropriate to return to the role of transfer-RNA (t-RNA) and the way in which it locates the amino acids on the messenger-RNA attached to the ribosomes. The problem was recognized by Crick as soon as it became apparent that the template was messenger-RNA and that each codon must be a triplet of bases. How does a particular amino acid recognize a specific codon? There had been a good many inspired guesses as to the way in which the three-dimensional structures of the codons and amino acids might be complementary in themselves but nothing convincing emerged. Then Crick suggested that the amino acids would be attached to a small RNA which would itself contain a triplet of bases, known as an anti-codon, which would be complementary to the codon of the messenger-RNA, the bases of the codon and anti-codon being held by the same type of hydrogen bonding as in native DNA. Crick named this hypothetical component adaptor RNA; later it became apparent that t-RNA fulfilled this role. (It is worth mentioning that this finding does not explain how a particular amino acid recognizes and comes to be linked to the 3' terminal adenosine of a specific t-RNA.)

Let us now look at the kind of experiment that can be done to illustrate the role of t-RNA. We can study the conditions under which [^{14}C]phenylalanine attached to t-RNA will bind to *E. coli* ribosomes. For this to occur the synthetic messenger-RNA (poly-U) must be present. We

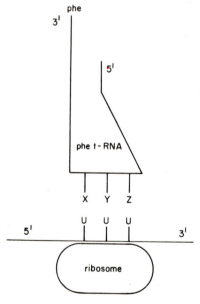

Fig. 12.15 A diagram to indicate the attachment of phenylalanine t-RNA to a ribosome from *E. coli* in the presence of UUU. It is assumed that a triplet of bases on the t-RNA(XYZ) hydrogen bond to the UUU.

find that binding takes place in the presence of UUU but not with either U or U-U. The binding of phenylalanine t-RNA to the ribosomes is completely specific to U-U-U so that under these conditions no other amino acid will bind. As shown, therefore, in Fig. 12.15, we conclude that there must be three bases XYZ in the t-RNA which hydrogen bond to the U-U-U.

In considering the nature of the anti-codon XYZ we know that all the different amino acids are attached to the 3′ terminal adenosine which occurs at the terminus of all the different t-RNA. Moreover, the 5′ terminus is usually guanosine. It follows that the triplet anti-codon cannot be at either terminus and must occur among the 70 internal nucleotides of the t-RNA.

Since t-RNA is a relatively small species of RNA it has been possible to determine the primary structure of many of the specific t-RNAs, and to search directly for the presence of the anti-codon. What might we hope to find? If we take alanine as an example there are four possible codons.

G-C-U

G-C-C

G-C-A

G-C-G

We might, therefore, expect to find four t-RNAs each with a different anti-codon. That this is not in fact so is explained as follows. For the first two codons we have:

	5′ → 3′	5′ → 3′
Codon Found	G-C-U	G-C-C
Anticodon Expected	C-G-A	C-G-G
	3′ ← 5′	3′ ← 5′

The expected anti-codons are assumed on the usual base pairing of G-C and A-U and that the strands are anti-parallel.

Holley was the first person to determine the primary structure of a t-RNA and it was in fact a t-RNA for alanine. The result is shown in Fig. 12.16. We have previously mentioned that although RNA is normally single-stranded it is known from physical measurements to form loops on itself which are held together by hydrogen bonding between bases. If one takes the known primary sequence of bases for alanine t-RNA and loops it round to form the maximum number of base pairs then one gets the clover leaf structure shown. This is true for all the t-RNA molecules for which the primary structure is known. It must be emphasized that this does not prove that the secondary structure of t-RNA is a clover leaf but it begins to look rather likely.

We would expect the anti-codon to be situated at the end of a loop which was not base-paired because the bases in the anti-codon must base pair with the codon on messenger-RNA and not with the bases in t-RNA. In fact we find the 3 residues C-G-I (3′ → 5′) at the top of a loop. One of the characteristics of t-RNA is that it has many unusual nucleosides and one of these is inosine (I) which is deaminated guanosine (G). Hence I will

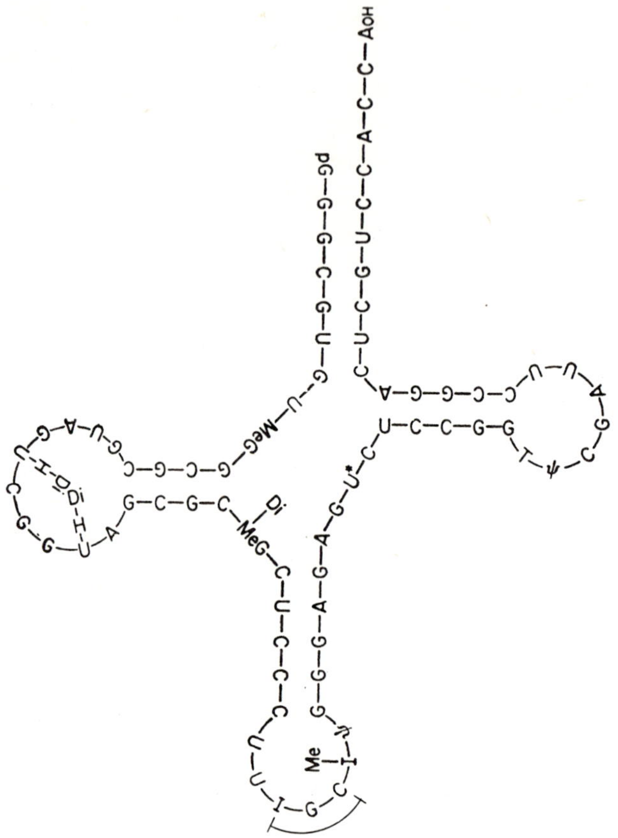

Fig. 12.16 The primary structure of alanine t-RNA. The triplet marked with a line is the anticodon.

base pair as if it were G. Hence we have located one of the predicted anti-codons for alanine which was CGG. How do we account for the other predicted anti-codon CGA?

If there were a different t-RNA for each amino acid codon there would, of course, be 61 different t-RNAs. Experimentally there were good reasons to doubt the existence of so many t-RNAs. Reference to the code shows that very often there is a change in the third base of the codon without a corresponding change in the nature of the amino acid. We saw this in alanine where all four possible changes were effected in the 3rd base. Crick suggested that perhaps it was not necessary in the binding of codon and anti-codon for the pairing of each of the three bases to be between, what we may regard as, the traditional pairs, i.e., A-U, G-C. Perhaps in the third position a less satisfactory pairing was possible and he suggested that G-U would be adequate. Thus for alanine one anti-codon CGI could pair with either GCC or GCU. In this way one particular t-RNA could serve for two different codons.

In fact this prediction has worked out very satisfactorily now that we know the primary structure of many different t-RNA. We have:

	Alanine $5' \to 3'$	Serine $5' \to 3'$	Tyrosine $5' \to 3'$	Valine $5' \to 3'$	Phenylalanine $5' \to 3'$
Codons	G-C-C G-C-U	U-C-C U-C-U	U-A-C U-A-U	G-U-C G-U-U	U-U-U U-U-C
Anti-codon	C-G-I $3' \leftarrow 5'$	A-G-I $3' \leftarrow 5'$	A-U-G $3' \leftarrow 5'$	C-A-I $3' \leftarrow 5'$	A-A-G $3' \leftarrow 5'$

The anti-codon CGI would not do for both the two alanine codons which we have not considered, namely GCA and GCG. We would expect a second alanine t-RNA with an anti-codon CGU. Thus in all there are probably about 40 different molecules of t-RNA. The prediction of Crick concerning the third base is known as the "Wobble hypothesis".

12.24. Punctuation in the Code—Initiation

We showed previously that it would be necessary for the code of messenger-RNA to be read from a fixed point. From the experiments concerning poly-U and phenylalanine it seemed possible that no particular mechanism was necessary to ensure this. It is now, however, clear that there is such a mechanism.

You will recall that polypeptides are synthesized in the direction $NH_2 \to COOH$. Reference to the code will show that for methionine there is only one codon, AUG. In spite of this Sanger and Marcker found two different t-RNA for methionine in extracts of *E. coli*. They found, moreover, that if methionine was attached to one type of t-RNA it could be formylated on the free NH_2 and in the other case it could not be. (Formylation is effected by an enzyme which utilizes formyltetrahydrofolic acid (THFA) and so we have $-NH_2 + HCOOH \to -NHCHO$.) The two t-RNAs are denoted t-RNA^{Met} and t-RNA^{fMet}. When both formyl methionine attached to t-RNA^{fMet} and methionine attached to t-RNA^{Met} were incubated with *E. coli* ribosomes in the presence of poly AUG the formylmethionine as expected was found at the N-terminus of the resulting polymethionine. More surprisingly methionine was never found at the N-terminus. It is now clear that all bacterial proteins are started by the insertion of N-formyl methionine at the N-terminus. In most cases the formyl group, and sometimes also the methionine, is cleaved from the polypeptide after completion of its synthesis.

For a long time the method of chain initiation with 80S ribosomes of eukaryotes was unknown but it is now clear that in this case methionine itself is the initiator. It is placed in position by the methionine t-RNA^{fMet} which allows the formylation of methionine but the transformylation enzyme is absent from animal cells. The methionine is removed before the completion of the polypeptide chain so that unless it is the genuine N-terminal amino acid it is not found in the synthesized polypeptide.

12.25. Synthesis of Mitochondrial Proteins

We should add a note at this point concerning the synthesis of proteins by mitochondria. Mitochondria contain a small amount of DNA and all the apparatus necessary for the synthesis of some protein. The protein

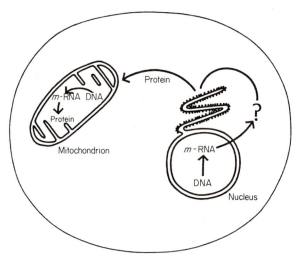

Fig. 12.17 The origin of mitochondrial protein. Some protein is synthesized in the mitochondrion itself through the transcription of mitochondrial DNA and the translation of the resulting m-RNA. The soluble mitochondrial proteins are transferred from the cytoplasm.

synthesized, however, is almost certainly the insoluble structural protein, the soluble protein of the mitochondria coming from the ribosomes in the cytoplasm. The position can, therefore, be summarized as in Fig. 12.17. The ribosomes in the mitochondria are smaller than the 80S ribosomes in the cytoplasm. Moreover, protein synthesis is initiated in mitochondria by formyl-methionine rather than methionine. It appears, therefore, that mitochondrial protein synthesis is more typical of bacteria than animal cells.

12.26. Punctuation in the Code—Full Stops

The length of messenger-RNA that codes for one complete polypeptide chain is known as a "cistron". If a single length of messenger-RNA contains information for more than one cistron it is polycistronic but in this case there must be a signal for the completion of a particular polypeptide chain. There is now good evidence that the three nonsense codons UAA, UAG, and UGA are employed to indicate termination or full stops.

12.27. Polyribosomes

Each of the ribosomal subunits has a different function in the synthesis of a polypeptide. Thus the small subunit binds the messenger-RNA whereas the larger subunit is the site for the formation of polypeptide. It is now apparent that ribosomes do not effect the synthesis of a polypeptide as single units but are linked together by messenger-RNA. These aggregates of ribosomes are known as polyribosomes or polysomes. Figure 12.18 shows

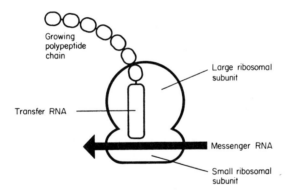

Fig. 12.18 A diagrammatic representation of the relationship of the growing polypeptide chain and m-RNA to the ribosomal subunits.

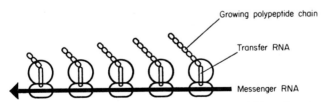

Fig. 12.19 The way in which ribosomes are linked by m-RNA to form polyribosomes. The arrow indicates that the m-RNA moves with respect to the ribosomes.

the relationship of the growing polypeptide chain and the messenger-RNA to the ribosomal subunits and Fig. 12.19 the way in which the ribosomes are aligned in a polyribosome. As the polypeptide chain is synthesized the ribosome moves with respect to the messenger-RNA. This arrangement means that there is great economy in the use of the information contained in the messenger-RNA and accounts for the fact that the latter represents only about 2% of the RNA of the cell.

We know now that the ribosomes dissociate into their subunits as the polypeptide chain is completed and the small subunit joins the messenger-RNA before the large subunit according to the scheme shown in Fig. 12.20.

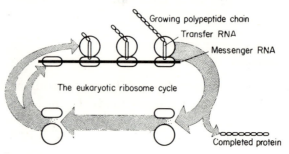

Fig. 12.20 The ribosome cycle. After the completion of the polypeptide chain the ribosomes split into their subunits. The small subunit then re-attaches to m-RNA followed by the large subunit.

12.28. The Formation of the Peptide Bond on the Large Subunit of the Ribosome

Figure 12.21 shows the arrival on the large ribosomal subunit of the t-RNA molecules bearing their amino acids. The subunit has two sites for the attachment of t-RNA. The central outline of the ribosome shows the P (peptidyl) site occupied by Ala t-RNA with the alanine linked through the

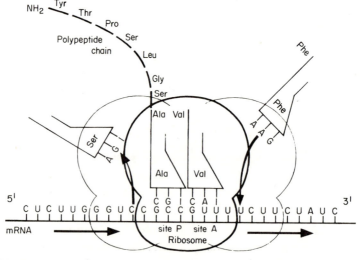

Fig. 12.21 The elongation of the polypeptide chain while attached to the large ribosomal subunit.

amino group to the polypeptide chain. The A (aminoacyl) site is occupied by a Val t-RNA that will provide the next amino acid in the peptide chain. The ribosome outlined on the left represents a slightly earlier state in which the Ser t-RNA is being ejected. The outline on the right indicates the next event, the arrival of the incoming Phe t-RNA as the ribosome moves. In our example valine is added to the peptide chain by the transfer of the peptide from the P site to the A site. This involves an enzyme, peptidyl transferase, and in chemical terms represents an attack by the free NH_2 group of the incoming amino acid on the ester link of the peptide t-RNA with the formation of an amide (peptide). For further elongation to occur the A site must be vacated for the incoming Phe t-RNA. This

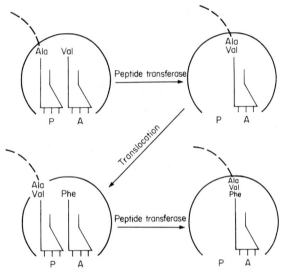

Fig. 12.22 The movement of polypeptide-t-RNA between the two sites on the large ribosomal subunit during chain elongation.

process is called "translocation" and is achieved as the ribosome migrates relative to the messenger-RNA. An enzyme "translocase" is involved in the process. The movement of the messenger-RNA brings the Phe codon UUC to the A site while the Val t-RNA bearing the peptide is now at the P site. The steps are illustrated in Fig. 12.22.

It will be noted that during chain elongation the incoming amino acid attached to its t-RNA arrives at the A site on the ribosomal subunit. For chain initiation it is necessary for the incoming t-RNA to attach to the P site. This change of site is, therefore, the major problem of chain initiation. It is the special property of the t-RNA^{fMet} either with formylmethionine (70S ribosomes) or methionine (80S ribosomes) attached to find its way into the P site. The sequence of events in chain initiation is shown in Fig. 12.23.

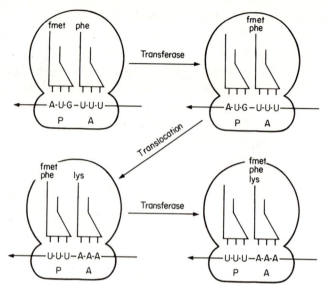

Fig. 12.23 The mechanism of chain initiation by fMet-t-RNA[FMet] and subsequent chain elongation.

In these processes of chain initiation, peptidyl transferase and translocation, many protein factors are required and GTP plays an essential role. Thus the requirement for GTP as well as ATP as originally discovered by the Zamecnik group in their studies on protein synthesis by the microsome fraction is explained.

The interrelationship of replication, transcription and translation is summarized in Fig. 12.24.

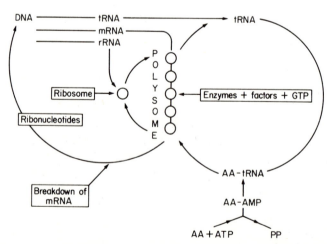

Fig. 12.24 The interrelationship between DNA replication, RNA transcription and translation and amino acid activation. (Modified from Watson, J. D. *Molecular Biology of the Gene*.)

12.29. The Mechanism of Action of Antibiotics

As we said at the outset, one of the objectives in attempting to elucidate the way in which nucleic acids and proteins are synthesized in the cell, is to enable drugs to be devised which will inhibit these processes in bacteria without affecting those in man. In fact certain useful drugs are already known to do just this, others have proved to be too toxic in man but have been useful in elucidating mechanisms, and yet others are still in the state of development. Certain antibiotics will now be considered in order to illustrate how they may inhibit some of the key reactions.

12.30. Actinomycin D

This is a member of a group of brightly coloured peptides obtained from *Actinomycetes*. They proved to be very potent anti-tumour agents but unfortunately are very toxic and so are not in general use therapeutically. Actinomycin stops the synthesis of RNA by DNA-dependent RNA polymerase. It does so by combining with the DNA template (actually deoxyguanosine) in such a way that it is no longer effective as a template. While it inhibits all RNA synthesis it is virtually specific for messenger-RNA when used over a short time since this RNA turns over very rapidly. It is, therefore, used to test whether the effect of a hormone, or other substance, on protein synthesis is dependent on the continued synthesis of messenger-RNA.

12.31. Rifampicin

A group of antibiotics, known as the Rifamycin complex, has been isolated from soil bacteria. The most useful of the range is known as Rifampicin and this is used clinically in the treatment of tuberculosis. This antibiotic inhibits DNA-dependent RNA polymerase by binding to the enzyme. It is very specific for a range of bacterial enzymes and is almost without effect on animals. It, therefore, has a quite different mode of action to Actinomycin.

Perhaps the most interesting action of this antibiotic is a more recent discovery that it is also active against viral infections. It is for example very useful in the treatment of Trachoma which is a serious eye disease prevalent in the Middle East and which is caused by a DNA virus. Rifampicin is the first effective anti-viral antibiotic and in view of the possible role of viruses in cancer it is easy to see the reason for the widespread interest in this antibiotic.

12.32. Puromycin

Of all the antibiotics, puromycin has probably been the most useful to the biochemist for the elucidation of the mechanism of protein synthesis.

Unfortunately it is too toxic for therapeutic use. Soon after its structure was known, de la Haba and Yarmolinsky pointed out its close structural resemblance to the terminal portion of t-RNA bearing an amino acid. The two structures are compared in Fig. 12.25. The amino group of

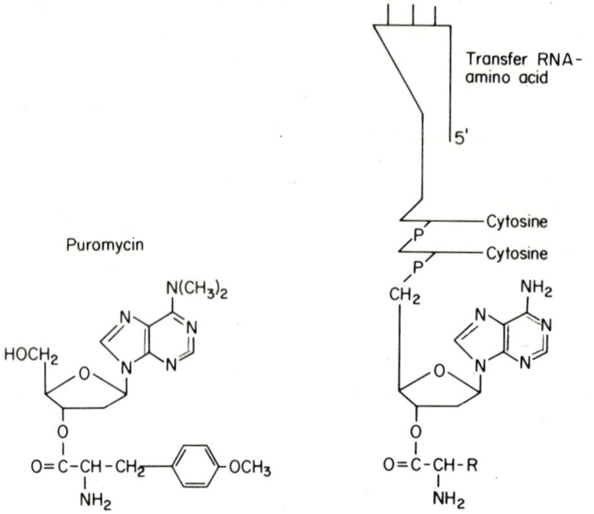

Transfer RNA–
amino acid

5'

Cytosine

P

Cytosine

Puromycin

P

N(CH$_3$)$_2$

CH$_2$

NH$_2$

HOCH$_2$

O

O

O=C−CH−CH$_2$⟨ ⟩−OCH$_3$
NH$_2$

O=C−CH−R
NH$_2$

Fig. 12.25 A comparison of the structure of puromycin (left) and the 3' terminal adenosine of t-RNA with amino acid attached (right).

Puromycin forms a peptide linkage with the growing peptide chain, as shown in Fig. 12.26, but as the peptide-Puromycin bears no anti-codon it is released without further elongation of the nascent polypeptide chain.

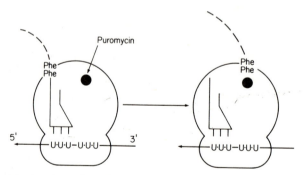

Puromycin

Phe
Phe

Phe
Phe

5'   3'
U·U·U—U·U·U   U·U·U—U·U·U

Fig. 12.26 The mechanism of action of puromycin on the polypeptide chain attached to the large ribosomal subunit. A peptidyl puromycin is formed and released.

Not surprisingly it is a powerful inhibitor of protein synthesis and has been used to demonstrate that an experimental observation, e.g., a hormonal action, depends on the synthesis of new protein.

12.33. Chloramphenicol and Cycloheximide

Chloramphenicol, which is a therapeutically useful antibiotic, inhibits protein synthesis by 70S ribosomes, but is inactive against 80S ribosomes. It also inhibits mitochondrial protein synthesis thus providing further evidence for the similarity of this process to bacterial protein synthesis.

Chloramphenicol is used in typhoid infections but can cause Aplastic Anaemia so that it is now only used when essential. The present evidence is that Chloramphenicol acts by inhibiting the transferase reaction, as shown in Fig. 12.27. In contrast to Chloramphenicol, Cycloheximide inhibits the synthesis of protein by 80S ribosomes but is without effect on 70S ribosomes. Cycloheximide inhibits the "translocase" reaction as shown in Fig. 12.27b. It also appears to inhibit initiation.

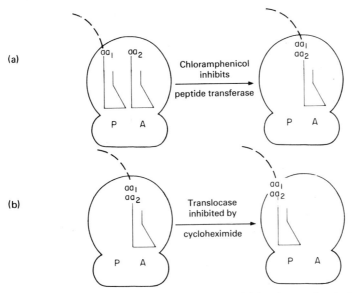

Fig. 12.27 The probable mechanism of action of (a) chloramphenicol on bacterial and mitochondrial ribosomes, and (b) cycloheximide on the ribosomes of eukaryotic cells.

12.34. Streptomycin and Neomycin

Streptomycin is effective in the treatment of tuberculosis whereas Neomycin is effective for the treatment of a wide range of bacterial infections of the gut.

The sensitivity of a bacterium to Streptomycin depends on the presence of one particular protein associated with the small ribosomal subunit. In the susceptible strains this protein is modified. Under *in vitro* conditions Streptomycin and Neomycin cause a misreading of the code so that instead of U-U-U coding only for phenylalanine it also codes for leucine.

Hence the wrong protein is synthesized in the presence of these anti-biotics. It is assumed that their activity *in vivo* is associated with this misreading.

12.35. Tetracyclines

These antibiotics are important clinically. The basis for their antibiotic activity is their direct inhibitory action on protein biosynthesis. In systems containing isolated ribosomes the Tetracyclines inhibit protein synthesis by both 70S and 80S ribosomes although they are rather more active against 70S ribosomes. In intact cells the Tetracyclines are much more active against bacteria than animal cells and this is no doubt due to the greater permeability of the bacterial cells to the antibiotic. The Tetracyclines inhibit the binding of aminoacyl t-RNA into the ribosomal A site.

12.36. Penicillin

It is worth noting that Penicillin inhibits the synthesis of a bacterial cell-wall hexapeptide but its synthesis does not involve ribosomes. Small peptides (including the antibiotic peptides such as Gramicidin), are synthesized by another mechanism. Penicillin, therefore, has no effect on protein synthesis.

12.37. The Control of Protein Synthesis in Bacteria

In bacteria a fundamental form of control of protein synthesis concerns the induction and repression of enzyme synthesis. By this means the bacterial cell avoids synthesizing enzymes which it does not need, yet retains the ability to start the synthesis when conditions make this necessary, e.g., when a change of nutrient occurs. The mechanism of enzyme induction has been much studied in bacteria. Although bacterial cells do not undergo differentiation to the degree encountered in higher forms of life nevertheless it is clear that the mechanisms of enzyme induction are relevant to the study of differentiation in animal cells.

The scheme in Fig. 12.28 shows the mechanism in the induction of a group of enzymes concerned with lactose utilization by *E. coli*. If lactose (a galactoside) rather than glucose is supplied in the nutrient medium for these bacteria, they cannot utilize the new energy source until the appropriate enzymes have been synthesized.

The transcription of a cluster of genes (known as an operon) for three proteins (G_1, G_2, G_3) necessary for the degradation of lactose, trans-acetylase, permease and β-galactosidase, is controlled by a repressor (R).

In the absence of lactose the repressor (shown as T), which is formed by transcription of a regulator gene (i), binds to the operator gene (o) and prevents the formation of the messenger-RNA for the three proteins. When lactose is present this combines with the repressor so that it is no longer able to combine with the operator gene. Since this is a negative form of control the removal of the repressor allows the genes for the three proteins to be transcribed by the polymerase which initiates at the Promoter (p).

The repressor has now been identified by Gilbert and Ptashne and shown to be a protein. Hence the combination of lactose with the protein may be presumed to change its tertiary structure such that the repressor no longer is able to combine with the operator gene. This is analogous to the binding of an allosteric effector to an enzyme, thereby affecting the combination of the enzyme with the substrate.

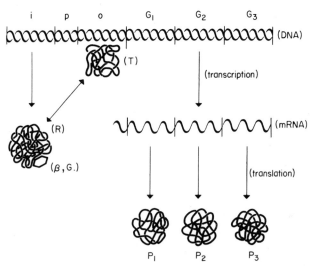

Fig. 12.28 The Jacob-Monod scheme for the induction of enzymes in bacteria. (From *Chance and Necessity*, J. Monod, trans. A. Wainhouse. Copyright © A. A. Knopf, New York, 1971. Reprinted by permission of the publisher.) Regulation of the synthesis of the enzymes in the "lactose system". R: repressor-protein, in state of association with the galactoside inducer shown by the hexagon. T: repressor-protein in state of association with operator segment (o) of DNA. i: "regulator gene" governing synthesis of the repressor. p: "promoter" segment, point of initiation for synthesis of messenger RNA (m-RNA). G_1, G_2, G_3: "structure" genes governing synthesis of the three proteins in the system, marked P_1, P_2, P_3.

12.38. Differentiation in Animal Cells

The DNA present in all the somatic cells of a given individual is believed to be identical with respect to base sequence. Thus each somatic cell has the information to enable it to synthesize any and all of the proteins typical of the body of the whole individual. We do not yet understand the basis whereby the nature of the proteins synthesized in a particular cell is controlled. Why is it that the liver cell does not make milk proteins and how is it that at parturition the mammary gland starts to synthesize these proteins?

While we do not at present know the answer to these questions three possible mechanisms come to mind.

1. One possibility is that the control takes place at the level of transcription as in the case of enzyme induction in bacteria. According to this hypothesis the cell would have gene regulators having a similar function to the repressors of the bacterial cell. A characteristic of the eukaryotic cell is the presence in it of small basic proteins called histones which are associated with the DNA. It has been argued that histones function as gene regulators. The truth is that the evidence for their role in this respect is extremely slim and that today we have no firm evidence for the presence of repressors in eukaryotic cells.

2. All the various messenger-RNA molecules for the whole range of proteins could be synthesized but they could be selectively degraded. We know that only a small proportion of the RNA made in the nucleus passes to the cytoplasm, the remainder is degraded in the nucleus. The real trouble about this hypothesis is that we cannot at present conceive of a way in which the breakdown of messenger-RNA could be adequately selective.

3. Perhaps all the possible messenger-RNA molecules are synthesized and leave the nucleus, but translation is controlled by the combination of repressors with selected messenger-RNA molecules. There is some evidence, mainly from the work on *Acetabularia* by Henry Harris that the various messenger-RNA molecules are present in the cytoplasm for a long time before they are translated.

Only time will show whether any or none of the above hypotheses is correct.

12.39. Possible Mechanism of Action of Hormones

Many hormones are able to affect the rate of synthesis of nucleic acids and proteins. Thus for example growth hormone and thyroid hormone increase the rate of protein synthesis of various cells. We have seen that the processes of protein synthesis and nucleic acid synthesis are so closely related that it is difficult to decide at which point the hormone has its primary effect. Thus an increase in protein synthesizing activity could result either from an increase in the availability of messenger-RNA or in the rate of translation of the already available messenger-RNA. The effect on the actions of the hormone of the administration of Actinomycin may give an indication of the role of RNA synthesis in the process and the use of Puromycin may indicate whether the synthesis of new protein is involved. These are, however, as we have seen, rather crude tools and many problems remain in the elucidation of the mode of action of hormones at the molecular level.

Recent work has indicated the presence in the cytoplasm of proteins which specifically bind hormones. If these proteins serve as repressors then the binding to them of a hormone might have an allosteric effect similar to the case of enzyme induction in bacteria.

13

The Regulation of Metabolism

13.1. Introduction

If metabolism were restricted to a single pathway consuming only one substrate in a solitary cell, control would be as pointless as signals on a single-track, one-locomotive railway. But given a complex, multi-branched metabolic pathway, with many starting points, many end products and numerous alternative routes, some means of regulation is essential, even in an isolated cell, if chaos is to be prevented. In higher organisms there is a further need to harmonize the various requirements of several specialized tissues and organs. Now that biochemistry has reached a stage in its development where all the main pathways of metabolism are known, the emphasis has shifted to gaining some insight into the ingenious mechanisms of metabolic control and the inter-relationships of different tissues. In this chapter we shall be concerned with some of the basic principles of the regulation of metabolism.

Since enzymes are the indispensible catalysts for all reactions in the cell, the control of metabolism is ultimately concerned with the regulation of enzyme activity. Indeed this is the most basic aspect of the problem and the one we must examine first. An alteration in enzyme activity can be achieved in two fundamental ways, either by effects on pre-existing enzyme molecules or by changing the rate of enzyme synthesis. Both types of regulation occur—the first provides for rapid, fine control, while the latter is appropriate for long term, coarse adjustments. It has also been suggested that changes in the rate of irreversible degradation of the enzyme protein may be a third way in which activity is controlled, but there is little evidence that this is an important mode of regulation. Some of the means of regulating enzyme activity have already been outlined in Chapter 5 and the concepts of induction and repression have been mentioned in respect of bacteria in Section 12.37.

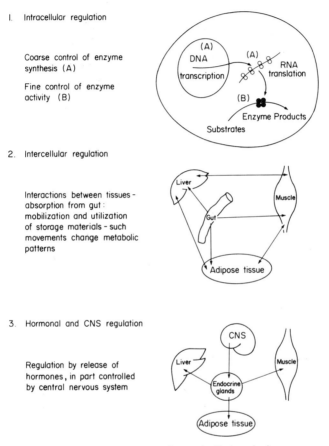

Fig. 13.1 Three levels of metabolic regulation.

A second means of regulating metabolism is through the supply of substrates for the enzymes. Such regulation operates, for example, during the absorption of foodstuffs or in the mobilization of stored energy reserves. Finally, superimposed on both these mechanisms (and influencing both) there is a communication system involving hormones as chemical messengers as well as the central and peripheral nervous systems.

The investigation of control processes is a more complicated and difficult task than simply describing the pathways themselves. Implicit in the concept of control is the facility to adapt to change. Control mechanisms can, therefore, only be revealed experimentally by pushing the system off balance. In an isolated tissue or cell fraction one might attempt to localize the regulatory sites by removal of oxygen or by poisoning with an inhibitor, followed by a determination of the various enzyme activities and concentrations of metabolites. The same idea applies to experiments in animals. Control mechanisms might be revealed by a nutritional disturbance—a prolonged fast, for example, or perhaps by creating a hormonal deficiency by surgical removal of an endocrine gland. The difficulty is to decide between cause and effect when interpreting the often rather complex changes that follow such experiments. Only by rigorous experimental controls and careful checking of hypotheses by alternative experimental approaches can progress be made. Fortunately, although there are still many areas of obscurity, the last ten years has seen great strides and we now have a fair understanding of a number of basic regulatory systems.

13.2. Key Enzymes in Metabolic Pathways

(a) General considerations

If a single pathway is catalyzed by, say, six consecutive enzymes it would be unnecessary and uneconomic to arrange for each individual enzyme activity to be regulated when control at a single site would be sufficient. What seems to happen is that control is usually exerted on the enzyme with the lowest activity. Thus if we measure the activity of each individual enzyme, assaying it at an optimal concentration of substrate, we find a wide variation—some activities are superabundant compared with others. An analogy picturing the pathway as flowing through a pipe is shown in

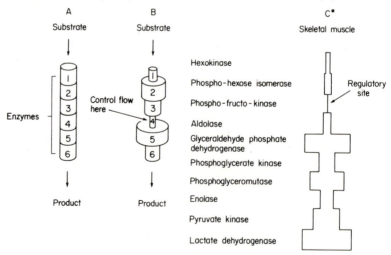

Fig. 13.2 The control point in a pathway. In A the pipe is of constant bore. In B it is of variable diameter and the logical point at which to regulate flow is 4. C shows glycolytic enzyme activities in skeletal muscle—width of column being proportional to activity—(μmole of substrate/min/g muscle). *Taken from Hales, C. N. *Essays in Biochemistry*, vol. 3.

Fig. 13.2. In A the pipe is of uniform bore—each enzyme has the same activity. In B the bore varies—in section 4 it narrows and forms a bottle neck—this would be the obvious point at which to exert control. The validity of the analogy can be seen in C, which shows the pattern of glycolytic enzyme activities in skeletal muscle. The actual range of activities is even greater than in our analogy. The bottle-neck in the pathway is obvious and we can identify phosphofructokinase as the key enzyme in regulating glycolysis. The other enzymes in this sequence may be limited by the availability of substrate and the reactions they catalyze may, therefore, approach equilibrium. But for phosphofructokinase, equilibrium is not achieved because the limiting factor is the enzyme itself. If we measure glycolytic substrates we find that the concentration of fructose 6-phosphate is high relative to the others. If we then exclude

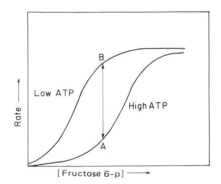

Fig. 13.3 Phosphofructokinase—an example of an allosteric enzyme.

oxygen from the tissue we find that, within a few seconds, fructose 6-phosphate is depleted due to a very rapid activation of phosphofructo-kinase. The result is an increase in glycolysis by the activation of this key enzyme. Generally speaking, key or pace-maker enzymes show allosteric properties. ATP is not only a substrate for phosphofructokinase but also acts as an allosteric inhibitor (Fig. 13.3). In high concentration it inhibits enzyme activity, i.e., it shifts the sigmoid shaped curve to the right. It is easy to see that when ATP is depleted in the cell, by anaerobiosis, phosphofructokinase activity immediately rises. This is the basis of the Pasteur effect (see Chapter 6.12).

We can now use these general principles to try to identify the important regulatory enzymes in the main pathways. These are set out in Fig. 13.4. Some of the allosteric effectors, inhibitors or activators, are given in Table 13.1. Nearly all of the key enzymes are activated or inhibited by one or more of a variety of nucleotides and metabolites. We must now try to discern the logic behind these allosteric effects.

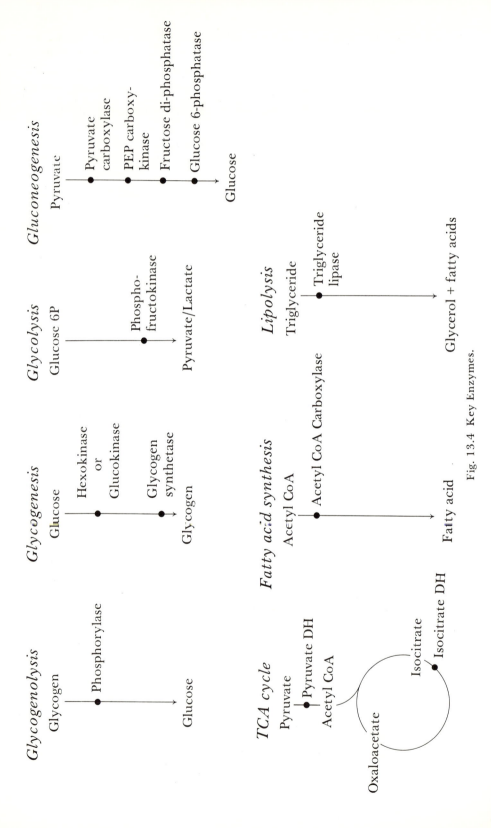

Fig. 13.4 Key Enzymes.

TABLE 13.1. Some activators and inhibitors of key enzymes

Enzyme	Activators	Inhibitors
Phosphorylase	cAMP (indirectly)	
Glycogen synthetase		cAMP (indirectly)
Hexokinase (muscle)		Glucose 6-phosphate
Phosphofructokinase	ADP AMP Phosphate cAMP	ATP Citrate
Fructose 1,6-diphosphatase		AMP Fructose 1,6-diphosphate
Pyruvate kinase	Fructose 1,6 diphosphate	ATP
Pyruvate dehydrogenase		ATP Acetyl CoA
Isocitrate dehydrogenase	ADP	ATP
Acetyl CoA carboxylase	Citrate (*in vitro*)	Palmitoyl CoA
Mobilizing lipase (adipose tissue)	cAMP (indirectly)	

(b) Some important components

(i) ADENINE NUCLEOTIDES

ATP is produced by glycolysis and by oxidation in the mitochondria. If production exceeds demand so that the ratio ATP : ADP + AMP rises, three control mechanisms begin to operate to curtail production. The respiratory chain is impeded through lack of ADP (see Chapter 9.8); secondly, glycolysis is reduced by inhibition of phosphofructokinase and pyruvate kinase and thirdly the TCA cycle oxidations are reduced by inhibition of pyruvate and isocitrate dehydrogenases. Alternatively, when utilization of ATP in biosyntheses, muscular contraction or other energy-consuming processes exceeds production, the same control mechanisms ensure that the production rate is stepped up to meet the demand.

(ii) CITRATE

This plays an analogous role to ATP. It, too, is a product of energy-producing catabolic pathways in mitochondria and it plays an essential role in supplying (mitochondrial) acetyl CoA for (cytoplasmic) fatty acid synthesis. Its accumulation reduces glycolysis (inhibition of phosphofructokinase) and stimulates fatty acid synthesis by an effect on acetyl CoA carboxylase.

(iii) NICOTINAMIDE ADENINE NUCLEOTIDES

These also play a part in regulation, but their role is rather different from that of the allosteric modifiers listed in Table 13.1. Normally the cellular concentration of the oxidized form NAD^+ greatly exceeds that of the reduced form, NADH. In liver the $NAD^+/NADH$ ratio is about 500:1 in the cytosol and 5:1 in the mitochondrion. The reactions catalyzed by dehydrogenases are reversible and the equilibrium of any pair of reactants, e.g., pyruvate-lactate, will depend on the $NAD^+/NADH$ ratio. A change in this ratio can have, therefore, widespread effects on metabolism. One example of this phenomenon is the effect produced by ethanol. The oxidation of ethanol in the liver by alcohol dehydrogenase requires NAD^+ and even a moderate dose of ethanol given to a rat (equivalent to 50 ml of whisky in man) causes a sharp fall in the $NAD^+/NADH$ ratio. This change is associated with a shift in the concentrations of numerous metabolites such that the reduced form is increased over the oxidized form. In contrast, the $NADP^+/NADPH$ ratio is normally predominantly in favour of the reduced form. In the liver the ratio is of the order of 0·004 : 1. The reduced form provides the reducing power for some biosynthetic reactions, in particular, fatty acid biosynthesis.

(iv) ACETYL CoA

This is an important regulator of pyruvate utilization. When its concentration rises (as a result of oxidizing fatty acids) it reduces pyruvate oxidation (pyruvate dehydrogenase) and stimulates the pyruvate carboxylase, which has an absolute requirement for acetyl CoA. The effect is to restrict carbohydrate utilization (because another energy source is being used) and favour gluconeogenesis (Fig. 13.5).

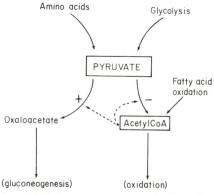

Fig. 13.5 The effect of acetyl CoA on pyruvate metabolism. An increased concentration of acetyl CoA effectively diverts pyruvate metabolism so as to restrict carbohydrate utilization and increase gluconeogenesis.

(v) PALMITOYL CoA

This together with other long chain fatty acyl CoA derivatives, inhibit the formation of malonyl CoA by the action of acetyl CoA carboxylase. Under conditions where fatty acids are synthesized at a rate greater than they can be esterified to triglyceride an accumulation of palmitoyl CoA may block this key enzyme in fatty acid synthesis.

(vi) CYCLIC 3',5'-ADENYLATE (cAMP)

This is a nucleotide of great importance in the control of metabolism. Its formation is largely dependent on hormone action and its effects are both widespread and variable in different tissues. It is primarily concerned with the phosphorylation of certain enzymes, a subject which is discussed in the next section. In regard to Table 13.1, we can note, at this stage, that cAMP promotes the mobilization of both glycogen and triglyceride while blocking the synthesis of glycogen.

13.3. Phosphorylation and Dephosphorylation of Enzymes—the Role of Cyclic AMP

The roles of kinases in transferring a phosphate group from ATP to a substrate and of phosphatases in hydrolyzing phosphorylated substrates, are already familiar in regard to simple substrates like glucose. Enzymes of both types are found which phosphorylate and dephosphorylate protein substrates. In this case it is a serine residue in the protein which is esterified with the phosphate group:

$$\text{Enzyme Protein} \quad \begin{array}{c} \text{NH} \\ | \\ \text{CH-CH}_2\text{O-P-O}^- \\ | \\ \text{CO} \end{array} \quad \begin{array}{c} \text{O}^- \\ | \\ \\ || \\ \text{O} \end{array} \quad \text{a phosphoserine residue}$$

This chemical change provides an important means of regulating enzyme activity. The best understood examples are the enzymes involved in glycogen metabolism and triglyceride hydrolysis (Table 13.2). When glycogen phosphorylase (see Chapter 6.4) and glycogen synthetase (see Chapter 6.5) are phosphorylated, the effect is to promote glycogen breakdown and prevent glycogen synthesis. Dephosphorylation of the enzymes reverses this. When the mobilizing lipase in adipose tissue is phosphorylated the effect is to mobilize fatty acids from the store of triglyceride in adipose tissue. The relevance of these processes to different nutritional rates will be clearer when we examine the part played by hormones.

TABLE 13.2. Some enzymes known to be regulated by phosphorylation
and dephosphorylation

Enzyme	Phosphorylated form	Dephosphorylated form
Glycogen phosphorylase	active	inactive
Glycogen synthetase	inactive	active
Mobilizing lipase (adipose tissue)	active	inactive
Pyruvate dehydrogenase	inactive	active

The balance between the two forms of the enzyme is the result of the
action of a pair of enzymes—a protein kinase and a protein phosphatase.

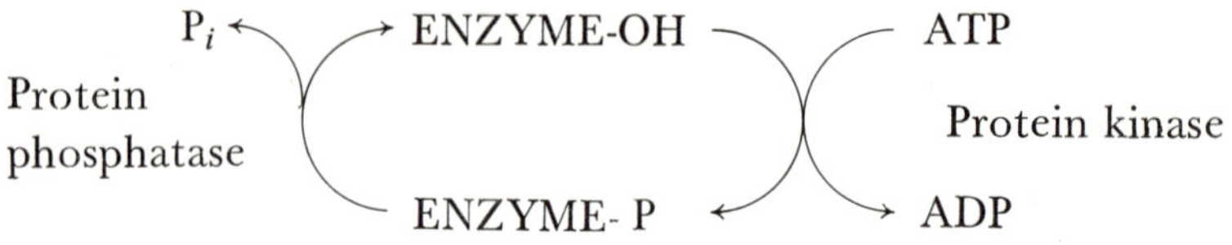

Depending on which of these two predominates at any instant the enzyme
will be activated or inactivated. This type of control differs fundamentally
from allosteric regulation because it involves an enzyme-catalyzed change
in the covalent structure of the enzyme. Whether the kinase or phosphatase predominates depends in turn on cyclic 3′,5′-adenylate.

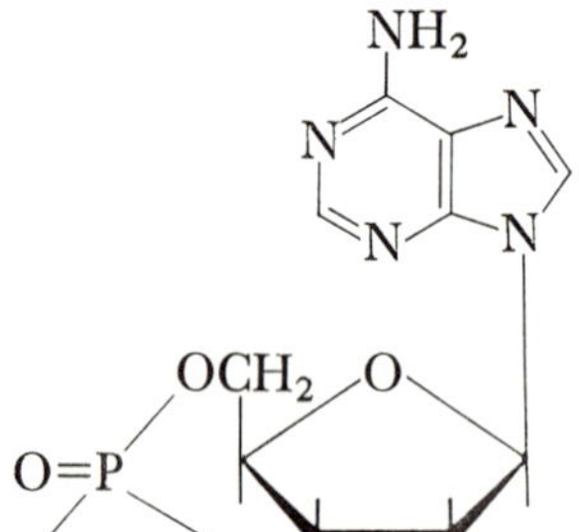

cyclic 3′5′-adenylate
(cAMP)

This nucleotide differs structurally from 5′-AMP only in the participation
of the C-3 hydroxyl in the formation of a phosphate-diester—a difference
which gives cAMP unique properties. cAMP is formed from ATP by an
enzyme—adenyl cyclase.

$$\text{ATP} \rightarrow \text{cAMP} + \text{PP}_i \, (\Delta G^{0'} = -5 \cdot 9 \text{ kJ})$$

It is destroyed by another enzyme cyclic nucleotide diesterase, which
specifically hydrolyzes it to 5′-AMP.

$$\text{cAMP} + \text{H}_2\text{O} \rightarrow 5'\text{-AMP} \, (\Delta G^{0'} = -11 \text{ kJ})$$

This enzyme is inhibited by certain purine derivatives—the methyl xanthines, e.g., caffeine:

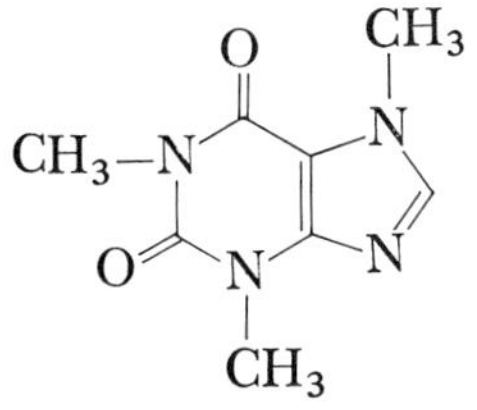

Caffeine (trimethylxanthine)

Theophylline and theobromine are dimethyl xanthines and exert the same effect. All three are natural products present in tea, coffee and cocoa. They have a variety of pharmacological effects, including stimulation of the central nervous system, heart and respiration, relaxation of smooth muscle, dilatation of the coronary arteries and renal diuresis. The only known biological action of the methylxanthines is to inhibit cyclic diesterase and thereby preserve cAMP from inactivation. All their pharmacological effects are therefore attributable to the actions of cAMP in various tissues.

Both adenyl cyclase and cyclic nucleotide diesterase are present in almost all living cells from mammals to bacteria. In mammals only the red cell lacks these enzymes. The brain contains the highest activities—about ten times as much as is found in liver or adipose tissue. Adenyl cyclase is always firmly bound to cell membranes, mostly to the plasma membrane—a location which is important in view of the influence of hormones on cyclase activity.

13.4. Hormones and Adenyl Cyclase

Cyclic AMP was discovered by Sutherland in the course of a long-term research programme on the mode of action of adrenaline and glucagon in liver. One effect of either hormone is to raise the blood glucose concentration by promoting glycogenolysis. Sutherland localized the effect to that on glycogen phosphorylase but found that the hormones required other cellular components for their action on this enzyme. The system (Fig. 13.6) was elucidated by some brilliant research between 1948 and 1962. The hormones stimulate liver adenyl cyclase (located in the plasma membrane) to produce cAMP. This activates a protein kinase which phosphorylates phosphorylase and tips the balance in favour of glycogen breakdown. The effect ceases when the hormones and cAMP have been inactivated and the balance then shifts in favour of inactive, dephospho-phosphorylase.

The importance of Sutherland's work was not that it provided an explanation for the action of one or two hormones on a single enzyme

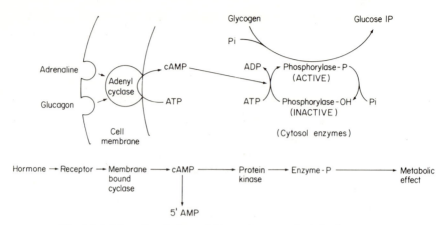

Fig. 13.6 The stimulation of Liver glycogenolysis by hormones.

system but that it revealed a general mechanism of the utmost significance, encompassing effects, not all hormonal, in many different cell
types. Many of these examples of cAMP-mediated effects are now
understood. Others, in particular the precise role of the highly active
adenyl cyclase in the central nervous system, still await elucidation.

We are concerned here with the role of adenyl cyclase in the control of
metabolism. The two most important areas are those mentioned earlier, in
Table 13.2, concerning glycogen and triglyceride metabolism. Control is
exerted, therefore, over the two principal energy reserves: the control of
glycogen metabolism is important in liver and muscle tissues, that of
triglyceride metabolism in adipose tissue. The scheme as regards glycogen
metabolism is shown in Fig. 13.7. Glucagon and adrenaline stimulate and

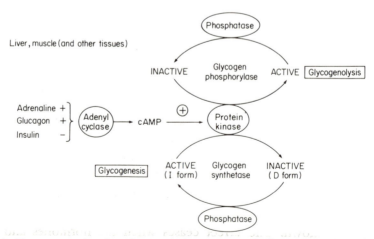

Fig. 13.7 Hormones, adenyl cyclase and glycogen metabolism. The scheme is slightly
over simplified—glucagon exerts an effect on liver but not on muscle. The hormones do
not act directly on the cyclase but bind to a specific receptor which then triggers the
effect on adenyl cyclase. Also phosphorylase kinase is not directly stimulated by cAMP
but is itself activated by the cAMP-sensitive protein kinase shown in the scheme.

insulin depresses adenyl cyclase. Insulin, therefore, reduces cAMP levels and promotes glycogen synthesis rather than glycogen breakdown. The other hormones, which increase cAMP, have the reverse effect.

Adipose tissue adenyl cyclase responds to many hormones (Fig. 13.8). Adrenaline*, glucagon, corticotropin, growth hormone, cortisol and thyroxine all stimulate adenyl cyclase while insulin depresses its activity. Those that stimulate promote lipolysis—the breakdown of triglyceride stores and the release of non-esterified fatty acids into the plasma. Insulin, therefore, tends to block this effect. Now another triglyceride lipase is known which hydrolyzes the neutral fat that has been absorbed from the intestine, via the lacteals, and which circulates in the blood in the form of chylomicrons. This enzyme, known as clearing factor lipase or lipoprotein lipase, is also present in adipose tissue and its function is to promote the uptake of the circulating lipid by adipose tissue and its deposition as

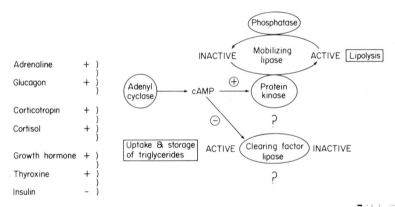

Fig. 13.8 Hormones, cyclase and triglyceride hydrolysis in adipose tissue and by clearing factor lipase. The opposing effects of cAMP on the two lipases are well established. There is also good evidence that a kinase is necessary for the activation of the adipose tissue mobilizing lipase, but, since the clearing factor lipase has not so far been purified the roles of the protein kinase and phosphatase in respect of this lipase are unknown.

stored triglyceride. This enzyme is required therefore in the period following a meal, but not during a fast. Insulin is secreted when food is absorbed from the intestine and this hormone brings about a lowering of cAMP in cells. Clearing factor lipase is inhibited by cAMP and this seems to be the means whereby insulin increases the activity of the enzyme. The opposing effects of cAMP on mobilizing and clearing factor lipases is well established and there is good evidence that the mobilizing lipase undergoes phosphorylation and dephosphorylation. It is possible that clearing factor lipase is regulated by the same mechanism but, as yet, this has not been demonstrated.

* Also known as epinephrine.

13.5. Hormones and the Synthesis of Enzymes

(a) Introduction

We have so far been concerned with the rapid adjustments of enzyme activity that result from allosteric effectors or from the effect of cAMP on enzyme phosphorylation. We must now consider the second way in which enzyme activity is controlled—by alteration in rates of synthesis of new enzyme molecules. Although induction or repression of enzyme synthesis may be observed within a few minutes of exposing a bacterial culture to the inducer or corepressor (see Section 12.37), the time scale in mammalian tissues is measured in hours rather than minutes. Moreover, the roles of inducer and corepressor seem more often to be played by hormones rather than by simple metabolites. Although the Jacob and Monod model of the operon is valuable in considering the mechanism of induction and repression of mammalian enzymes there is no direct evidence that the mammalian mechanism is identical with that proven to operate in the bacterial cell.

Most of the inducible mammalian enzymes have been demonstrated in liver; other tissues, muscle for instance, show less dramatic changes in rate of enzyme synthesis. Perhaps this is because the liver receives a very variable supply of nutrients from the intestine and needs to adjust to changing nutritional states. The experimental evidence that a change in enzyme activity is the result of induction of enzyme synthesis often depends on the use of inhibitors of protein synthesis. If the rise in an enzyme activity following hormone administration is prevented by Actinomycin D, it can be argued that the hormonal action involves transcription of DNA. If the effect is blocked by Puromycin but not by Actinomycin D, the hormone is probably affecting translation of the messenger RNA. Although such experiments provide convincing evidence that hormones may affect enzyme synthesis it has not been possible, so far, to demonstrate this action *in vitro*. In other words, the effect on protein synthesis is observed only in intact animals or in tissue culture of mammalian cells. Some examples of how hormones affect enzyme synthesis will serve to illustrate this aspect of metabolic regulation.

(b) Thyroid hormone

The principal thyroid hormone, thyroxine:

$$\text{HO}-\underset{\text{I}}{\overset{\text{I}}{\bigcirc}}-\text{O}-\underset{\text{I}}{\overset{\text{I}}{\bigcirc}}-\text{CH}_2\text{CHNH}_2\text{COOH}$$

exerts a general control over growth, metamorphosis (in amphibia) and oxidative metabolism. The amount of oxygen consumed at rest (the basal metabolic rate) is increased by thyroxine and depressed on removal of the thyroid. It is now established that this effect is achieved by increased synthesis of many of the mitochondrial dehydrogenases and cytochromes and that the primary effect is on the DNA-dependent RNA polymerase in the cell nucleus.

Figure 13.9 shows that, following a single small dose of thyroid hormone, RNA synthesis in the nucleus is increased three-fold within a few hours. Since this response precedes the other known effects of thyroid hormone it is believed to be the primary event.

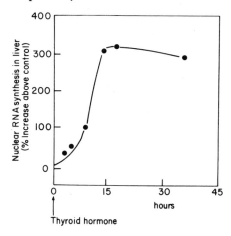

Fig. 13.9 Effect of Thyroid hormone on RNA synthesis in liver in rats. The experiment was performed in rats from which the thyroids had been previously removed. A single injection of thyroid hormone was given to each rat (except the controls) and the livers removed at various intervals thereafter. (After Tata.)

(c) Cortisol and glucagon—the defences against hypoglycaemia

The blood glucose concentration is maintained within narrow limits in health. Insulin is secreted by the pancreas after ingestion of a meal and serves to keep blood glucose from rising too high. During a period of fasting it is even more essential to prevent hypoglycaemia occurring. Although the nervous system utilizes ketone bodies during fasting, its ability to function normally depends on a supply of glucose via the blood. Unchecked hypoglycaemia will cause coma, convulsions and death. Glycogen stores in the liver are the first line of defence against hypoglycaemia and, as we have seen, glucagon, a polypeptide hormone from the pancreas, stimulates glycogenolysis and will help to maintain the blood glucose concentration. But glycogen stores are soon exhausted unless they can be replenished by the catabolism of protein. The conversion of glycogenic amino acids to carbohydrate is called gluconeogenesis and involves a number of key enzymes. These include several transaminases of broad specificity; some more specific deaminases, e.g., serine deaminase and threonine dehydrase; pyruvate carboxylase, phos-

amino acids

• (Transaminases
(Deaminases

• Pyruvate carboxylase

• PEP carboxykinase

• Fructose 1, 6-diphosphatase

Glucose 6-P ⟶ Glycogen

• Glucose 6-phosphatase

Glucose

Fig. 13.10 Key enzymes of gluconeogenesis in liver.

phoenol-pyruvate carboxykinase, fructose 1,6-diphosphatase and glucose 6-phosphatase (see Fig. 13.10). In part gluconeogenesis is encouraged by predominance of fatty acid oxidation as an energy source. The accumulation of acetyl CoA acts to restrict carbohydrate utilization and promote the pyruvate carboxylase reaction (see Section 13.2 above). Glucagon helps to mobilize triglyceride by stimulating lipase and so encourages gluconeogenesis. In some circumstances it may induce the synthesis of gluconeogenic enzymes.

(d) Cortisol

A more powerful hormone in this type of adaptation is cortisol—the main steroid produced by the adrenal cortex. Removal of the adrenal glands or their destruction by pathological processes (a condition known as

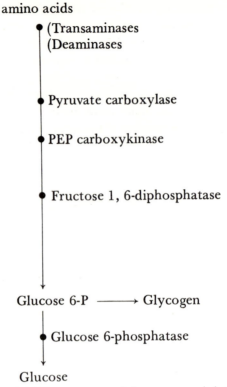

Cortisol

Addison's disease) leads to a fatal derangement of metabolism. One aspect of this disorder is the inability to tolerate fasting without developing the symptoms of hypoglycaemia. The way in which cortisol protects against hypoglycaemia is by inducing the synthesis of the key enzymes required for gluconeogenesis. All of the enzymes mentioned above are synthesized at an increased rate. An example—fructose 1,6-diphosphatase—is shown in Fig. 13.11 and is typical of this group of enzymes in its response to cortisol. The effect of Actinomycin D in preventing the induced enzyme synthesis confirms that nuclear RNA synthesis is a primary event in the process of induction.

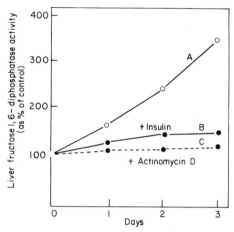

Fig. 13.11 Induction of a gluconeogenic enzyme in liver by administration of a cortisol derivative. A synthetic derivative of cortisol was administered daily to each of three groups of rats. One of the key gluconeogenesis enzymes in liver—fructose 1,6-diphosphatase is increased three-fold (A). The effect is blocked in (C) by an inhibitor of RNA synthesis—Actinomycin D or by the simultaneous administration of insulin in (B) which acts as a corepressor.

(e) Insulin

Insulin plays a role counter to the hormones, glucagon and cortisol. It is secreted following the ingestion of food and helps to adapt metabolism so as to assimilate the absorbed nutrients. Insulin seems to have several points of action. One has been mentioned already—insulin prevents the activation of adenyl cyclase in adipose tissue and liver. By lowering the concentration of cAMP it shifts the balance of glycogen and triglyceride metabolism away from catabolism and towards the synthesis of these storage molecules. Insulin also exerts effects on the plasma membrane of many cells including skeletal and heart muscle, adipose tissue, but not the liver. It promotes the transport of glucose and amino acids into the cells and hence increases the uptake of these nutrients from the blood. It also exerts effects on the induction and repression of enzyme synthesis.

Insulin opposes the action of cortisol by causing repression of the gluconeogenesis enzymes. This effect can be seen in the experiment shown

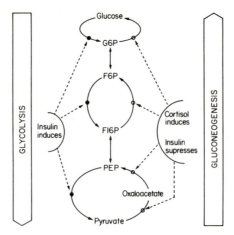

Fig. 13.12 Inducible enzymes in glycolysis and gluconeogenesis.

in Fig. 13.11, in which it prevented the induction of fructose 1,6-diphosphatase by the adrenal cortical steroid. It suppresses the other key gluconeogenesis enzymes in the same way. The positive role of insulin in inducing the synthesis of enzymes in liver is manifested on three glycolytic enzymes—phosphofructokinase, glucokinase and pyruvate kinase. These three enzymes are the irreversible steps of glycolysis, all kinases, which, in the reverse pathway of gluconeogenesis, are replaced by alternative reactions, see Fig. 13.12.

The relationship of hexokinase and glucokinase deserves a special word. In many cells, those of muscle and adipose tissue for example, the plasma membrane imposes a permeability barrier to glucose entering the cell and insulin specifically stimulates glucose entry in these tissues. Phosphorylation of glucose is effected by a non-specific hexokinase which is present in abundant amounts and which is not rate-limiting unless the product, glucose 6-phosphate accumulates. In contrast the liver cell membrane is freely permeable to glucose but the formation of glucose 6-phosphate depends on a specific glucokinase. This enzyme has a high K_m (low affinity) for glucose and is a rate-limiting step in glucose utilization. It is this enzyme rather than hexokinase that is induced by insulin.

13.6. Cyclic AMP and the Induction of Enzyme Synthesis

In our examples of hormonal regulation we have seen that several hormones appear to have more than one site of action—they may, for example, influence adenyl cyclase and also affect the rate of enzyme synthesis. The response of adenyl cyclase is rapid—usually manifested within a few minutes, while changes in enzyme synthesis require many hours. It would be logical to ask whether cAMP may be the mediator for both types of effect—not only the short-term effect resulting from enzyme phosphorylation but the long-term effect stemming from enzyme synthesis. Quite recently there have been a number of reports which suggest that the answer may be yes.

The lac operon in *E. coli* is one of the most studied examples of induction and repression. When these bacteria are presented with lactose as a sole carbon source for growth they quickly synthesize a group of three enzymes, including a β-galactosidase, necessary to utilize this disaccharide. If glucose is also provided this sugar is metabolized in preference to lactose and the synthesis of β-galactosidase is repressed. Now if cAMP is also added to the medium the induced enzyme synthesis continues and glucose no longer acts as a corepressor (Fig. 13.13). Many mutant forms of *E. coli* are known and their investigation has provided the experimental proof of the Jacob and Monod hypothesis. Recently a mutant lacking adenyl cyclase has been isolated. This organism not only

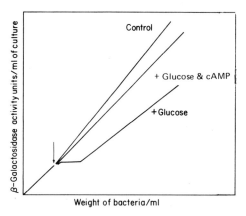

Fig. 13.13 Induction and repression of β-galactosidase synthesis in *E. coli*. A galactoside was added to a culture of *E. coli* in order to induce the synthesis of β-galactosidase. Ten minutes after adding the inducer the culture was divided into three parts (indicated by ↓). No additions were made to the control, which continued to produce β-galactosidase. To the second portion glucose was added: this repressed the induced enzyme. The third portion received glucose and cAMP: in this case β-galactosidase synthesis was not repressed. (After Pastan and Perlman.)

failed to produce cAMP but also refused to synthesize β-galactosidase when presented with inducers—a refusal that was reversed if cAMP was also added to the medium. These two observations show that cAMP is in some way implicated in the induction of enzyme synthesis: if added to the culture it prevents repression; if genetically absent, induction cannot occur. The role of cAMP in induction is now becoming clearer. RNA polymerase binds to the promoter site (P in Fig. 12.28). The binding requires a protein which binds cAMP (known as CRP, cAMP receptor protein) without which it is ineffective in binding RNA polymerase. Hence the induction of β-galactosidase requires cAMP. The effect of glucose (the corepressor) is to lower the cellular concentration of cAMP.

This explains why the addition of cAMP to the culture can overcome the effect of glucose in repressing β-galactosidase synthesis.

There is evidence, too, that cAMP is implicated in the synthesis of some liver enzymes which are induced by hormones. Cyclic AMP has been shown to mimic the effect of glucagon on the induction of gluconeogenesis enzymes in tissue cultures of liver cells and in intact animals. The mechanism of this effect is unknown, but one possibility is that histones are involved. These very basic nuclear proteins are not present in bacteria and their role in regulating gene action in eukaryotes has been the subject of much speculation. It is known that histones can be phosphorylated by a histone kinase in liver and that cAMP stimulates this reaction. If this effect is related to the process of enzyme induction it would help considerably to unify our theories of how hormones act.

13.7. Metabolic Regulation Before and After Birth

The alteration in rates of enzyme synthesis by which cells adapt to changing conditions have, in our examples, been concerned with adaptations in adult animals. Before leaving this subject we must consider the need for metabolic regulation at a unique moment in a mammal's life. This is the moment, at birth, when the foetus ceases to be dependent on the placental circulation for nutrients. Until birth the foetus exists in an environment of plenty, but the abundant supply of nutrients is suddenly

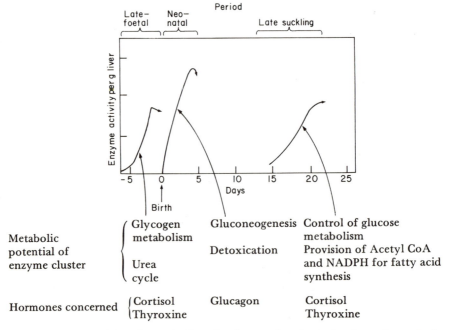

Fig. 13.14 Regulation of metabolism in the rat by the induction of synthesis of "clusters" of enzymes before and after birth. (After Greengard.)

terminated and is followed by a period of starvation before the maternal
milk secretion is adequate. It is now that the newborn must draw upon its
own resources if it is to survive. The means by which it defends itself
against hypoglycaemia and other hazards of an independent life are
beginning to be understood.

Just before birth a group of enzymes, known as the "late foetal cluster"
is synthesized in the liver (Fig. 13.14). Glycogen synthetase is in this
group and it leads to a rapid accumulation of liver glycogen. The urea
cycle enzymes also appear, in preparation for the time when the newborn
must provide its own excretory system. Both cortisol and thyroxine are
concerned in the induction of these enzymes. Almost immediately after
birth another functional group of enzymes appears—the "neonatal
cluster". The stimulus for this to happen is the onset of hypoglycaemia,
which is inevitable once the maternal circulation stops supplying the
foetus with glucose. Glucagon is secreted and this exerts its rapid effect on
phosphorylase (to mobilize the extensive glycogen stores) and its slower
effect in evoking the synthesis of the gluconeogenesis enzymes, including
an upsurge of glucose 6-phosphatase (Fig. 13.15). The adrenal cortex
plays little or no part in inducing these enzymes, indeed it seems to be
temporarily dormant. Glucagon alone is adequate to evoke the neonatal
cluster of enzymes either normally, by the hypoglycaemic trigger, or
prematurely if either glucagon or cAMP is administered to the foetus a
day or two before term. The late suckling cluster of enzymes is concerned
with the process of weaning, involving a switch to a diet richer in
carbohydrate and lower in fat and protein than that supplied by suckling.
This new phase demands a closer control of carbohydrate utilization.
Among the enzymes now appearing are glucokinase, pyruvate kinase as
well as citrate cleavage enzyme and NADP-linked dehydrogenases which

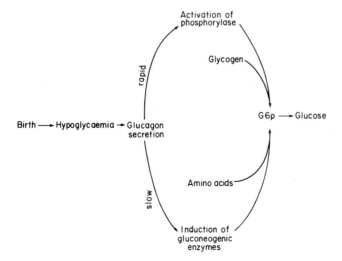

Fig. 13.15 How the newborn defends itself against hypoglycaemia.

are necessary for effective synthesis of fatty acids. By now the adrenal cortex is active again and the cortical steroids, together with thyroxine, are more important than glucagon as the hormonal inducers of the late suckling cluster.

13.8. An Attempt at Integration

The object of this chapter was not to provide a comprehensive survey of metabolic regulation but rather to convey a basic understanding of the mechanisms involved and the physiological sense behind these mechanisms. We began by specifying the key, pacemaker enzymes and identifying some of the allosteric effectors which influenced their activity. We progressed to the role of cAMP in phosphorylation of enzymes and saw how some hormones control the formation of this nucleotide. And lastly we observed how hormones can lead to the induction and repression of groups of enzymes. It was necessary to dissect these components involved in metabolic regulation in order to examine the operation of each separately. But to do so is unreal, for in the whole animal, each component interacts with the others and the overall adjustment of metabolism is the result of many co-operative effects. As an illustration of some of these interrelationships, let us examine the metabolic effects that occur in diabetes mellitus.

A condition resembling human diabetes can be produced by destroying or removing the insulin secreting cells of the pancreas. This may be

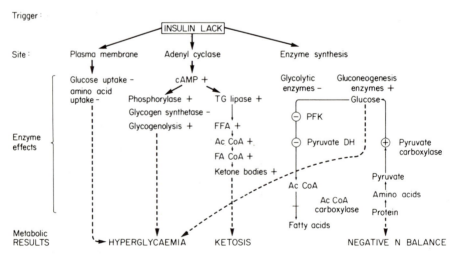

Fig. 13.16 Metabolic effects of Diabetes. The scheme summarizes the main sequence of metabolic events in diabetes mellitus. See text for detailed explanations. For simplicity, the tissues involved in these events have not been identified. + = increased — = decreased. TG lipase = adipose tissue triglyceride lipase. PFK = phosphofructokinase. Pyruvate DH = pyruvate dehydrogenase.

achieved with a specific cytotoxic agent streptozotocin, which destroys the β-cells of the islets of Langerhans or by surgical removal of the entire pancreas. Assuming that we produced a diabetic state by such means, what is the sequence of events which would be triggered off by this hormonal imbalance?

First (Fig. 13.16) consider the sites at which insulin exerts a direct effect. These are:

(i) The plasma membrane of many tissues, including muscle and adipose tissue which will be less permeable to glucose and amino acids in the absence of insulin.

(ii) The stimulation of adenyl cyclase in various tissues—liver and adipose tissue for example—by various hormones will no longer be opposed by insulin. Hence the concentration of cAMP will rise in these cells and the breakdown of triglyceride and glycogen will be promoted.

(iii) Certain liver enzymes—glucokinase, phosphofructokinase and pyruvate kinase—normally induced by insulin will be synthesized at a slower rate and others—those concerned with gluconeo-genesis—and normally suppressed by insulin, will now increase in amount.

Next we must consider the secondary effects of insulin deficiency. The crucial alteration is the increased hydrolysis of triglyceride which leads to the export of FFA from the adipose tissue cells into the plasma and its uptake by other tissues—muscle and liver. Fatty acids then become the major energy substrate. Increased concentrations of FFA, long chain fatty acyl CoA and acetyl CoA inhibit glycolysis and fatty acid synthesis while leaving gluconeogenesis unimpaired. Triglyceride synthesis is further restricted by a lack of α-glycerophosphate normally derived from glycolysis. Acetyl CoA production (by fatty acid oxidation) exceeds the liver's capacity to utilize it in the TCA cycle with the result that the output of acetoacetate increases.

Finally, there are the general metabolic effects that follow these changes.

(i) Glucose utilization is reduced and at the same time glucose formation from glycogenic amino acids is increased. The blood glucose concentration therefore rises. When it exceeds the capacity of the renal proximal tubule to reabsorb it, glucose begins to be excreted in the urine.

(ii) The mobilization of tissue protein—predominantly from skeletal muscle—for gluconeogenesis leads to a negative nitrogen balance.

(iii) The over-production of ketone bodies—acetoacetate, β-hydroxy-butyrate and acetone—by the liver may exceed the capacity of other tissues to utilize them. They accumulate leading to a condition known as keto-acidosis. The excretion of these substances, together with glucose, in the urine causes an excessive loss of water and salts. A state known as diabetic coma may ensue, characterized by drowsiness or unconsciousness, deep (acidotic) breathing, dehydration, ketosis and hyperglycaemia. It will prove fatal unless insulin is injected and the water and electrolyte deficiencies made good.

Further Reading

Chapter 2—Acid-base Dissociations and Their Relevance to Biological Systems

CHRISTENSEN, H. N. *Body Fluids and the Acid-Base Balance*. A learning program for students of the Biological and Medical Sciences. W. B. Saunders Co., Philadelphia (1964).

MASORO, E. J. and SIEGEL, P. D. *Acid-Base Regulations: Its Physiology and Pathophysiology*. W. B. Saunders Co., Philadelphia (1971).

MORRIS, J. G. *A Biologist's Physical Chemistry* (2nd Edition) Edward Arnold, London (1974). Chapter 5 and 6.

Chapter 3—Structure and Properties of Carbohydrates and Nucleotides

DAVIDSON, J. N. *Biochemistry of the Nucleic Acids*. (7th Edition) Chapman and Hall, London (1972).

Chapter 4—Structure of Proteins

CAVILLI-SFORZA, L. L. The Genetics of human populations. *Scientific American*. p. 80 (September, 1974).

DICKERSON, R. E., and GEIS, I. *The Structure and Action of Proteins*. Harper and Row, New York, Evanston, London (1969).

GIVOL, D. Affinity labeling and topology of the antibody combining site. *Essays in Biochemistry*, Vol. 10, p. 1 (1974).

GRANT, P. T. and COOMBS, T. L. Proinsulin, a biosynthetic precursor of insulin. *Essays in Biochemistry*, Vol. 6, p. 69 (1970).

GUILLEMIN, R., and BURGINS, R. The hormones of the hypothalamus. *Scientific American*, p. 24 (November, 1972).

LOEWY, A. G., and SIEKEVITZ, P. *Cell Structure and Function*. Holt, Rinehart & Winston, New York. (1969).

PHILLIPS, D. C., and NORTH, A. C. T. *Protein Structure*. Oxford Biology Readers, Ed. by J. J. Head. O.U.P. (1973).

ROITT, I. M. *Essential Immunology*. Blackwell Scientific Publications Oxford (1977).

Chapter 5—Enzymes

BERNHARD, S. *The Structure and Function of Enzymes*. W. A. Benjamin, Inc. New York (1968).
Organic Chemistry of Life—Readings from the Scientific American. W. H. Freeman & Co., San Francisco (1973). The following chapters:
OLD, L. J., BOYSE, E. A., and CAMPBELL, N. A. L-Asparagine and Leukemia.
NEURATH, H. Protein-digesting Enzymes.
PHILLIPS, D. C. The three-dimensional Structure of an Enzyme.

Chapter 6—Carbohydrate Metabolism

Carbohydrate Metabolism and its Disorders. Ed. by F. Dickens, P. J. Randle & W. J. Whelan. 2 Vols. Academic Press London (1968).

Chapter 7—Nitrogen Metabolism

MEISTER, A. *Biochemistry of the Amino Acids* (2nd Edition) 2 Vols. Academic Press, New York (1965). A useful source book for information on the occurrence, isolation, structure and metabolism of amino acids.

Chapter 8—Structure and metabolism of Lipids

MASORO, E. J. *Physiological Chemistry of Lipids in Mammals*. W. B. Saunders Co., Philadelphia (1968).

Chapter 9—Bioenergetics of Mitochondria

GREVILLE, G. D. A scrutiny of Mitchell's chemiosmotic hypothesis of respiratory chain and photosynthetic phosphorylation in Current Topics. In *Bioenergetics*. Ed. R. Sanadi. Vol. 3. Academic Press, New York (1969). A very clear and penetrating analysis of one of the main hypotheses.
KLINGENBERG, M. Metabolite transport in mitochondria: an example for intracellular membrane function. *Essays in Biochemistry*, Vol. 6, p. 119. (1970).
LEHNINGER, A. L. *Bioenergetics*. W. A. Benjamin, Inc. New York (1965). A very readable and elementary explanation.
MORRIS, J. G. *A Biologist's Physical Chemistry* (2nd Edition) Edward Arnold, London (1974). Includes a clear account of thermodynamics as applied to biochemistry including redox potential.
RACKER, E. The two faces of the inner mitochondrial membrane. *Essays in Biochemistry*, Vol. 6, p. 1. (1970).
Membranes of Mitochondria and Chloroplasts. Ed. E. Racker. Van Nostrand Reinhold Co., New York (1970). Contains reviews of many aspects, including chemical structure, electron microscopy, function and biogenesis.
ROODYN, D. B., and WILKIE, D. *The biogenesis of mitochondria*. Methuen & Co., Ltd., London (1968).

Chapter 11—Membranes and Transport of Materials

The Movement of Molecules across Cell Membranes. Ed. W. D. Stein, Academic Press, London (1967).

Membrane Molecular Biology. Ed. C. F. Fox and A. D. Keith. Sinauer Assoc. Inc. Stamford, Conn. (1972).
The Structure and Function of Biological Membranes. Ed. L. I. Rothfield. Academic Press, New York (1971).
Four articles of interest are:
 CAPALDI, R. Scientific American, pp. 26–33 (March, 1974).
 CRANE, R. K. Federation Proceedings, 24, pp. 1000–1006 (1965).
 GUIDOTTI, G. Annual Reviews of Biochemistry, 41, pp. 731–750 (1972).
 SINGER, S. J., and NICOLSON, G. L. Science, 175, pp. 720–731 (1972).

Chapter 12—Nucleic Acid and Protein Synthesis

DAVIDSON, J. N. *The Biochemistry of Nucleic Acids* (7th Edition) Chapman and Hall, London (1972).
INGRAM, V. M. *Biosynthesis of Macromolecules* (2nd Edition) W. A. Benjamin, New York (1972).
KORNBERG, A. *DNA Synthesis*. W. H. Freeman & Co., San Francisco (1974).
MATHEWS, M. B. Mammalian messenger RNA. *Essays in Biochemistry*, Vol. 9, p. 59 (1973).
NOVIKOFF, A. B., and HOLTZMANN, E. *Cells and Organelles*. Holt, Rinehart & Winston, New York. (1970).
WATSON, J. D. *Molecular Biology of the Gene* (2nd Edition). W. A. Benjamin, New York (1970).
The Chemical basis of life and Organic Chemistry of life. Readings from Scientific American. W. H. Freeman & Co. San Francisco (1973).
In *Companion to Biochemistry*. Ed. A. T. Bull *et al.* Protein biosynthesis. B. F. C. Clark, p. 1. Longmans, London (1974).

Chapter 13—Regulation of Metabolism

ASHMORE, J., and WEBER, G. *Hormonal control of Carbohydrate Metabolism in liver, in Carbohydrate Metabolism and its disorders*. Ed. by F. Dickens, P. J. Randle & W. J. Whelan. Academic Press, London (1968).
GREENGARD, O. Enzymic Differentiation in Mammalian Tissues. *Essays in Biochemistry*, Vol. 7, p. 159 (1971).
HALES, C. N. Some actions of hormones in the regulation of glucose metabolism. *Essays in Biochemistry*, Vol. 3, p. 73 (1967).
STADTMAN, E. R. Allosteric Regulation of Enzyme Activity. *Advances in Enzymology*, Vol. 28 (1966).

Subject Index